P DATE DUE, BOTH BELOW AND ON CARD

DATE DUE	DATE DUE	DATE DUE

Synthesis of the Caledonian Rocks of Britain

NATO ASI Series
Advanced Science Institutes Series

A series presenting the results of activities sponsored by the NATO Science Committee, which aims at the dissemination of advanced scientific and technological knowledge, with a view to strengthening links between scientific communities.

The series is published by an international board of publishers in conjunction with the NATO Scientific Affairs Division

A	Life Sciences	Plenum Publishing Corporation
B	Physics	London and New York
C	Mathematical and Physical Sciences	D. Reidel Publishing Company Dordrecht, Boston, Lancaster and Tokyo
D	Behavioural and Social Sciences	Martinus Nijhoff Publishers
E	Engineering and Materials Sciences	The Hague, Boston and Lancaster
F	Computer and Systems Sciences	Springer-Verlag
G	Ecological Sciences	Berlin, Heidelberg, New York and Tokyo

Series C: Mathematical and Physical Sciences Vol. 175

Synthesis of the Caledonian Rocks of Britain

edited by

D. J. Fettes
British Geological Survey, Edinburgh, U.K.

and

A. L. Harris
Department of Geological Sciences,
University of Liverpool, Liverpool, U.K.

D. Reidel Publishing Company

Dordrecht / Boston / Lancaster / Tokyo

Published in cooperation with NATO Scientific Affairs Division

Proceedings of the NATO Advanced Study Institute on
Synthesis of the Caledonian Rocks of Britain
United Kingdom
August 19-September 2, 1984

Library of Congress Cataloging in Publication Data

NATO Advanced Science Institute on Synthesis of the Caledonian Rocks of Britain, U.K.
  (1984 : Cardiff, South Glamoren, etc.)
  Synthesis of the Caledonian rocks of Britain.

  (NATO ASI series. Series C, Mathematical and physical sciences; vol. 175)
  "Proceedings of the NATO Advanced Study Institute on Synthesis of the Caledonian
Rocks of Britain, U.K., August 19-September 2, 1984"—T.p. verso.
  Includes index.
  1. Geology, Stratigraphic—Paleozoic—Congresses. 2. Geology—Great Britain—Congresses. I. Fettes, D. J. II. Harris, A. L. (Anthony Leonard) III. Title. IV. Series:
NATO ASI series. Series C, Mathematical and physical sciences; vol. 175.
  QE654.N37 1984   551.7'2'0941   86-4972
  ISBN 90-277-2235-8

Published by D. Reidel Publishing Company
P.O. Box 17, 3300 AA Dordrecht, Holland

Sold and distributed in the U.S.A. and Canada
by Kluwer Academic Publishers,
190 Old Derby Street, Hingham, MA 02043, U.S.A.

In all other countries, sold and distributed
by Kluwer Academic Publishers Group,
P.O. Box 322, 3300 AH Dordrecht, Holland

D. Reidel Publishing Company is a member of the Kluwer Academic Publishers Group

All Rights Reserved
© 1986 by D. Reidel Publishing Company, Dordrecht, Holland.
No part of the material protected by this copyright notice may be reproduced or utilized
in any form or by any means, electronic or mechanical, including photocopying, recording
or by any information storage and retrieval system, without written permission from the
copyright owner.

Printed in The Netherlands

CONTENTS

Preface	vii
List of Contributors	xi
List of Participants	xiii
The Caledonian Rocks of Britain / R.E. Bevins, W. Gibbons, A.L. Harris and G. Kelling	1
A Geotraverse through the Caledonides of Wales / M.G. Bassett, R.E. Bevins, W. Gibbons and M.F. Howells	29
Field Guide to the Lake District and Southern Uplands / G. Kelling, W.S. McKerrow and P. Stone	77
The Caledonian Geology of the Scottish Highlands / A.L. Harris, D.J. Fettes, W.G. Henderson, J.L. Roberts, J.E. Treagus, A.J. Barber, M.P. Coward and R. Strachan	113
A Comparison of the Lower Palaeozoic Volcanic Rocks on Either Side of the Caledonian Suture in the British Isles / C.J. Stillman	187
The Tectonic Setting of the Southern Uplands / W.S. McKerrow	207
The Contribution of the Finnmarkian Orogeny to the Framework of the Scandinavian Caledonides / D.M. Ramsay and B.A. Sturt	221
The Moine Thrust Zone: A Comparison with Appalachian Faults and the Structure of Orogenic Belts / R.D. Hatcher Jr.	247
The Moine Thrust Structures / M.P. Coward	259
Some Aspects of Geophysics in the Caledonides of the UK / R.T. Haworth	281
The Caledonian Geology of the Scottish Highlands / D.J. Fettes, A.L. Harris and L.M. Hall	303
Subject Index	335

PREFACE

    The Advanced Science Institute on which this publication is based took the somewhat unusual form of a geological field symposium held during late August 1984. It was designed to demonstrate to experienced earth scientists from the North Atlantic area the full range of geological phenomena encountered in the British Caledonian rocks. The ASl travelled from South Wales to the far northwest of Scotland by the route shown on the map and in doing so examined sedimentary, igneous and metamorphic rocks from Pembrokeshire (Dyfed), Cardigan (Ceridigian), Snowdonia, Anglesey, the English Lake District and the Southern Uplands and Highlands of Scotland. Thus the fifty or so participants in the ASl studied the geological history and major structures of rocks exposed on either side of the supposed Lower Palaeozoic Iapetus Ocean the British sector of which closed to the south of the present Southern Uplands.
    Wales (1-5) afforded insight into the nature of the late Precambrian basement of England and Wales and the relationship of sedimentary and volcanic cover sequences to this basement. The Ordovician sequence in Wales is a sample of the volcanic rocks typical of a marginal basin, and were examined in Pembrokeshire and Snowdonia. The English Lake District (6) displays rocks from an island arc also of Ordovician age.
    To the north of the suture where the two sides of the ocean came together is an accretionary prism of turbiditic and pelagic sediments locally resting against slivers of the ocean crust on which they were laid down in the late Ordovician and early-mid Silurian. These rocks now compirse the Scottish Southern Uplands (7-8) which on their north-west side are flanked by the Ballantrae Complex which occurs near Girvan and which is thought by some workers to be a fragment of an obducted ophiolite complex.
    Whereas the rocks of Wales, the Lake District and the Southern Uplands comprise the paratectonic Caledonides, those to the north of the Highland Boundary Fault comprise the orthotectonic zone. The metamorphic rocks of the Scottish Highlands (9-14) offer a strong contrast to the rocks in the paratectonic zone although many of the geological events affecting the opposite sides of the ocean were

approximately contemporary.

The orthotectonic zone comprises strongly metamorphosed and deformed rocks of the Dalradian (late Precambrian to mid Cambrian) and Moine (mid to late Proterozoic) which were deposited on a basement of Archaean (Lewisian) >2500 Ma rocks, comparable with the ancient basement of Greenland and Laurentian North America.

To the northwest the Caledonian rocks are limited by a major thrust zone the study of which formed the final stage of the field symposium. The traverse made by the symposium across the Highlands towards the thrust zone extended from the Highland Boundary Fault, a major fracture, across classic outcrops of metamorphic Dalradian rocks and major granites and beyond the Great Glen Fault into the Northern Highlands, where complexely deformed and highly metamorphosed Moine rocks are splendidly exposed.

The field symposium was particularly appropriate because it brought together earth scientists from the USA, Canada, France, the British Isles, Spain, Belgium, Norway and Sweden - all countries in which Caledonian rocks are of major importance. Thus an opportunity was offered for discussions between, say, geologists skilled and experienced in Appalachian problems and those conversant with similar problems in Europe. They came together in Britain where Caledonian rocks were first described and were exposed to the stimulus of type localities and classic relationships.

Part of this publication comprises the guide to the various localities visited by the field symposium while the remainder, in the form of written papers, embodies the several lectures and discussions which were a feature of every evening during the symposium. All the field guides and the papers have been extensively revised in the light of the experience and discussions during the field symposium, a process which to some extent accounts for the delay in the publication of this work.

All the participants and particularly the director of the symposium and the editors of this volume are appreciative of the unique opportunity offered by the NATO Scientific Committee which enabled this meeting to take place. The editors would like to record their appreciation of the help and support not only of the contributors but of all the participants. It is a matter of great sadness that they record the death of Professor David Wones of the Virginia Polytechnic Institute whose friendship and contributions during the meeting did much to make the symposium a success.

D.J. Fettes   A.L. Harris

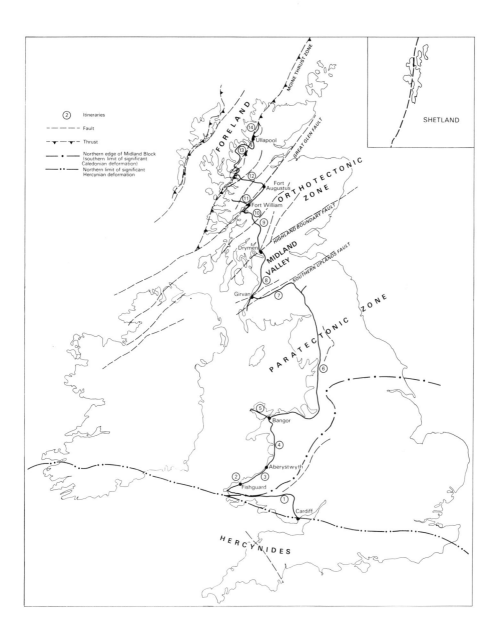

Map showing the main divisions of the British Caledonides and the generalised route of the field itineraries.

LIST OF CONTRIBUTORS

Dr. A.J. Barber, Department of Geology, Royal Holloway and Bedford Colleges, Egham Hill, Egham, Surrey, U.K.
Dr. M.G. Bassett, Department of Geology, National Museum of Wales, Cathays Park, Cardiff CF1 3NP, U.K.
Dr. R.E. Bevins, Department of Geology, National Museum of Wales, Cathays Park, Cardiff CF1 3NP, U.K.
Professor M.P. Coward, Department of Earth Sciences, University of Leeds, Leeds LS2 9JT, U.K.
Dr. D.J. Fettes, British Geological Survey, Murchison House, West Mains Road, Edinburgh EH9 3LA, Scotland.
Dr. W. Gibbons, Department of Geology, University College of Wales, Cardiff CF1 3NP, Wales.
Professor L.M. Hall, Department of Geology and Geography, University Of Massachusetts, Amherst, Mass. 01003, U.S.A.
Dr. A.L. Harris, Department of Geology, University of Liverpool, P.O. Box 147, Liverpool L69 3BX, U.K.
Professor R.D. Hatcher, Dept. of Geology, University of South Carolina, Columbia, South Carolina 29208, U.S.A.
Dr. R.T. Haworth, British Geological Sciences, Nicker Hill, Keyworth, Nottingham NG12 5GG, U.K.
Dr. M.F. Howells, British Geological Survey, Bryn Eithyn Hall, Llanfarian, Aberystwyth, Dyfed SY23 4BY, Wales.
Professor G. Kelling, Department of Geology, University of Keele, Keele, Staffs St5 5BG, U.K.
Dr. W.S. McKerrow, Department of Geology & Mineralogy, University of Oxford, Parks Road, Oxford OX1 3PR, U.K.
Professor D.M. Ramsay, Department of Geology, University of Dundee, Dundee DD1 4HN, Scotland.
Dr. J.L. Roberts, Department of Geology, University of Newcastle, Newcastle upon Tyne NE1 7RU, U.K.
Professor C.J. Stillman, Department of Geology, Trinity College, Dublin 2, Eire.
Dr. R. Strachan, Department of Geology, Oxford Polytechnic, Oxford, UK
Professor B.A. Sturt, Geologisk Institutt Avd. A., Bergen University Allegt 41, 5014 Bergen, Norway.
Dr. J. Treagus, Department of Geology, University of Manchester, Manchester, U.K.
Dr. W.G. Henderson, British Geological Survey, Murchison House, West Mains Road, Edinburgh EH9 3LA, Scotland.
Dr. P. Stone, British Geological Survey, Murchison House, West Mains Road, Edinburgh EH9 3LA, Scotland.

LIST OF PARTICIPANTS

Dr. T.B. Anderson, Department of Geology, Queen's University of Belfast Belfast BT7 1NN, Northern Ireland.
Dr. P.G. Andreasson, Department of Mineralogy & Petrology, University of Lund, Solvegatan 13, S-233 62 Lund, Sweden.
Dr. A.J. Barber, Department of Geology, Royal Holloway and Bedford Colleges, Egham Hill, Egham, Surrey TW20 OEX, U.K.
Dr. M.G. Bassett, Department of Geology, National Museum of Wales, Cathays Park, Cardiff CF1 3NP, Wales.
Dr. D. Bates, Department of Geology, University College of Wales, Aberystwyth, Dyfed, Wales
Dr. R.E. Bevins, Department of Geology, National Museum of Wales, Cathays Park, Cardiff CF1 3NP, U.K.
Dr. D. Bruton, Palaeontologisk Museum, Universitetet i Oslo, Sars Gate 1, Oslo 5, Norway.
Dr. I. Bryhni, Mineralogisk-Geologisk Museum, Universitetet i Oslo, Sars Gate 1, N-Oslo 5, Norway.
Dr. B. Cabanis, Universite Paris VI, Lab de Geochimie Comparee et Systematique, Tour 16-26 3 Etage, 4 Place jussieu, 75320 Paris Cedex 05, France.
Professor M.P. Coward, Department of Earth Sciences, University of Leeds, Leeds LS2 9JT, U.K.
Dr. D. Dallmeyer, Department of Geology, University of Georgia, Athens 30334, Georgia, U.S.A.
Dr. A.A. Drake Jr., United States Geological Survey, Mail Stop 926, Reston, Virginia 22092, U.S.A.
Dr. D.J. Fettes, British Geological Survey, Murchison House, West Mains Road, Edinburgh EH9 3LA, Scotland.
Dr. L. Fyffe, Dept. of Natural Resources, Mineral Resources Division, College Hill Road, P.O. Box 6000, Fredericton, N.B., Canada E3B5H1
Dr. D.G. Gee, Geological Survey of Sweden, Box 670, S-751 28 Uppsala, Sweden.
Professor F. Geukens, Katholieke Universiteit Leuven, Redingenstraat 16, 3000 Leuven, Belgium.
Dr. W. Gibbons, Department of Geology, University College of Wales, Cardiff CF1 3NP, Wales.
Dr. P.L. Guillot, Dept des Sciences de la Terre, University d'Orleans, 45046 Orleans cedex, France.
Dr. L. Gunderson, United States Geological Survey, Denver, Colorado, U.S.A.
Professor L.M. Hall, Department of Geology and Geography, University of Massachusetts, Amherst, Mass. 01003, U.S.A.
Dr. A.L. Harris, Department of Geology, University of Liverpool, P.O. Box 147, Liverpool L69 3BX, U.K.
Professor R.D. Hatcher, Dept. of Geology, University of South Carolina, Columbia, South Carolina 29208, U.S.A.

LIST OF PARTICIPANTS

Dr. R.T. Haworth, British Geological Sciences, Nicker Hill, Keyworth, Nottingham NG12 5GG, U.K.

Dr. W.G. Henderson, British Geological Survey, Murchison House, West Mains Road, Edinburgh, Scotland.

Dr. M.F. Howells, British Geological Survey, Bryn Eithyn Hall, Llanfarian, Aberystwyth, Dyfed SY23 4BY, Wales.

Professor G. Kelling, Department of Geology, University of Keele, Keele, Staffs ST5 5BG, U.K.

Dr. D. Keppie, Department of Mines & Energy, Box 1097, Halifax, Nova Scotia, Canada.

Dr. R. Kumpulainen, Mineralogisk-petrologiska avdelningen, Uppsala Universitet, Box 555, S-751-22 Uppsala, Sweden.

Dr. J-P. Lecorche, Lab de Geologie Dynamique, Univ de Droit, d'economie et des Sciences d'Aix-Marseille, 13397 Marseille Cedex 13, France.

Dr. J.P. Lefort, Las de Geol dynamique, Centre Armoricain d'Etude Structurale des Soc., Campus de Rennes-Beaulieu, 35042 Rennes Cedex, France.

Dr. B. Long, Department of Energy, Geological Survey of Ireland, 14 Hume Street, Dublin 3, Eire.

Dr. W.S. McKerrow, Department of Geology & Mineralogy, University of Oxford, Parks Road, Oxford OX1 3PR, U.K.

Dr. J.R. Mendum, British Geological Survey, Murchison House, West Mains Road, Edinburgh, Scotland.

Dr. D.P. Murray, Department of Geology, University of Rhode Island, Kingston, R.I. 02881, U.S.A.

Dr. R.B. Neuman, E-501, U.S. National Museum of National History, Washington DC 20560, U.S.A.

Dr. C.Q. Ochoa, Instituto Geologico y Minero de Espana, Rios Rosas 23, Madrid 3, Spain.

Professor P.H. Osberg, Department of Geological Sciences, University of Maine at Orono, 110 Boardman Hall, Orono, Maine 04469, U.S.A.

Dr. A. Ploquin, Centre National de la Recherche Scientifique, B.P. 20, 54 501 Vandoeuvre, Les Nancy Cedex, France.

Dr. W.H. Poole, Energe, Mines & Resources Canada, Earth Sciences, Geol. Surv. of Canada, 601 Booth Street, Ottawa, Ontario K1A OE8, Canada.

Professor D.M. Ramsay, Department of Geology, University of Dundee, Dundee DD1 4HN, Scotland.

Dr. D.R. Rankin, United States Geological Survey, Mail Stop 962a A, Reston, Virginia 22092, U.S.A.

Dr. P. Rathbone, British Geological Survey, Keyworth, Nottingham NG12 5GQ, U.K.

Dr. A.J. Reedman, British Geological Survey, Bryn Eithyn Hall, Llanfarian, Aberystwyth, Dyfed Sy23 4BY, Wales.

Ms. K. Reksten, Mineralogical-Geological Museum, Sars Gate 1, Oslo 5, Norway.

Dr. J.L. Roberts, Department of Geology, University of Newcastle, Newcastle upon Tyne NE1 7RU, U.K.

Dr. P. Robinson, Department of Geology & Geography, University of Massachusetts, Amherst, Massachusetts 01003, U.S.A.

LIST OF PARTICIPANTS

Dr. A.A. Ruitenberg, Dept of Natural Resources, Geological Survey
   Branch, P.O. Box 1519, Sussex, New Brunswick, Canada EOE 1P0.
Dr. D. Santallier, Universite de Lyon 1, Lab de Petrographie,
   27-43 Bd. du 11 Novembre, F-69622 Villeurbanne cedex, France.
Dr. M.B. Stephens, Instituut fur Geologi, Universitetet i Oslo,
   Postboks 1047, Blindern, Oslo 3, Norway.
Professor R.K. Stevens, Department of Earth Sciences, Memorial
   University of Newfoundland, St. Johns, Newfoundland, Canada A1B3X5
Professor C.J. Stillman, Department of Geology, Trinity College,
   Dublin 2, Eire.
Dr. P. Stone, British Geological Survey, Murchison House, West Mains
   Road, Edinburgh, Scotland.
Dr. R. Strachan, Department of Geology, Oxford Polytechnic, Oxford, UK
Professor B.A. Sturt, Geologisk Institutt Avd. A., Bergen University
   Allegt 41, 5014 Bergen, Norway.
Dr. W.A. Thomas, Department of Geology, Alabahama State University, 915
   S. Jackson St., P.O. Box 271, Montgomery, Alabahama 36195, U.S.A.
Dr. B.O. Torudbakken, Institutt for Geologi, Universitetet i Oslo,
   Postboks 1047, Blindern, Oslo 3, Norway.
Dr. J. Treagus, Department of Geology, University of Manchester,
   Manchester, U.K.
Dr. H. Williams, Dept of Earth Sciences, Memorial University of
   Newfoundland, St. Johns, Newfoundland, Canada.
Professor D.R. Wones, Department of Geology, Virginia Polytechnic
   Institute & State University, Blacksburg, Virginia 24061-0796, USA

THE CALEDONIAN ROCKS OF BRITAIN

R. E. Bevins
Dept. of Geology
National Museum of Wales
Cardiff, UK

W. Gibbons
Dept. of Geology
University College
Cardiff, UK

A. L. Harris
Dept. of Geology
The University
Liverpool, UK

G. Kelling
Dept. of Geology
The University
Keele, UK

ABSTRACT. The Caledonian rocks of Britain are described in terms of four geologically and geographically distinct areas. (1) Wales and the adjacent English midlands may be described in terms of three terranes - the Monian, Welsh Basin and S Britain Platform terranes. These are separated by major, long-lived faults or lineaments. The Welsh Basin characterised by Ordovician bi-modal volcanism and Silurian turbidites probably comprised a basin marginal to the Iapetus Ocean and related to SE-directed oceanic lithospheric subduction throughout the Ordovician. (2) Also related to this subduction is the Lake District which, from the calc-alkaline nature of its volcanics, is probably a fragment of a mid-Ordovician island arc, now partially thrust below the Southern Uplands. (3) The Southern Uplands itself comprises an Ordovician-Silurian accretionary prism accumulated at the NW margin of Iapetus Ocean and related to NW directed thrusting. The suture marking the site of the closure of the former ocean probably lies along the S edge of the Southern Uplands. (4) To the north of the Highland Boundary Fault, two major units of metamorphic rocks are recognised - the Proterozoic Moine and the late Precambrian-to-Cambrian Dalradian. Both units are floored by Archaean basement - the Lewisian complex. Dalradian and Moine suffered polyphase deformation and metamorphism but at different times. The Moine having undergone 1000 Ma orogenesis was reworked by Caledonian thrust-related deformation during the mid-Ordovician to Silurian, while the Dalradian acquired structures and metamorphic assemblages during the late Cambrian to early Ordovician Grampian orogeny. Although some plutons of mid-Proterozoic to early Ordovician age are recognised, most granitic plutons in the Scottish Highlands are 440-400 Ma.

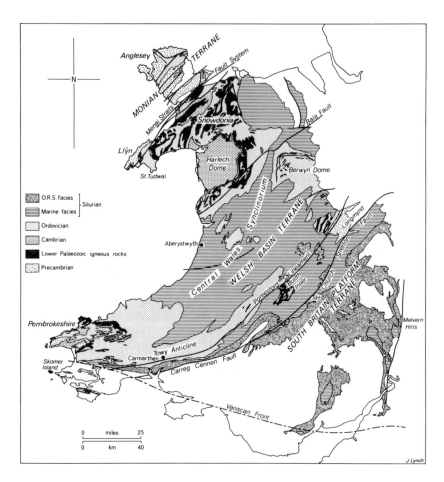

Fig 1.1 Geological map of the Welsh Caledonides.

WALES AND THE WELSH BORDERLAND

Exposures of pre-Devonian rocks south of the Lake District are chiefly confined to Wales and the Welsh Borderland, with relatively small inliers occurring elsewhere in the English Midlands. The area may be subdivided into three distinct regions: a northwestern "Monian Terrane", a central "Welsh Basin Terrane", and a southeastern "South Britain Platform Terrane" (Figure 1.1). The boundaries between these three units are major, steep fault systems and the terranes are therefore suspect in nature, ie. the possibility exists for significant strike-slip displacements having occurred along them. The NW boundary of the Welsh Basin Terrane is defined by the long-active Menai Straits Fault System (which includes the Berw, Dinorwic, and Aber-Dinlle Faults), movements along which appear to have been essentially vertical since late Lower Cambrian times. Any major transcurrent faulting along this lineament is likely to be late Precambrian or early Cambrian in age. The SE boundary of the Welsh Basin Terrane is defined by the Pontesford-Church Stretton Fault System, another major lineament with an equally long and complex movement history. The involvement of Precambrian basement along this fault system, combined with gravity and magnetic data, indicate a deep-rooted, steep shear zone. The lineament, also probably initiated in late Precambrian times, as a transcurrent fault, was intermittently active throughout the Palaeozoic era. Both Palaeozoic dip slip and strike slip displacements have been identified, although as yet no major ($>$100km) Phanerozoic transcurrent movements have been proven.

The S Britain Platform Terrane preserves a veneer of mainly shallow water marine Lower Palaeozoic sediments (with a major early-mid Ordovician hiatus) resting unconformably upon a dominantly igneous and volcaniclastic late Precambrian basement. By contrast the Welsh Basin Terrane exposes a much thicker, more complete, and generally deeper water marine Lower Palaeozoic sequence, the sialic basement to which is not exposed. The Monian Terrane, quite different again, is characterised by fault-bounded strips of probable late Precambrian basement overlain in places by a markedly incomplete Lower Palaeozoic (mostly Ordovician) sedimentary succession. The oldest radiometric date obtained so far from England and Wales is 702 ± 8 Ma, from the Stanner-Hunter Rocks in the Welsh Borderland. Unlike northern Britain, there is no proof of the existence of a really ancient gneissose basement in Britain south of the Iapetus Suture until the Pentevrian basement to the Armorican Massif is reached in the Channel Islands. Instead all dates obtained so far from basement rocks (mostly plutonic rocks) exposed on either side of the Welsh Basin Terrane cluster around the late Precambrian-early Cambrian boundary. Despite this similarity in age, however, there are marked differences between the Precambrian rocks of the Monian and S Britain terranes. The latter exposures are mostly intermediate-to-acid plutonic and volcanic rocks (in Pembrokeshire, the Welsh Borderland, Malvern Hills and the English Midlands) associated with marine and fluvial, sometimes volcaniclastic sediments, notably the Longmyndian succession of the Welsh Borderland and the Charnian of Charnwood Forest. Broad correlations have been made between these rocks and the Avalonian sequences of SE Newfoundland. The igneous geochemistry of the basement has been shown to be calc-alkaline and the sequence therefore interpreted as preserving the remnants of a late Precambrian island arc once active along the margins of an Avalonian plate.

Exposures of Monian rocks in Anglesey, Llŷn, and SE Ireland display an extraordinary range of lithologies. There are three broad Monian subdivisions, namely:

(a) plutonic rocks (granitic to dioritic) and gneisses (basic amphibolites, granitic migmatites, sillimanitic pelites, calc-silicates). Rb-Sr whole-rock dates from these exposures in Anglesey and Llŷn range from 542 ± 17 to 603 ± 34. The effect of shearing deformation on these rocks is ubiquitous with cataclasite and mylonite belts being particularly common along contacts with other lithologies.

(b) linear belts of fine grained, intensely foliated schists, mostly derived from sedimentary and volcanic rocks. The metamorphic grade is variable, with greenschist, lower amphibolite and, in SE Anglesey, blueschist conditions being represented.

(c) lower grade metasediments (mostly turbidites and mudstones) and volcanic rocks belonging to the Monian Supergroup. Widespread exposures of the famous olistostromal Gwna mélange are normally incorporated as the highest formation within the supergroup, although the exact nature of the stratigraphic relationship between this mélange and non-chaotic supergroup strata remains uncertain.

Dating control on both schists and supergroup rocks is poor, although all Monian units appear to predate the late Lower Cambrian initiation of deposition in the adjacent Welsh Basin Terrane. With such poor dating control, little modern work, and most of the original stratigraphic relationships destroyed by shearing, the tectonic significance of the Mona Complex remains obscure. The presence of blueschists and mélange clearly suggests an active plate margin and a connection between these rocks and the postulated late Precambrian arc to the SE seems reasonable. However, recent work has emphasised the likely importance of late Precambrian - early Cambrian strike-slip fault movement displacing the Monian Terrane. It is therefore likely that neither blueschist nor mélange originally lay in their present position with respect to Southern Britain.

The subsidence of the Welsh Basin began in Lower Cambrian times. In NW Wales this subsidence heralded the eruption of thick, fault-ponded, acidic ash-flows tuffs which now form the Padarn Tuff Formation at the base of the Arfon Group. The eruption of these tuffs was immediately followed by the deposition of locally derived tuffaceous clastic sediments and tuffs. These in turn were covered by marine mudstones (the Llanberis Slates Formation) dated by the late Lower Cambrian trilobite Pseudatops viola, the oldest fossil known from the Welsh Basin Terrane. Although for long assumed to be of Precambrian age, the negligible time gap between the eruption of the Padarn Tuff Formation and the deposition of the overlying sediments makes a Lower Cambrian age for these volcanic rocks more likely.

A long period of Lower Palaeozoic sedimentation occurred subsequent to the Arfon Group eruptions, and produced the classic

"Welsh Basin" sequence of marine sandstones and mudstones. Sedimentological interpretations of this succession have been dominated for many years by the development and application of the turbidite concept. However, any palaeogeographic model of a single, rapidly subsiding NE-SW oriented Welsh Basin is something of an oversimplification; water depths within the basin varied greatly both in space and time. A major marine transgression in the early Cambrian for example, was followed by regression in later Cambrian times, with deposition of non-marine sediments in some areas (eg. Pembrokeshire). Evidence for local, mild end-Cambrian folding and erosion prior to another important phase of marine deposition is provided by the sub-Arenig angular unconformity at St. Tudwal's. The Arenig transgression is recorded within all three terranes in Southern Britain although there are significant differences in facies and faunas between the Monian and Welsh Basin Terranes. An anomalous "Celtic" brachiopod fauna occurs in the shallow marine Ordovician of Anglesey and may be a result of either geographic separation or sedimentary facies differences.

Shallow water and emergent conditions are also in evidence for much of Ordovician times, from Snowdonia across the Welsh Basin Terrane to the east and southeast. These areas were characterised by volcanic islands, particularly during Caradoc times, and the passage of ash-flows can be traced from subaerial across intertidal and into submarine environments. Unlike in North Wales, early to mid Ordovician sedimentation in the southern part of the Welsh Basin Terrane was dominated by a series of basins generally starved of terrigenous detritus, although early Ordovician age turbidite sandstones are present in the Carmarthen area. The quiet accumulation of black mud in these basins contrasted with the often violent volcanicity which occurred at various times in the Ordovician throughout the Welsh Basin Terrane. Another contrast is provided by the almost non-volcanic, shallow water or emergent conditions recorded by Ordovician rocks of the S Britain terrane. The Pontesford-Church Stretton Fault System separating these two terranes underwent an important period of probable strike-slip movement during Ashgill times, although neither the amount nor the direction of movement are established.
Silurian deep water sedimentation produced many of the classic Welsh turbidite deposits such as the Aberystwyth Grits. These turbidites were deposited in several contiguous basins such as the Denbigh and Montgomery Troughs in the northeast part of the terrane. Certain so-called turbidite sequences of Silurian age have recently been re-interpreted in the context of a shallow marine environment, with reworking of sediments by traction currents. A final shallowing of the basin occurred in late Silurian (Ludlow-Přídolí) times.

There are many dissimilarities in Silurian geology between the Welsh Basin Terrane and the platform to the southeast, although this most probably only reflects the slope-shelf transition rather than any major transcurrent fault movement. After a late Ordovician marine regression from the platform the sea returned with the worldwide transgression during Middle to Upper Llandovery times. The Silurian platform succession records a reduced terrigenous sediment supply with

Llandovery sandstones and mudstones being replaced by Wenlock and
Ludlow mudstones and limestones. A final regression with the onset of
red bed facies began as early as late Wenlock times in South Wales.

The Welsh Basin Terrane may be regarded as an ensialic marginal
basin subsiding behind an active arc probably situated somewhere to the
NW. Study of the abundant volcanic rocks in this terrane, chiefly of
Ordovician age, has provided a major stimulus in plate tectonic
reconstructions of explaining the Lower Palaeozoic evolution of the
southern British Caledonides. These studies have also generated
insights into the physical processes of volcanic eruptions,
particularly those in the subaqueous environment. An early plate
tectonic reconstruction, based on the petrochemistry of Ordovician and
Silurian volcanic rocks, suggested that most of the volcanic rocks were
calc-alkaline in character and erupted above a SE dipping subduction
zone. A change towards a more alkaline character in the south was
attributed to increasing distance from a trench present to the NW.
Recently, however, more detailed work has modified this earlier
interpretation. Following an early island arc-like volcanic episode in
Tremadoc times (the Rhobell Fawr Volcanic Group) igneous activity for
much of Ordovician times was characterized by bimodal acidic-basic
(chiefly tholeiitic not calc-alkaline) magmatism. Basic magmas were
erupted and intruded at a high level to produce pillowed flows and
massive sheets, with only minor pyroclastic products. In contrast
acidic volcanism was commonly explosive, and resulted in the widespread
distribution of pyroclastic ejecta in ash-flows, particularly well seen
today in the tuff sequences of Snowdonia. However, rhyolitic lavas,
domes and high-level intrusions were also developed. Petrochemical
studies suggest that eruptions occurred in a back-arc marginal basin
environment, with the rhyolites most probably being generated chiefly
by crustal melting. As in other marginal basins, however, the periodic
eruption of calc-alkaline magmas occurred. The later stages of
Ordovician volcanism, in Caradoc times, witnessed a change in the magma
type being erupted, becoming transitional towards alkaline in
character. Such an evolution in Wales most probably reflects the
decreasing influence with time of a subduction zone under the Welsh
Basin Terrane. Whether volcanic rocks of Silurian age, exposed on
Skomer Island and in Somerset, are related to this phase of Ordovician
volcanism is doubtful, whilst the source of the numerous bentonites
present in Silurian strata over much of Wales is not known.

The igneous rock of Wales are also of considerable significance in
terms of their associated mineral deposits. The volcanogenic nature
of the Coed-y-Brenin porphyry copper and the Parys Mountain copper-
lead-zinc deposits is well established, whilst the British Geological
Survey is currently engaged in investigations concerning possible
relationships between base metal mineralization in Snowdonia and the
closely associated, commonly host Ordovician volcanic rocks.

Many accounts describe the Lower Palaeozoic rocks of the Welsh
sector of the paratectonic Caledonides as being unmetamorphosed.
However, recent studies, based on metabasite mineral assemblages, clay
mineral assemblages, illite crystallinity studies, conodont colour
alteration and graptolite reflectance, show that this is not the case.

The findings illustrate a range in metamorphic grade, from probable zeolite facies (or 'diagenesis' grade, based on illite crystallinity studies) in the S Britain Platform Terrane, through prehnite-pumpellyite facies (or 'anchizone' grade) over much of the Welsh Basin Terrane and the Monian Terrane, with greenschist facies (or 'epizone' grade) in the more central parts of the basin. This pattern has been attributed to the effects of burial metamorphism, the greater depth of burial of sequences in the basin resulting in higher metamorphic grades. However, it is possible that in some areas, such as Llŷn and Snowdonia, metamorphic grade may have been influenced in part by tectonism.

The structure of the Welsh Basin Terrane is dominated by relatively open, upright to SE-verging folds, characterized generally by a single, near axial planar cleavage. The orientation of the folds is typically NE-SW, although in the extreme SW and NE axes and cleavage swing towards an E-W trend, possibly related to basement structures. Locally an early low angle cleavage is developed in parts of North Wales, whilst in Central and SW Wales a late crenulation cleavage is present. It has been suggested that much of the main cleavage development owes its origin to wet-sediment deformation. This view has generally found little favour and the important episode of folding and cleavage development most probably occurred at the end of Lower Devonian times. However, the role of active tectonism throughout Lower Palaeozoic times is well established, particularly the localisation of volcanic activity and deformation along active, vertical faults. More recently it has been suggested that strike-slip faults might have played an important part in the evolution of the Welsh area. Based on a thin-skinned thrust tectonic interpretation, 43km of crustal shortening across Wales has been estimated, although this has not found universal acceptance. Support for the existence of major thrusts at depth is provided by the South Irish Sea Lineament, identified by the BIRPS offshore seismic reflection studies, which dips WNW at $\underline{c}.$ 25° to a depth of 15 - 18 km. However the depth of focus and focal mechanism of the 19 July 1984 Llŷn earthquake suggest that this movement was related to slip along a steeply-dipping ($\underline{c}.45^{o}$) N-S oriented fault, at a depth of 18 - 22 km. The fact that brittle failure can still occur at such depths clearly has implications for any models involving décollement.

THE LAKE DISTRICT

Situated on the basis of Ordovician faunal evidence to the south of the Iapetus suture-zone, the Lake District region of northwest England forms part of the Leinster-Isle of Man-Lake District geotectonic zone. The Caledonide succession of the Lake District spans the early Ordovician to late Silurian time-interval and includes a variety of sedimentary and volcanic rocks. The thick Lower and Middle Ordovician sequences are mainly exposed in the Cumbrian Mountains of the northern and central Lake District while Upper Ordovician and Silurian rocks form extensive outcrops in the Coniston-Howgill Fells region of the southern Lake District (Fig. 1.2).

The oldest rocks are the Skiddaw Group, several kilometres of fine-grained, 'slaty' and graptolitic sediments of Arenig and basal Llanvirn age. The base of these silt- and mud-turbidites is not exposed within the Lake District and the four formations recognised within the Group comprise alternating argillaceous and arenaceous assemblages. The Skiddaw succession commences with the thin Hope Beck Slates, which pass gradationally up into the Loweswater Flags, up to 900 metres of thin-bedded, fine, well-graded turbidites, alternating with thin, greenish mudstones. Lenticular pebbly greywackes occur locally, and are best developed near the base of the Flags on Watch Hill, where they rest upon a thin, impersistent felsic lava, the Watch Hill 'Felsite'. The overlying Kirk Stile Slates Formation comprises several hundreds of metres of dark 'striped' silty mudstones, with a few thin pebbly mudstones. Similar rocks are known in the Cross Fell Inlier, east of the Lake District and in the Black Combe area of southwest Cumbria (Fig. 1.2). In the latter district, important tuffs and tholeiitic lavas of Arenig age occur around Millom Park. Silty mudstones equated with the uppermost Kirk Stile Slates have yielded Llanvirn ($\underline{D}$. bifidus) faunas in the tunnel-sections around Ullswater, in the east-central Lake District. These may be partly coeval with the topmost Skiddaw Formation, the Latterbarrow Sandstone, 2-300m of green to brown quartzitic arenites that probably represent a northwards-thinning wedge intercalated within the topmost Skiddaw mudstones.

The Eycott Group gradationally succeeds the Skiddaws and in the northern Lake District comprises at least 2 km of lavas, tuffs and agglomerates of transitional calc-alkaline to tholeiitic affinities, with subordinate mudstone interbeds which become dominant near the top of the Eycott succession. Volcanic elements are less abundant in coeval sequences of the eastern and central Lake District (Cross Fell, Ullswater, etc). Faunal evidence indicates that the Eycott Group is largely of Llanvirn age, although uppermost Arenig microfloras have been recorded from basal Eycott tuffs and mudstones in the northeast Lake District.

Rocks of the Borrowdale Volcanic Group form most of the mountains and crags of the central Lake District. The Borrowdales comprise up to 6000 metres of andesitic, rhyolitic and basaltic lavas, tuffs and agglomerates of calc-alkaline type, with very minor intercalated mudstones. The group rests with erosional unconformity on both the Skiddaw and Eycott Groups, and lenticular rudites (Bampton Conglomerate and equivalents) are extensively developed above this surface. The Borrowdale Group has yet to yield stratigraphically diagnostic faunas, but it falls between dated upper Llanvirn and middle Caradoc rocks.

The unconformably succeeding late Ordovician Coniston Limestone Group is a sequence (with an aggregate thickness exceeding 1000 m) of terrigenous clastics, calcareous mudstones, thin limestones and volcanics. Locally the sediments contain rich shelly faunas of mid-Caradoc and Ashgill age. The oldest formation in this Group is the Longvillian Drygill Shales, which occur within a small faulted outlier in the Carrock Fell igneous complex, and their age-equivalents in the Cross Fell area. Within the central Lake District the overlying

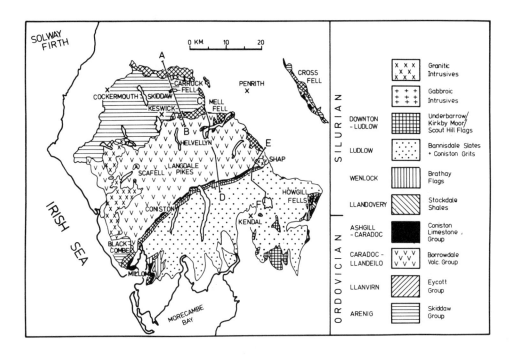

Fig. 1.2. Generalised geological sketch-map of the Lake District, after Moseley, 'Geology of the Lake District', <u>Yorks. Geol. Soc.</u>, 1978. Post Silurian areas left blank. A-B, C-D, and E-F indicate location of structural profiles shown in Figure 1.3.

Ashgill sequence commences with the Stile End Beds (Cautleyan) a fluviatile and shallow marine series of conglomerates, sandstones and calcareous mudstones, followed by the 70m Yarlside Rhyolite. The succeeding Applethwaite Beds comprise up to 100 m of muddy limestones and nodular calcarous mudstones, again of Cautleyan age. Traced to the southwest the upper part of the Applethwaite Beds becomes more silty and the basal limestone thickens substantially. During late Cautleyan or early Rawtheyan times, the central Lake District was emergent or subjected to erosion since the next rock-unit, the White Limestone, is of late Rawtheyan age and rests disconformably on the upper Applethwaite strata. The White Limestone and the late Ashgill (Hirnantian) Mucronata Beds, a few metres of grey-green calcareous shales. These pass up into the Ashgill Shale Formation, 20 m or so of grey, calcareous mudstones, sandy in the basal few metres and carrying a restricted Hirnantia brachiopod assemblage.

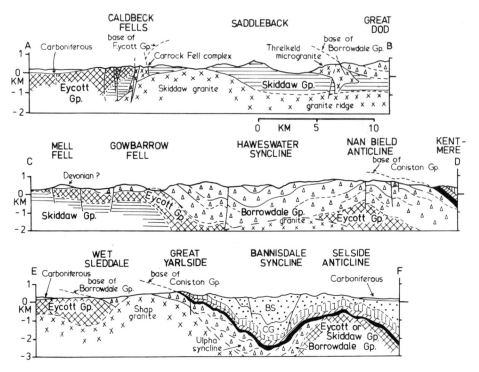

Fig. 1.3. Schematised structural sections across the Lake District along the lines shown in Figure 1.2. After Johnson et al., In: The Caledonides of the British Isles Reviewed, 1979.

The fault-disrupted Upper Ordovician successions of the Howgill Fells and Cross Fell Inlier (Fig. 1.2) appear to be thicker, more complete and more lithologically uniform than those of the central Lake District, with locally developed Ashgill reef-limestones (Keisley and Swinsdale Limestones).

The early Silurian sequences of the Lake District are traditionally assigned to the Stockdale Shales Group and reflect gradual deepening of the marine environment. The Lower and Middle Llandovery Skelgill Beds (persculptus - sedgwickii graptolite zones) commence with impersistent basal limestones, laterallly variable in thickness and lithological character, which are succeeded by 20 - 30 m of black, pyritous graptolitic shales, thickening westwards. The upper Llandovery is represented by the Browgill Beds, some 20 -70 m of grey mudstones with thin black graptolitic shale layers and occasional sandstones, which are more abundant towards the southeast. This formation is capped by thin impersistent red mudstones, overlain by more laterally continuous grey mudstones.

Most of the Wenlock is represented by the Brathay Flags Formation, about 400 m of blue-grey, laminated graptolitic mudstones, containing a few thin sandstones and bentonite horizons. To the west an eastwards-thinning lenticular body of silt- and sand-turbidites (Lower Coldwell Beds) reflects the localised influx of coarser sediment from the west or northwest. The uppermost Wenlock is represented by the Middle Coldwell Beds, about 100m of grey bioturbated mudstones, thin limestones and graptolitic shales.

The base of the Ludlow-Downton succession in the Lake District is marked by the Upper Coldwell Beds, a thin formation comprising alternations of grey graptolitic mudstones and laminated siltstones. These are abruptly succeeded by the Coniston Grits, up to 1800 m of fine-sand turbidites, with occasional coarser units and a widespread medial shale member. The overlying Bannisdale Slates (Middle Ludlow) comprise around 600 m of silty mudstones, thin turbiditic arenites and dark graptolitic mudstone films. Rich in situ faunas of trilobites and bivalves are known from several horizons. The succeeding Underbarrow Flags and Kirby Moor Flags (Upper Ludlow) are composed of thick, well-sorted siltstones, often calcareous and with an abundant shelly benthos, while the Scout Hill Flags (Downton) are similar but include several groups of red beds.

The Lower Palaeozoic stratigraphic record in the Lake District thus can be assigned to three stages:
(i) During the early Ordovician a thick sequence of hemipelagic muds and distal turbidites was deposited on or near a N-sloping continental margin forming the southern border of Iapetus. Coarser units yield evidence of southerly provenance and of predominant current flow towards N and NNE. Infrequent tholeiitic magmatism gradually became more calc-alkaline and abundant, leading to the major volcanic outburst recorded in the Eycott Group.
(ii) The Borrowdale Group was formed during an important phase of Middle Ordovician andesitic volcanicity, within a calc-alkaline island arc, and the laterally variable volcanic components appear to represent a series of over-lapping submarine edifices, which were locally, and ephemerally, emergent. This episode was followed by general compressive deformation.
(iii) The late Ordovician and Silurian sequences record a complex history of post-arc evolution, commencing with pronounced uplift and erosion, followed by a prolonged phase of southward marine transgression that eventually resulted in discordant deposition of late Caradoc and Ashgill shelf-carbonates directly upon pre-Borrowdale strata.
However, gradual deepening of this Lake District platform (perhaps locally accentuated by contemporaneous faulting) occurred during much of the Silurian, culminating in deposition of considerable thicknesses of turbidites during the latest Wenlock and early Ludlow. Moreover, the Ludlovian Coniston Grits record the arrival in this area of north-derived sands and mark the effective elimination of the Iapetus Ocean and the merging of the Southern Uplands and Lake District sedimentary

environments.

The gross structure of the Lake District involves a series of ENE-WSW trending broad and upright folds (Fig. 1.3) with well-developed cleavage of arcuate trend. However, this simple pattern is deceptive, concealing considerable complexity within each of the major stratigraphic groups. Thus the three major unconformities (pre-Borrowdale, pre-Ashgill, and end-Silurian) resulted from erosion and uplift associated with an important period of Caledonian diastrophism.

The NNE-trending pre-cleavage folds in the Skiddaw and Eycott Groups have been assigned to an early orogenic episode, but recent studies suggest that this anomalous trend and westward extension indicated by the fold-styles are more compatible with their origin as massive slumps down a W-facing early Ordovician palaeoslope. Folds within the Borrowdale Group are mainly of E or ENE trend and some of these are observed to pass below the regional unconformity at the base of the Coniston Limestone Group, indicating a pre-Ashgill phase of regional deformation. The complex relationships between the Skiddaw-Eycott and Borrowdale Groups in the northern Lake District suggest that the main Lake District Anticline may have been initiated during the accumulation of the Borrowdales, and the inception of some of these E-W structures thus may date from the Caradoc.

The major tectonic episode in the evolution of the Lake District occurred between deposition of the Downtonian Scout Hill Flags and the formation of the molassic Mell Fell Conglomerate (? Lower Devonian). Polyphase deformation resulted in the formation, accentuation or modification of major E-W and ENE-WSW trending upright or northward-facing folds and parasitic, sideways-closing minor folds. Sub-greenschist-facies metamorphism is widespread and is associated with the regional Caledonoid (NE-SW) cleavage, which in the Silurian rocks consistently strikes $5^\circ - 10^\circ$ clockwise with respect to the local fold traces. Emplacement of granite plutons is also associated with this post-Silurian episode but faulting (mainly thrusting and strike-slip), while present, is not the major control on outcrop-pattern, as it is in the Southern Uplands. Thus there is little evidence in Lake District structure or stratigraphy for subductive accretion, and south-eastwards subduction probably ceased beneath this region in the late Ordovician. The bulk of the Lake District deformation is thus ascribed to early Devonian continental collision, while the earlier episodes may relate to re-orientation of convergent stresses under the influence of structural controls exerted by the local basement. During later collision it appears that the Lake District sedimentary cover became detached from its basement and geophysical evidence suggests that the Lake District basement, possibly accompanied by an accretionary wedge, may have been conveyed beneath the Southern Uplands during collisional suturing.

THE SOUTHERN UPLANDS

Distinct faunal provinces recognised in the Lower Palaeozoic sequences of England and Wales (including the Lake District) and of southern

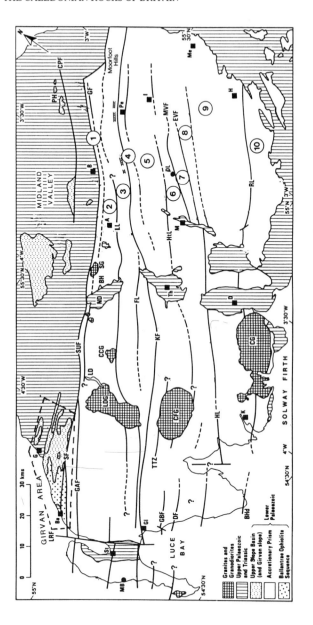

Fig. 1.4. Generalised structural map of the Southern Uplands, illustrating distribution of major strike-faults. After Leggett, et al., J. Geol. Soc. London, Vol. 136, 1979.

Place-names: A - Abington; B - Biggar; Ba - Ballantrae; D - Dumfries; DL - Dobbs Linn; G - Girvan; Gl - Glenluce; H - Hawick; I - Innerleithen; K - Kirkcudbright; L - Moffat; MB - Morroch Bay; Me - Melrose; Pe - Peebles; PH - Pentland Hills; St - Stranraer; Th - Thornhill.

Faults: CPF - Carmichael-Pentland Fault; EVF - Ettrick Valley Fault; FL - Fardingmullach Line; DF - Drumblair Fault; GAF - Glen App Fault; GBF - Gillespie Burn Fault; GF - Grassfield Fault; HL - Hawick Line; HtL - Hartfell Line; KF - Kingledores Fault; LL - Leadhills Line; LRF - Loch Ryan Fault; MVF - Moffat Valley Fault; RL - Riccarton Line; SF ; Stinchar Fault; SUF - Southern Upland Fault; TTZ - Talnotry Thrust Zone.

Devonian Granites: CG - Criffel Granites; CCG - Carsphairn Granite; CFG - Cairnsmore of Fleet Granite; LDG - Loch Doon Granite; SG - Spango Granite.

Scotland (including the Girvan area) indicate that the Ordovician and Silurian rocks now found in the Southern Uplands of Scotland were deposited on the northwest margin of the Iapetus Ocean. During the late Ordovician this ocean still was sufficiently wide to separate benthic faunas with pelagic larval stages and it appears to have closed by northwesterly subduction, initiated during the Llandeilo or Caradoc.

For the present purpose, it is appropriate to deal first with the main part of the Southern Uplands and then to outline the principal features of the Girvan area, north of the Stinchar Valley Fault, since the latter region provides evidence critical to interpretation of the nature and evolution of the northwest margin of Iapetus.

Since the pioneer work of Lapworth and of Peach and Horne, the Lower Palaeozoic rocks of the Southern Uplands have been assigned to three main strike-parallel sections: the Northern Belt, where early Ordovician spilites and cherts pass up into thick greywacke turbidites and rudites; the Central Belt, where the greywackes commonly are underlain by thick late Ordovician/early Silurian graptolitic shales; and the Southern Belt, dominated by greywacke-turbidites, interbedded with thin, later Silurian graptolitic shales.

Modern work has demonstrated that this model is oversimplified and that at least 10 distinct stratigraphic sequences can be recognised within the Southern Uplands on the basis of lithological, petrographical and palaeontological criteria (Fig. 1.5). These sequences may be several hundreds to several thousands of metres in thickness and some can be traced for 100km or more along strike. Some sequences are repeated across strike and this repetition, formerly attributed to isoclinal folding, is now explained by means of powerful strike faults with reverse throws, which usually truncate the steeply dipping sequences on their southeastern (basal) boundaries. Stratigraphical polarity within individual sequences is predominantly towards the northwest, although successive fault-bounded slices are generally younger towards the southeast, accounting for the traditional distribution of Ordovician and Silurian 'belts'.

In addition to the basal assemblages of early to mid-Ordovician spilites, cherts and black shales, several of the northwestern slices include significant thicknesses of boulder- and pebble-rudites. These form laterally lenticular bodies within the thick sequences of late Ordovician turbidites, and evidently represent proximal portions of laterally supplied submarine fans. Local bodies of volcanic rocks (mainly Caradoc in age) also occur within the greywacke sequences of the northern sector of the Southern Uplands. These bodies include submarine lavas, mass-flow agglomerates and occasional air-fall tuffs in varying proportions and in aggregate thicknesses ranging from a few metres to several hundreds of metres. Compositionally these rocks are alkaline basalts and trachy-andesites. They appear to represent remnants of sea-mounts and smaller ocean-floor volcanic edifices, scraped off during subductive accretion.

Another distinctive sequence occurs in structural slices 6 and 7 (Fig. 1.4) where an exceptionally thick development of graptolitic mudstones and siltstones (with thin bentonitic clays) spans the interval from Upper Llandeilo to Middle Llandovery. This Moffat

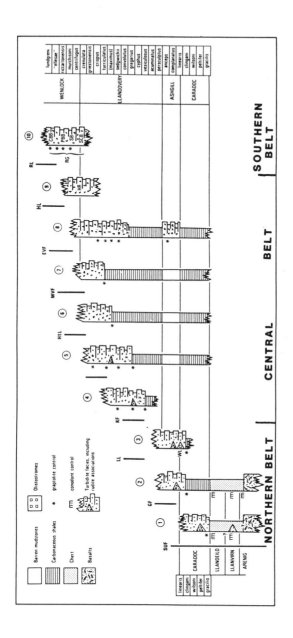

Fig. 1.5. Stratigraphic sequences within the major fault-bounded sectors shown in Figure 1.4. After Leggett et al.; J. Geol. Soc. London, 136, 1979. Fault names as in Figure 1.4.

Shales Group evidently resulted from a prolonged episode of pelagic and hemipelagic sedimentation on an elevated portion of the Iapetus ocean floor, protected from the activity of bottom-hugging turbidity currents which deposited coarse sediment on either flank of this elongate 'axial rise', possibly an outer trench swell or high.

It is noteworthy that volcanic rocks are entirely absent both from the basal assemblages and from the main turbidite successions in the central and southern regions (slices 4 - 10). Moreover, south of the Ettrick Valley (Fig. 1.4: fault-slice 8), the basal pelagic mudstones are no longer seen. The late Llandovery and Wenlock sequences thus comprise great thicknesses of thin- to medium-bedded greywacke-turbidites, axially transported (mainly from NE to SW), with only local occurrences of coarser sandstones and channel-rudites, eg. Raeberry Castle Beds displaying anomalous flow directions (to E and SE).

Another noteworthy feature of the Southern Uplands sedimentary sequences is the petrographic diversity of the greywacke-turbidites. The late Ordovician and early Silurian greywackes are replete with ophiolitic and andesitic arc-derived detritus, enabling recognition of laterally extensive bodies of greywacke with distinctive petrographic attributes, each probably related to a lateral point-source on the northwest margin of Iapetus. Later Silurian turbidites, however, are dominated by siliceous components and are less diverse in composition.

Turning to structural aspects, the pattern of major strike-parallel high-angle reverse faults with southward translation, commonly replacing the southern limb of anticlines, and the prevalence of early formed, imbricate SE-verging folds and listric thrusts within the major fault slices are features highly reminiscent of modern accretionary wedges. Later, more upright folds and more E-W trending, steeply plunging sinistral buckles and major shears appear to be genetically related to the principal cleavage. These are more controversial in origin, being variously attributed to continued, perhaps oblique, accretion, later continent-continent collision or to a combination of both mechanisms. Emplacement of the major granodiorite plutons appears to post-date this last phase of (? collisional) deformation and is dated as early to mid-Devonian.

In summary, the stratigraphical, sedimentological and structural evidence derived from the rocks of the Southern Uplands is broadly consistent with a model involving evolution of an accretionary prism along the northwestern margin of the Iapetus Ocean, during the Ordovician and Silurian. The major fault-bounded tracts are interpreted as packets of ocean-floor rocks and trench sediments, accreted in order from NW to SE. The sedimentological and petrographical data have been interpreted as indicating a late Ordovician to early Silurian phase of trench-fill sedimentation, succeeded by a post-late Llandovery phase of abyssal plain or open ocean-floor sedimentation, dominated by axially derived turbidite sands.

In the Girvan-Ballantrae area (Fig. 1.4) the preserved succession provides a tantalising glimpse of an early Ordovician magmatic association that is more extensively exposed in Ireland and may underlie much of the (southern) Midland Valley of Scotland. The

Ballantrae Volcanic Group includes a disrupted tectonic mélange and a thick pile of pillow-basalts and volcaniclastic rocks, intruded by serpentinites, gabbros and trondhjemites. Thin interbedded cherts and graptolitic shales are associated with conglomerates and sandstones. Recent graptolite finds strongly suggest tectonic repetition and juxtaposition of the Lower to Middle Arenig lavas and sediments (of possible hot-spot or seamount origin) now exposed in the southern coastal sector, which may have been obducted rapidly northwards (or transferred laterally by transform faulting) and then covered by late Arenig island-arc volcanics and sediments. This sequence of events appears to have been accomplished prior to the inception of the main phase of accretion in the Southern Uplands, dated as Llandeilo-early Caradoc.

The late Llanvirn-Ashgill sediments Barr and Ardmillan Groups, of the Girvan area record a northward-transgressive phase of shelf and slope sedimentation, involving conglomerates, sandstones, shales and thin limestones with abundant 'North American' shelly faunas. These display rapid variation in lithology and thickness, attributed to a series of ENE-WSW syndepositional faults, stepping down to the south. Sediment transport appears to be dominantly towards the south and south-east. Disconformably succeeding Llandovery-early Wenlock conglomerates, sandstones and shales also record deposition on an intermittently shoaling shelf or clastic platform. Early Silurian sediment transport was southwards but this pattern was reversed during the Wenlock.

Above the Ballantrae Complex the structural arrangement of the Girvan sequence is comparatively simple and involves a series of SE-verging major asymmetrical folds and high-angle reverse faults of NE-SW trend. Both the style and the polyphase deformation history of the Lower Palaeozoic sequence in this region are comparable with the tectonic pattern already described from the Southern Uplands proper, although the overall stratigraphic polarity is to the northwest, rather than to the southeast.

The tectonic environments of the Girvan area and of the Silurian inliers elsewhere in the Scottish Midland Valley is critical to interpretations of the geotectonic status and structural relationships of the Southern Uplands and the Midland Valley during the Lower Palaeozoic. The most popular current view is that, following the Arenig, the Midland Valley, including the Girvan area, was a fore-arc basin, separating the Southern Uplands accretionary prism from a volcanic arc, probably sited on uplifting Dalradian basement to the north. However, the clast-size, composition, depositional environment and palaeoflow evidence of Ordovician and Silurian rudites in the Midland Valley inliers and the northern part of the Southern Uplands are incompatible with the magnitude of the arc-trench gap required by such a model. A crystalline basement source with associated arc-rocks is required in close proximity, to provide the boulders and cobbles fed southwards in fan-deltas, on to the marginal slopes and into deep submarine fans. Utilising provenance and palaeocurrent evidence, recently has suggested that the original relationships have been telescoped. On this view the Southern Uplands accretionary prism has been thrust northwards in 'thin-skinned' style over a now concealed fore-arc basin, to rest upon a remnant of the proximal fore-arc and the adjacent magmatic arc, originally formed on sub-Midland Valley continental basement. Alternatively, significant oblique strike-slip displacements may account for anomalous juxtaposition and cross-strike foreshortening. At present, although the evidence appears to favour the concept of significant allochthony, it is inadequate to decide whether large-scale thrusting or transcurrent movements are the prime cause of this displacement.

SCOTTISH HIGHLANDS

The Scottish Highlands comprise the Metamorphic Caledonides of Britain and are commonly referred to as the orthotectonic zone (Fig.1.6). The events recorded by the metamorphic rocks and igneous bodies which make up the Highlands to some extent preceded the opening of Iapetus Ocean although much of the deformation, metamorphism and plutonism took place at a time when the ocean was probably open; the later stages of plutonism and the Old Red Sandstone molasse sedimentation overlapped with the closure of the ocean at the end of Silurian.

It is convenient to discuss the metamorphic rocks of Scottish Highlands in terms of three areas: - the NW Foreland (Hebridean craton), the Highlands to the NW of the Great Glen Fault and the Highlands to the SE of the Great Glen Fault.

1. Hebridean craton

This zone comprises the foreland of the Caledonides and forms the NW part of the Scottish Mainland and much of the Hebrides. Much of the area consists of late Archaean-to-Proterozoic gneisses and granulites which make up the Lewisian complex. Large parts of this gneiss and granulite complex is or has been at the granulite facies of metamorphism although much of it has been retrograded to amphibolite facies. Most of the material comprising the Lewisian came into existence between 2900 and 2800 Ma and may originally have been calc-alkaline volcanics and/or sub-volcanic plutons of similar composition. Basic igneous material, including an anorthositic body, occurs locally and part of the complex consists of metasedimentary rocks including mica schist and marble. The major gneiss-forming event took place shortly after the formation of the Lewisian protolith ($\sim$ 2800 Ma) and is referred to as Badcallian or early Scourian. Those parts of the early Scourian which escaped later reworking are markedly depleted in heat-forming elements, and are characterised by sharply discordant NW-trending dykes - the Scourian dykes - which were emplaced $\sim$ 2400 - 2200 Ma. The central part of the Lewisian complex of NW Scotland has largely escaped the imprint of later orogenesis during the late Scourian (Inverian) ($\sim$ 2400 Ma) and Laxfordian (post 2400 Ma) episodes. To the NE and SW of this early Scourian complex there is a strong reworking on NW-trending shear-zones on all scales such that the

THE CALEDONIAN ROCKS OF BRITAIN

Table 1.1

	Hebridean Craton	NW of Great Glen		SE of Great Glen	
Ordovician				Highland Boundary complex	
	Cambro-Ordovician sequence				
Cambrian				Dalradian groups	Southern Highlands Argyll Appin Grampian
Proterozoic	Torridonian	Moine (divisions) groups	Locheil Glenfinnan Morar	Central Highland Division	
Archaean	Lewisian	Lewisian			

regional Scourie dykes and the early gneissic foliation are transposed into the foliation imposed by reworking. Major vein complexes of granitic and pegmatitic material were emplaced (1750 - 1700 Ma) and the heat-forming elements replenished. Much of the Lewisian basement which is inferred to underlie most if not all of the Metamorphic Caledonides to the E and SE of the Hebridean craton probably underwent Laxfordian reworking because many of the zircons found in Caledonian granite yield Laxfordian ages.

The Lewisian rocks of the Hebridean craton became stable about 1800 - 1700 Ma age and were subsequently overlain by largely undeformed mid - to - late Proterozoic arkoses, shales and conglomerates, the Torridonian, and a Cambro-Ordovician sequence comprising mainly arenites and carbonates (Eriboll Quartzite, including pipe rock, and Durness Limestones). The Torridonian comprises two units - the Stoer Group (c. 1000 Ma) and the Torridon Group (c. 810 Ma). The Stoer Group rests with profound unconformity on the Lewisian Complex and suffered tilting and erosion before being overstepped by the Torridon Group. Both Torridonian units are thought to have been deposited in fault-bounded basins. The Torridonian is overstepped by the basal conglomeratic quartzite of the Eriboll Quartzite which itself commonly rests unconformably on the Lewisian basement. This feature is well displayed at Loch Assynt (Itinerary 14) where a famous "double unconformity" crops out. Here the Torridonian rests characteristically on a hilly landscape made of Lewisian gneisses and this unconformity is overstepped in spectacular fashion by the almost planar base of the Eriboll Quartzite.

The youngest member of the overlying Cambro-Ordovician carbonate sequence is Arenig-Llanvirn in age. The Llanvirn age of this carbonate places a maximum age on the displacements and mylonitisation associated with the Moine thrust zone because the carbonates are clearly involved in the thrust zone. A major thrust, of probable

Caledonian age, the Outer Hebrides thrust, lies immediately to the east of the Outer Hebrides, in many places cropping out just on shore.

Scottish Highlands NW of the Great Glen Fault

Almost all the metamorphic rocks of the N Highlands comprise units of the Moine series, but the Moine is locally interleaved with slices of Lewisian basement and intruded by pre-syn - and - late-orogenic igneous rocks, acid and basic. To the west the Moine series is separated from the Hebridean craton by the Moine thrust zone which intensely interleaved and mylonitised Lewisian, Torridonian, Cambro-Ordovician and Moine in a complex zone of displacement. It is now understood that the Moine thrust zone is only one of a number of Caledonian thrusts, two of which, the Knoydart and Sgurr Beag thrusts, lie to the east within the Moine tract. These have been traditionally described by Scottish geologists as 'slides' in recognition of the extremely ductile nature of the deformation zones associated with them. Within the Moine thrust zone itself a sequence of thrusting has been recognised; and a foreland-propagating system of thrusts has been recognised with the Moine thrust (<u>sensu stricto</u>) being the highest and oldest thrust and the Sole thrust being the youngest, on which the earlier thrusts were carried 'piggy-back' as a duplex. The direction of transport of these thrusts is regarded as WNW on the basis of shape fabrics within the mylonite while a maximum displacement of 77 km for the Moine thrust itself has been proposed. Some late brittle displacement in the thrust may well be extensional. Deep seismic experiments suggest that the Moine thrust extends eastwards through a crust <u>c</u>. 27km thick at an angle of <u>c</u>. $30°$. The thrusts exposed to the east of the Moine thrust zone, within the Moines are structurally higher and probably older than the Moine thrust and display more ductile phenomena than the Moine thrust; they have, moreover, been intensely folded. The most important of these thrusts - the Sgurr Beag - has been traced northwards from Ardnamurchan (Fig 1.6) for about 160 km and probably continues to the N coast of Scotland where it forms the Navar/Swordly slide (ductile thrust) system. Over much of its outcrop the Sgurr Beag thrust carries a thin, 1 m to several kilometres, slice of Lewisian basement. To the south, however, this slice is absent, possibly as a result of a regionally developed lateral hangwall ramp of the ductile thrust. The Sgurr Beag thrust separates the rocks of the Morar Division from those of the Glenfinnan Division. For more detailed information about these divisions and for the arguments which suggest that they are true lithostratigraphic groups, the reader is referred to the paper by Fettes and Harris in this volume.

In brief, three major units of Moine rocks are recognised in the N Highlands - the Morar, Glenfinnan and Locheil groups (Table 1.1). Each has its own distinctive lithostratigraphic sequence consisting of rather monotonous formations dominated by psammite or pelite or heterogeneously striped psammite and pelite. The abundance of minor but distinctive lithologies such as calc-silicates, heavy mineral laminae and mafic garnetiferous amphibolites serves to refine the gross

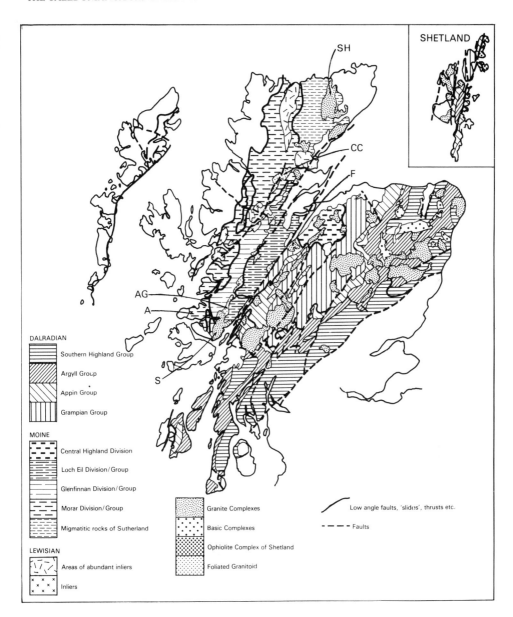

Fig. 1.6. Geological map of the Caledonides of the Scottish Highlands (from Fettes and Harris, this volume). A= Ardnamurchan; AG= Ardgour granitic orthogneiss; CC= Carn Chuinneag granite; F= Foyers granite; S= Strontian granite; SH= Strath Halladale granite.

lithostratigraphical characteristics while locally well preserved sedimentation structures indicate original way-up. It is believed by many Moine geologists that the Morar Group was deposited in one basin, floored by Lewisian basement, while the Glenfinnan Group, now known to be conformably overlain by the Locheil Group, was deposited in another. Dislacements of many tens or hundreds of kilometres are envisaged for the Sgurr Beag thrust while a sense of tectonic transport from ESE to WNW is demonstrated by shape fabrics in the blastomylonites which mark the ductile thrust zone.

Much of the Moine tract is at mid- to -upper amphibolite facies with moderate pressures being implied by Barrovian index minerals such as almandine and kyanite. Grade rises eastwards into the N Highlands steep belt where sillimanite (fibrolite) in pelites and diopside-bytownite in calc-silicates suggests high amphibolite facies. The formation of the gneisses, however, does not involve partial melting, and subsolidus reactions are invoked for the origin of essentially trondjhemitic leucosomes in the pelitic gneisses. Eastwards across the Loch Quoich line (Fig. 1.7)(Fettes and Harris, this volume) the grade falls to mid-amphibolite facies which persists as far as the Great Glen.

The main deformation features so far described in this account of the Moine have been indicated as Caledonian. The Sgurr Beag ductile thrust has been suggested as $\underline{c}$. 25 Ma earlier than the Moine thrusts while the age of structures which fold the ductile thrusts are constrained by the fact that they overprint a 456 ± 5 Ma syenite and a suite of $\underline{c}$. 440 Ma pegmatites. Interference patterns, the transposition of gneissosity and radiometric evidence, however, make it clear that the Caledonian structures in the Moine of the N Highlands have been superimposed on a much older complex of which perhaps only a small fragment remains, severely foreshortened by Caledonian thrusting. The metamorphism of this complex probably occurred at $\underline{c}$. 1004 Ma age, while deformed pegmatites dated at $\sim 776 \pm 15$ Ma intruded Morar Group Moine rocks which had already undergone two episodes of deformation. Arguably the strongest evidence for the pre-Caledonian age of the Moine derives from the Ardgour granitic orthogneiss (1030 ± 45 Ma) (Fig. 1.6), which, cutting the boundary between Glenfinnan and Locheil groups, seems to have been emplaced during the peak of regional deformation and metamorphism (Fettes and Harris,this volume). The present distribution of metamorphic isograds within the Moine of the N Highlands reflects grade established at this time (ie. $\underline{c}$. 1000 Ma) albeit modified by Caledonian deformation and thermal activity. Precambrian isograds are believe to have been subhorizontal or gently westward dipping prior to Caledonian orogenesis (Fettes and Harris,this volume); hence later Caledonian structures are parallel to these isograds irrespective of their orientation. Limits to the severe reworking of the Moine during the Caledonian are inferred to coincide with the Loch Quoich line (Fig 1.7) which defines the eastern edge of the N Highlands steep belt where Caledonian reworking is severe; to the east of the line, in the flat belt, Precambrian structures probably largely retain their pre-Caledonian attitude, ie. subhorizontal. Elsewhere, the Carn Chuinneag granite ($\underline{c}$. 550 ± 10 Ma)(Fig. 1.6) has

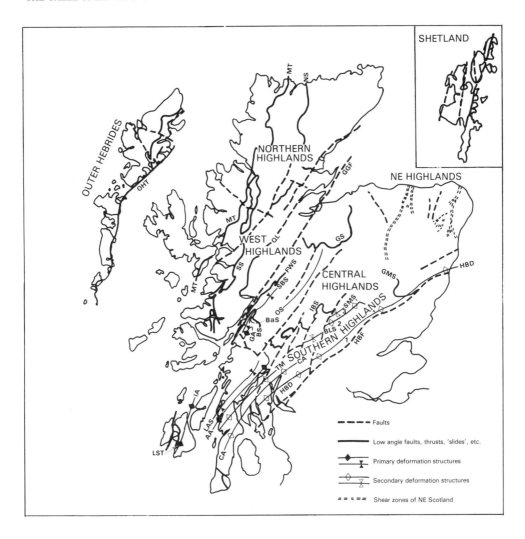

Fig. 1.7. Diagram illustrating structures referred to in text (after Ashcroft et al., 1984; Harris, 1983; Harte, 1979; Bradbury et al., 1979; Harte et al., 1984; Thomas, 1979; Roberts and Treagus, 1979; Piasecki et al., 1981). AA - Ardrishaig anticline; BAS - Ballachulish syncline; BLS - Ben Lawers synform; BS - Ballachulish slide; CA - Cowal anticline; FWS - Fort William slide; GA - Glen Creran anticline; GGF - Great Glen fault; GMS - Glen Mark slide; GS - Grampian slide; HDB - Highland Border downbend; HBF - Highland Boundary fault; IA - Islay anticline; IBS - Iltay Boundary slide; LAS - Loch Awe syncline; LST - Loch Skerrols thrust; MT - Moine thrust; NS - Naver slide; OHT - Outer Hebrides thrust; OS - Ossian steep belt; QL - Loch Quoich line; SBS - Stob Ban synform; SMS - Sron Mhor syncline; SS - Sgurr Beag slide; TM - Tarbert monoform. 1 - Highland Border steep belt; 2 - Flat belt; 3 - Loch Tummel steep belt. (from Fettes and Harris, this volume).

been deformed during the Caledonian while the Strath Halladale granite (649 ± 30 Ma) (Fig. 1.6) emplaced in the Sutherland migmatite complex has not. A similar Caledonian deformation front must occur on the N coast of Scotland to the west of the Sutherland migmatites because severe Caledonian reworking of Moine and Lewisian between the Moine thrust and the probable continuation of the Sgurr Beag ductile thrust, the Navar/Swordly 'slide' system, has been reported.

Direct evidence for the presence of Lewisian basement below the Moine rocks of the N Highlands is forthcoming from the occurrence of Lewisian slices along the Sgurr Beag and analogous ductile thrusts and also as inliers within the Morar Group where Lewisian occurs in the core of early (pre-Caledonian) major isoclines. Inliers of Lewisian occur as far east as the Great Glen Fault. Deep seismic evidence suggests the persistence of Lewisian basement SE beyond the Great Glen Fault where it underlies the largely Dalradian tract of the Central Highlands.

Scottish Highlands to the SE of the Great Glen Fault

Although geophysical evidence suggests that basement of Lewisian type persists SE of the Great Glen at least as far as the Highland Boundary Fault, direct evidence of its presence is only forthcoming in the far west of this area, on the island of Islay where severely retrograded Lewisian rocks are interleaved with Torridonian rocks of uncertain affinity.

Rocks designated as Moine have an extensive outcrop within the Central Highlands. These Moine rocks, however, fall into two entirely different categories based on their age and geological history. The younger unit, designated Grampian Division by some workers, is much better referred to as the Grampian Group and included with the Dalradian Supergroup as its oldest unit. The older Central Highland Division (Fig. 1.6) has some affinities with the Glenfinnan Group of the N Highlands (Fettes and Harris, this volume). The use of the term 'Moine' in the Central Highlands should be restricted to the Central Highland Division. These rocks, comprising mid-high amphibolite-facies psammites and pelitic gneisses with minor basic bodies appear everywhere to be separated by a ductile shear zone from the Grampian Group.

The Dalradian Supergroup (Table 1.1) comprises four groups of lithologically diverse metamorphosed sediments among the youngest of which mafic volcanics were emplaced. The groups - Grampian, Appin, Argyll and Southern Highlands - range in age probably from about 750 Ma until the middle Cambrian. Thus they comprise Riphean, Vendian and Lower Palaeozoic rocks and form a sequence which is cumulatively between 25 and 30 km thick but which never attains that thickness at any given locality. Palaeontological evidence based on early acritarchs suggests that the base of the Cambrian lies within the Argyll Group and must lie below the lower-middle Cambrian Southern Highland Group where trilobites, acritarchs and chintinozoa have been used to determine the age of the sequence. The boundary between the Appin and Argyll groups is marked by a tillite correlated tentatively with the Varanger tillite of Finnmark which is Vendian in age.

Sedimentological evidence suggests the evolution of the Dalradian basin to have been marked by increasing instability with the Grampian Group being made up of shallow marine or estuarine sediments and the upper part of the Argyll Group and the whole Southern Highland Group comprising turbidite sediments associated with mafic volcanics. It is likely that the instability was related to block-faulting of the basement as it was extended during basin evolution. The upper part of the Argyll Group (Crinan and Tayvallich subgroups) are comparable lithologically to the arenaceous and calcareous parts respectively of the Durness sequence of the foreland. This coincidence tempts correlation especially as the rocks are of similar age.

During the Grampian orogeny the Dalradian rocks were disposed in a major nappe complex which includes the Tay nappe, involving regional inversion of stratigraphy across the strike of some 20 km. In the SW Highlands a central steep zone has been indentified. Within this zone early folds are upright and face upwards. To the NW early folds face NW, while to the SE in the flat inverted zone early folds face to the SE. This simple pattern can be traced NE albeit in greatly complicated form, into the Central Highlands (Fettes and Harris, this volume).

In the Central Highlands early structures (D1 and D2) are cut by a 514^{+6}_{-7} Ma granite which was itself deformed by later structures (D3). Textural and fabric evidence suggests that the peak of regional metamorphicms also coincides with the D2-D3 interval. In the NE Highlands the 489 ± 17 Ma gabbros have been emplaced at a similar D2-D3 interval and these radiometric dates together constrain the peak of orogenic activity in the Dalradian to this Tremadoc – early Ordovician period. Thus the Grampian orogeny which was the main orogenic episode to affect the Dalradian of the Highlands to the SE of the Great Glen fault is constrained to a period several tens of millions years earlier than that which affected the Moine of the N Highlands and may imply the existence of two terranes of different Caledonian history. No plate-tectonic reconstruction has yet satisfactorily explained the the crustal shortening associated with either the Grampian orogeny of the Central Highlands, or the later Caledonian reworking of the N Highlands.

Grade of metamorphism in Highlands to the SE of the Great Glen is largely a product of Grampian orogenesis, but the Central Highland Division gneisses probably owe their mid-high amphibolite facies assemblages to the c. 1000 Ma events which affected the N Highlands. Much of the Dalradian is at mid-amphibolite facies with the metamorphic peak intervening between D1 and D3. Peripheral areas such as the SW Highlands and the Highland Boundary zone are characterised by lower grades being largely at greenschist facies. High structural levels such as those in NE Scotland are also characterised by low metamorphic grade. Traditionally the metamorphic terrain of the Highlands SE of the Great Glen have divided into Buchan (cordierite, andalusite but with staurolite) and Barrovian (almandine, staurolite and kyanite). The boundary between the two is marked by the andalusite-kyanite inversion isograd.

Along the Highland Border metamorphic grade is uniformally low, early tectonic structures almost every where face downwards and the rocks are steeply inclined. It is inferred that this is due to early displacement in the Highland Border fault zone. In many places, adjacent to those steeply inclined rocks, and everywhere tectonically separated from them, there occurs the Highland Border complex which consists of serpentinites, greenstones (spilites), cherts, jasper, black shales and coarse-grained arenites. It is believed that this complex is, at least in part, a dismembered ophiolite complex. It is possibly the site of a former short-lived basin floored by oceanic crust. The latter is consistent with the palaeontological evidence which suggests that the rocks of the complex are too young to have been involved in the Dalradian (Grampian) nappe complex.

Plutonism

The pre-to-syn-tectonic granites and gabbros of the Highlands have been referred to in the foregoing text where their radiometric age has a bearing on the timing of orogenic events. The bulk of Highland plutonic rocks are, however, late-to-post-orogenic and were largely emplaced in the 440 Ma - 400 Ma interval. Many are associated with penecontemporaneous intermediate-basic precursors remnants of which are preserved as xenoliths, synplutonic dykes or high level stockwork of dykes and sheets. In this respect they are largely I-type granites. However, many have characteristics which render them intermediate between I-type and S-type. Notable among these are strong indicators of crustal involvement indicated by $^{86}Sr/^{87}Sr$ ratios rather higher than those typical of I-type granites and by the presence of abundant metasedimentary xenoliths which testify to the stopping mechanism of their emplacement. The later bodies were probably emplaced as ring complexes high in the crust and locally, formed sub-volcanic plutons which fed volcanics including ashflows now interlayered with Old Red Sandstone sediments on the contemporary Upper Silurian - Lower Devonian land surface.

It has been suggested that this plutonism is related to NW directed subduction at the NW margin of Iapetus Ocean while the accretionary prism of Ordovician - mid Silurian sediments accumulated. There are, however, significant time discrepancies between the subduction recorded by the accretionary prism and the magmatism recorded by the plutons. The latter continued to be emplaced after the (Wenlock) youngest slice of the wedge was emplaced and there is a significant lack of mid-late Ordovician granites in the Highlands. Increasing evidence of lateral accretion of the Southern Uplands is also forthcoming (see this volume) and positive correlation between events in the prism as now constituted and plutonism in the Highlands would probably be misguided.

Faulting

While the late Silurian to early Devonian plutonism occurred in the Highlands, clearly overlapping with sedimentation, major faults also breached the contemporary surface to complete an impression of extremely vigorous landscpae with a significant interaction of structure, volcanicity and sedimentation. Major faults trend NE – SW and were predominantly lateral-slip in displacement. Many, having small displacements (3-5 km), such as the Loch Tay Fault, are demonstrably sinistral lateral-slip. Most have had prolonged and complex histories and some like the Great Glen and Highland Boundary faults are still active. Most if not all of these faults, were sinistral lateral slip. Notable among these was the Great Glen Fault the displacement of which was believed to be indicated by the 100 km distance between the Foyers and Strontian granites (Fig. 1.6). These two granites are now believed to be quite separate bodies, of only superficial similarity, having quite different isotopic characteristics. In spite of detailed knowledge of the Great Glen zone, it is becoming clear that no pre-Devonian feature is unambiguously common to both sides of the fault, although displacements of the order of thousands of kilometres are improbable. The Highland Boundary Fault had a major influence on thickness and facies of the Old Red Sandstone of the adjacent Midland Valley. Lower Old Red Sandstone sediments and volcanics were deposited unconformably on already downward-facing Dalradian rocks in the Highland Border zone to the north of the fault.

References

References relevant to this article will be found in the appropriate parts of the Fieldguide.

A GEOTRAVERSE THROUGH THE CALEDONIDES OF WALES

Coordinators: M.G. Bassett and R.E. Bevins
Contributors: M.G. Bassett, R.E. Bevins, W. Gibbons
and M.F. Howells

A. Introduction - The Proterozoic to Lower Palaeozoic marginal basin of Wales (MGB and REB)

B. Field itineraries

   1. The southern platform: Precambrian basement and Lower Palaeozoic marginal marine sedimentation (MGB)

   2. Southwest Wales: Ordovician (Arenig-Caradoc) volcanic, sedimentary and tectonic processes (REB)

   3. Ordovician (Llanvirn) volcanic processes of the southwest margin of the basin; and Silurian (Llandovery) trough sedimentation (REB and MGB)

   4. Central Snowdonia: Ordovician (Caradoc) volcanic processes and environments (MFH)

   5. An eastward traverse across Anglesey: Precambrian basement and the Lower Palaeozoic cover (WG)

C. References

M.G. Bassett, Department of Geology, National Museum of Wales, Cardiff, South Glam., CF1 3NP
R.E. Bevins, Department of Geology, National Museum of Wales, Cardiff, South Glam., CF1 3NP
W. Gibbons, Department of Geology, University College, Cardiff, South Glam., CF1 1XL
M.F. Howells, British Geological Survey, Regional Office for Wales, Bryn Eithyn Hall, Llanfarian, Aberystwyth, Dyfed, SY23 4BY

INTRODUCTION - THE PROTEROZOIC TO LOWER PALAEOZOIC MARGINAL BASIN
OF WALES

This field guide describes critical exposures to illustrate aspects of
the late Proterozoic to early Palaeozoic evolution of the marginal basin
in Wales; the full route is shown in Figure 2.1.

During Lower Palaeozoic times, Wales was underlain by continental crust
(Watson and Dunning, 1979) which, from the evidence of peripheral
outcrops, was composed principally of Upper Precambrian calc-alkaline
igneous rocks (Thorpe, 1979). This basement, which also forms the
Midland Platform (Fig.2.2) can be traced through Newfoundland into the
northeast Appalachians and constitutes the Avalonian terrane of the
Caledonides (Kennedy, 1979). The Mona Complex of Anglesey, bounding the
Avalonian terrane to the northwest, is distinctly different, however, and
the association of flyschoid sediments, mélanges, basic pillow lavas,
jaspers, gabbros, serpentinized ultramafics and lawsonite-glaucophane
schists has been modelled in terms of a subduction system (Dewey, 1969;
Baker, 1973; Thorpe, 1974; Wood, 1974), although Gibbons (1983b, 1984)
has emphasised the additional role of major strike-slip movements.
    Basinal conditions in Wales commenced with early Cambrian volcanism
and sedimentation controlled dominantly by active NE-SW trending faults
(Fig.2.2). Those along and defining the northwest margin of the basin
constitute the Menai Straits Fault System (Kokelaar et al., 1984b; see
also Reedman et al., 1984), and those defining the southeast margin
against the Midland Platform were the Pontesford-Linley and Church
Stretton Fault systems and their extensions to the southwest. These
fault systems, together with similarly oriented and N-S faults within the
basin, are thought to be basement-controlled (Kokelaar et al., 1984b) and
remained active throughout the Lower Palaeozoic, exerting a particularly
strong influence over the igneous and sedimentary history of Wales during
the Ordovician.
    Volcanic activity appears to have been of only very restricted
extent during the Cambrian Period, with evidence restricted to a few
horizons of tuffs and tuffaceous sediments in North and South Wales.
From late Tremadoc through to late Caradoc times Wales was the site of
major volcanic activity. Recent investigations (Kokelaar et al., 1984b;
Bevins et al., 1984) suggest that the marked petrological and geochemical
differences between Tremadoc volcanics and those of Arenig to Caradoc age
reflect an arc to marginal basin transition. The earlier volcanics,
represented by the Rhobell Fawr Volcanic Group of southern Snowdonia, and
the Trefgarn Volcanic Group exposed in north Pembrokeshire, comprise a
fractionation continuum from basalt through andesite to dacite, with
arc-like chemical characteristics. The later suites, characterized by
the Fishguard Volcanic Group, also exposed in north Pembrokeshire, and
the Aran Volcanic Group of southern Snowdonia, are represented by bimodal
basalt-rhyolite sequences. The basic rocks are typical of those
generated in ensialic marginal basin environments, with the rhyolites
probably related to crustal melting. The basin developed on the
southeast flank of the Iapetus Ocean, most probably behind the arc of the
Leinster-Lake District zone. Thin bentonites indicate the persistence of
volcanism during Silurian times, although the eruptive centres may have

A GEOTRAVERSE THROUGH THE CALEDONIDES OF WALES

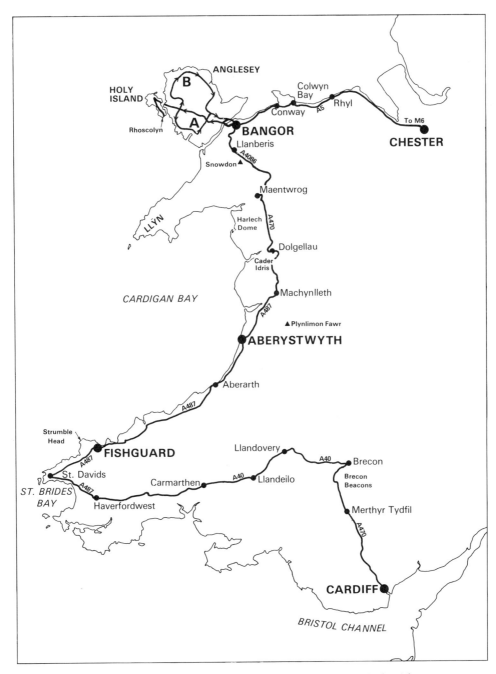

Fig 2.1 Generalised route map through the Welsh Caledonides. Itineraries 1-5.

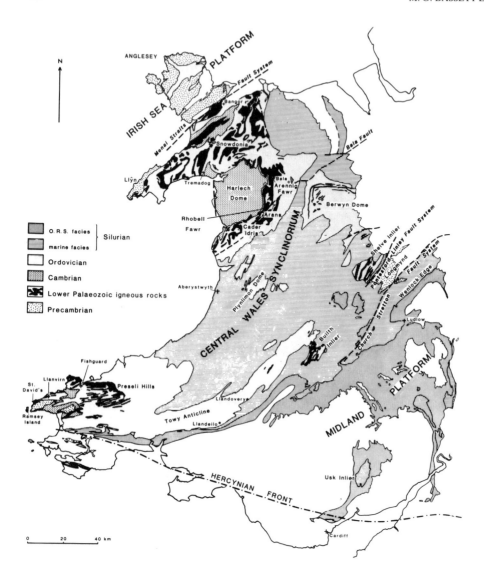

Figure 2.2. Generalised Precambrian and Lower Palaeozoic geology of Wales and the Welsh Borderland showing main structural elements and major features of palaeogeographical importance.

been distal, possibly even lying outside the Welsh Basin area. However, the thick sequence of alkali basalts, hawaiites, mugearites and rhyolites, of probable Llandovery age, exposed on Skomer Island and the neighbouring mainland are clearly proximal to the site of eruption.

The sedimentary sequences of the depositional basin across Wales and adjacent regions reflect the persistent nature of instability, with a marked contrast in facies between turbidites and fan deposits of the trough and its margins and the shallow marine sediments of the bounding platform areas. Finally, the basin was infilled by the spread of Old Red Sandstone molasse facies, which commenced as early as late Wenlock to early Ludlow times in southwest Wales (Walmsley and Bassett, 1976) and was completed by early in the Přídolí.

Pulsatory and cumulative phases of deformation extended at least until early-mid Devonian times to produce the present structural geometry, which appears partly to reflect basement control of a plastic cover. The structure of Wales is dominated by a strongly sigmoidal pattern of mainly upright to steep, SE-verging major folds, with an attendant, broadly axial planar, cleavage. Coward and Siddans (1979) estimated that a maximum shortening of 43 km has taken place across the basin between the Menai Straits and the Welsh Borderland. All of the Caledonian rocks of the Welsh Basin have suffered low-grade regional metamorphism, ranging from approximately zeolite facies in the platform region to greenschist facies in the trough area (Bevins and Rowbotham, 1983).

Figure 2.3 presents a chronostratigraphical classification of the Lower Palaeozoic strata exposed in Wales and the Welsh Borderland, together with the main bases for calibration and correlation; the stratigraphical nomenclature used in this part of the excursion guide is based on this classification. The excursion is described as commencing in Cardiff.

ITINERARY 1.   THE SOUTHERN PLATFORM

Precambrian basement and Lower Palaeozoic marginal marine sedimentation

In a westward traverse across South Wales (Fig.2.1), this itinerary demonstrates the persistent nature of shallow-water marine sedimentation along the southern margin of the Welsh Basin throughout Cambrian-late Silurian times, together with end-phases of infilling by Old Red Sandstone molasse facies. Precambrian basement is exposed only intermittently in this region, but it appears to have been a close southerly source of sediment throughout the early Palaeozoic.

Central Cardiff is situated on flat-lying Triassic red beds of lacustrine, fluviatile and marginal marine facies, but towards the northeast outskirts of the city there is a small Silurian inlier that includes beach and shoreface deposits of Wenlock age, indicative of the early spread of Old Red facies from a southerly source. To the north the route descends firstly onto Devonian terrestrial deposits and then, in a short distance, onto Carboniferous rocks forming the basin of the South Wales Coalfield. North of Merthyr Tydfil, across the Brecon Beacons, the rocks are entirely in Devonian Old Red Sandstone facies; as throughout

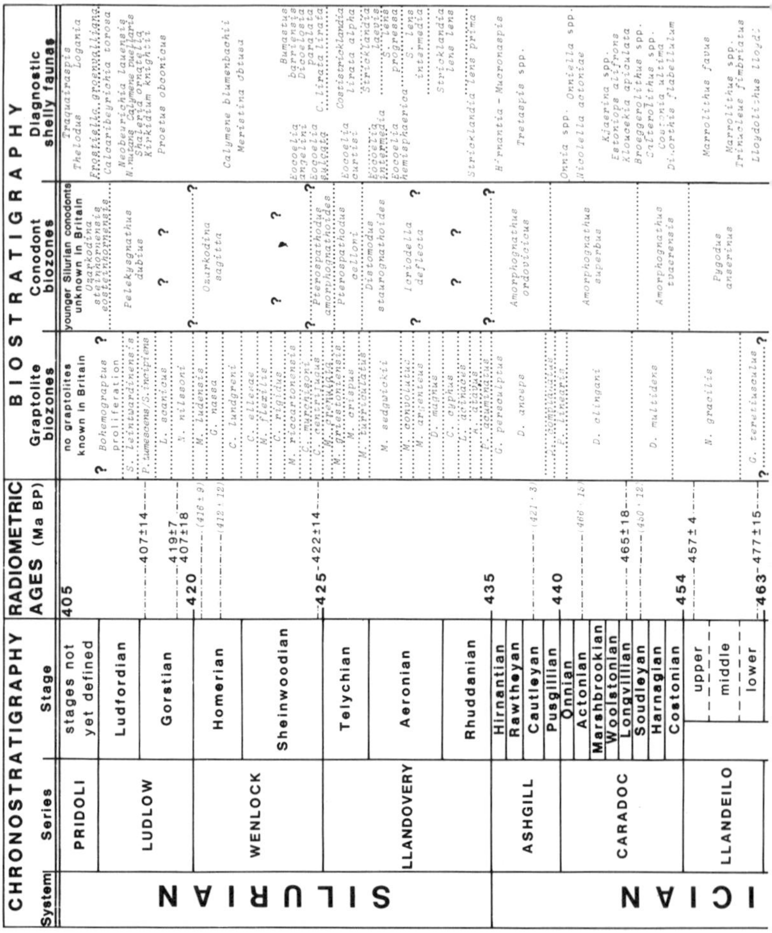

Figure 2.3. Summary of chronostratigraphical classification of Lower Palaeozoic rocks in Wales and the Welsh Borderland, with data for geochronological calibration and the biostratigraphical bases for correlation. Chronometric ages for the base of each System and Series are interpolations using data from many sources worldwide (eg. see Snelling, in press for summary); ages in the centre of this column are restricted to samples from Wales and England whose stratigraphical positions are known within close limits and whose 2σ errors are 2% or less; bracketed ages to the right also refer to stratigraphically well-localised samples from Wales and England but whose analytical errors are considered to be outside an acceptable range (see also McKerrow

et al., in press; these latter ages are shown here to indicate the order of magnitude of all stratigraphically accurate samples from the region). Graptolite and conodont biozones are those proved to be present in Wales and England; most are recognised internationally apart from a few of the Silurian conodont 'zones' that are diagnostic local ranges. The shelly faunas listed (trilobites, brachiopods, ostracodes) are also all taxa that occur in the region; apart from some of the Cambrian trilobites, their distributions are not considered to be of biozonal status but they are used here as examples of the reasonably diagnostic correlative value of such faunas within varying limits of accuracy.

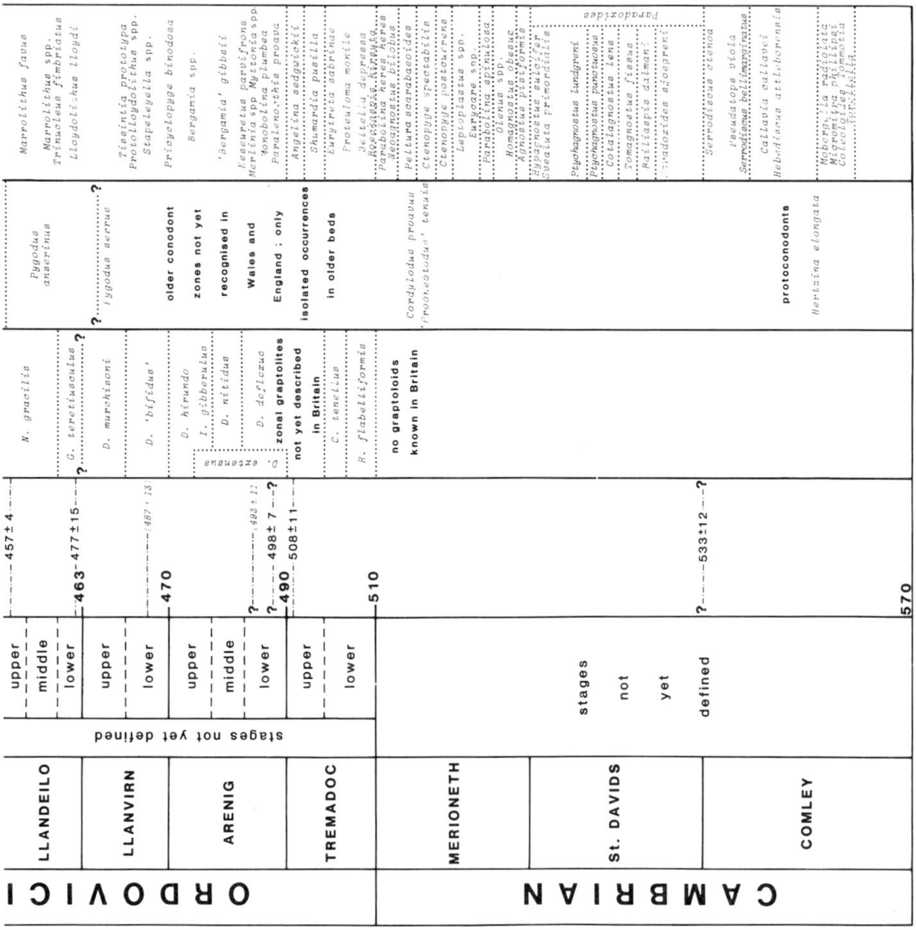

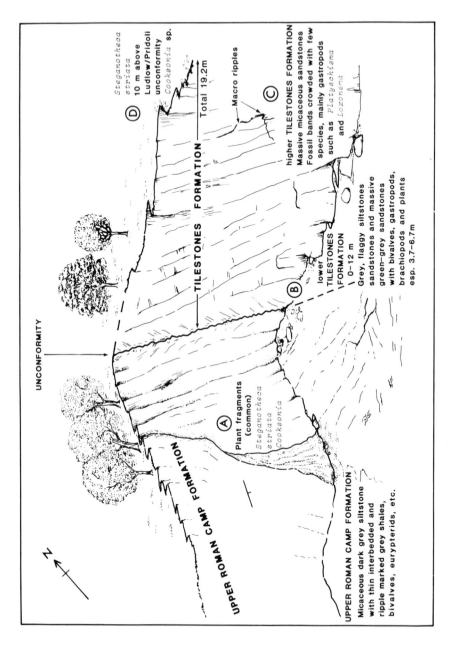

Figure 2.4. Geological sketch of Main Quarry at Capel Horeb (locality 1.1, showing position of unconformity between Ludlow and Přídolí beds (modified from Williams, 1978).

A GEOTRAVERSE THROUGH THE CALEDONIDES OF WALES

the whole of South Wales, middle Devonian beds are missing. From Brecon westwards towards Llandovery the route follows a descending sequence onto the Lower Palaeozoic.

Locality 1.1. Capel Horeb (SN 8444 3238) (Fig.2.4)

This quarry is on the north side of the A40 road 5 km west of Trecastle, Powys. Vehicles must be left on the road. The coarsening-upward succession records a regression from late Ludlow shallow subtidal sediments through unconformably overlying Pridoli shoreface and barrier sand environments to muddy intertidal and supratidal flats.

The Ludlow (Gorstian) Upper Roman Camp Formation forms the main face at the western end of the quarry (locality 1.1A); the rippled siltstones and shales contain a restricted inshore fauna of bivalves, brachiopods, gastropods and eurypterids. The unconformity exposed in the prominent cleft at the eastern end of the main face cuts out uppermost Ludlow beds, with the overlying Tilestones Formation being in Old Red Sandstone facies of Přídolí age. The lowest 12 m of the Tilestones Formation (locality 1.1B) are fine-medium micaceous and quartzitic sandstones, again with a restricted marine fauna in some beds; parallel laminated, rippled and hummocky laminated tabular sandstones represent storm-induced sand sheets of the lower shoreface. Higher beds (locality 1.1C) are increasingly rippled and trough cross laminated. Transport vectors mainly indicate a southerly or southeasterly source.

One of the chief interests in this quarry is the presence in both the Ludlow and Přídolí beds of fragmentary plant axes, including some of the earliest known evidence for vascular tissue, indicative of land plants (Edwards and Davies, 1976; Edwards and Rogerson, 1979) (localities 1.1A, 1.1D). The flora consists mainly of the rhyniophytes Cooksonia and Steganotheca, the former genus being a particularly distinctive member of late Silurian floras over wide areas of the northern hemisphere. The marginal marine to terrestrial environments created by the spread of Old Red Sandstone molasse in response to Caledonian tectonism were among the first to be colonised by higher plants.

Route to locality 1.2

From Capel Horeb the route continues westward through Llandovery and Llandeilo along the southern flank of the Towy 'Anticline' (Fig.2.2) - a persistent lineament of folding and associated faulting which was probably active through much of early Palaeozoic time and appears to have controlled the position of the southeastern margin of the Welsh trough (eg. George, 1963). South of Carmarthen, Precambrian rhyolites are overlain by tuffs that have yielded an Ediacaran fauna of medusoids and trace fossils (Cope, 1977). Arenig rocks in this region show rapid lateral facies changes suggestive of control by basement faults. Continuing west into the former county of Pembrokeshire, magnificent coastal sections expose a wide range of Precambrian to Silurian rocks. From Haverfordwest the route turns north around St Brides Bay to St David's (Figs 2.1, 2.2).

Locality 1.2. Porth Clais (SM 7395 2427) (Fig.2.5)

Small vehicles can be parked in the small car park adjacent to the bridge, but larger vehicles should be left further up the road towards St David's.

Exposures in crags on the hillside above the east side of the quay expose part of the most extensive Precambrian intrusion in the area, the St David's Granophyre (Baker, 1982; Williams and Stead, 1982). This is a highly siliceous, leucocratic granophyre containing orthoclase, perthite and soda-plagioclase, and it is often referred to as 'alaskite'. A small basic dyke cuts the granophyre close to the foot of the cliff path. This episode of intrusion is dated at 650-570 Ma (Patchett and Jocelyn, 1979).

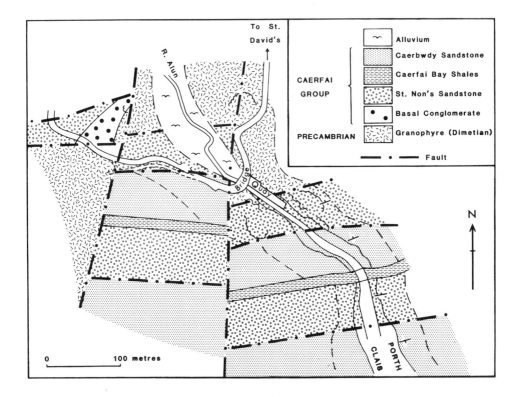

Figure 2.5. Geology of Porth Clais (locality 1.2) (modified after Williams and Stead, 1982).

The Cambrian sequence of Porth Clais is downfaulted against the Precambrian by E-W trending normal faults and displaced by N-S dextral wrench faults. Most of the Cambrian succession is exposed in the harbour, with the uppermost beds of the Caerbwdy Sandstone displayed on magnificent high-angle bedding planes on the southeast headland (SM 7468 2395). These beds are purple and green highly bioturbated micaceous and

A GEOTRAVERSE THROUGH THE CALEDONIDES OF WALES

feldspathic sandstones, with some units showing grading; deposition below wave base is inferred (Crimes, 1970a).

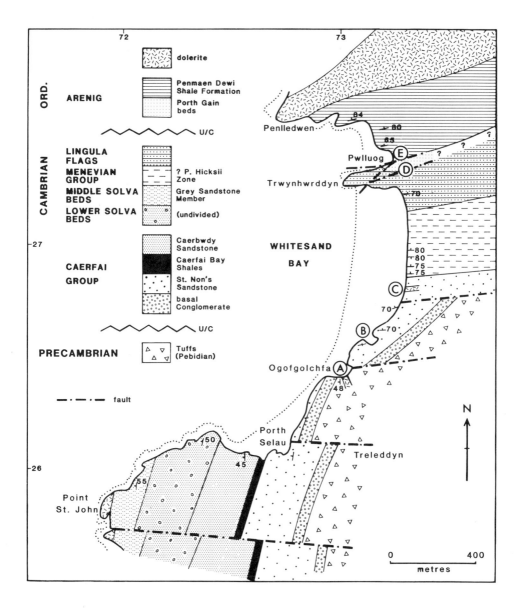

Figure 2.6. Geology of Whitesand Bay (locality 1.3) (modified after Williams and Stead, 1982).

Locality 1.3. Whitesand Bay (Fig.2.6)

There is a large car park above the beach in the north centre of the bay. The southern cliffs expose excellent sections through Precambrian and Cambrian units.

At Ogofgolchfa (locality 1.3A), Pebidian (Precambrian) tuffs and agglomerates are overlain unconformably by basal conglomerates of the Caerfai Group; the contact is locally faulted. Clasts of pebble and cobble grade include vein quartz, quartzites, schistose quartzites, acid tuffs and red argillites set in a red-purple sandstone matrix; the clasts are locally imbricated and become finer upwards, where rippled, scoured and burrowed sandstones become more prominent, consistent with deposition intertidally (Crimes, 1970a).

A few metres to the east, the Precambrian volcanics are intruded by a massive, dark green dolerite.

Green and buff shallow marine sandstones in the cliffs and foreshore at locality 1.3B are intruded locally by thin dolerites. The sediments are strongly bioturbated and occasionally ripple laminated. The beds are strongly disturbed, with bedding/cleavage relationships suggesting the presence of a large, downward-facing and SW-plunging syncline (Williams and Stead, 1982).

Alongside the steps at locality 1.3C (SM 7334 2688) is a well-exposed dolerite intrusion.

The Lingula Flags exposed on Trwynhwrddyn (locality 1.3D) comprise two facies of intertidal and shallow subtidal sediments (Turner, 1977). The first are rhythmically alternating fine, light-coloured sandstones and shales or mudstones, with wavy, rippled and lenticular bedding; these are interpreted as deposits of tidal mud flats. The second are well sorted, whitish quartzose sandstones, in which cross-lamination, convolution, and bedding structures suggest deposition on subtidal sand shoals or in channels to the seaward side of the tidal flats.

The contact between the Lingula Flags and the overlying sandstones of Ordovician age has been the subject of considerable controversy (Jones, 1940; Evans, 1948). Although for the most part it is a faulted contact, the presence of a thin pebble conglomerate at one locality suggests an unconformable relationship between the two rock sequences; sedimentary channelling is common throughout this sequence, including within the contact zone. Faunal evidence for the age of the Ordovician sandstones is not present, but based on lithological correlations with rocks exposed at Abercastle they are probably equivalent to the Porth Gain beds, of mid Arenig age.

On the north side of Trwynhwrddyn, these sandy beds are in sharp fault contact with black, strongly cleaved mudstones of the Penmaen Dewi Shale Formation which are exposed in the cliffs of Pwlluog (locality 1.3E). Trilobites and graptolites can be collected from this sequence, which indicate an upper Arenig age. Interbedded with these black mudstones are fine turbiditic sandstones and occasional white weathering rhyolitic tuffs, along with thin, highly irregular rhyolitic intrusions, indicating the onset of volcanism, which was to become very important during Llanvirn times. The turbidites increase upwards in number, thickness and grain size, suggesting accelerated clastic input that was probably related to the tectonic instability accompanying volcanism.

A GEOTRAVERSE THROUGH THE CALEDONIDES OF WALES   41

ITINERARY 2.   SOUTHWEST WALES

Ordovician (Arenig-Caradoc) volcanic, sedimentary and tectonic processes.

Coastal outcrops in north Pembrokeshire (now part of the county of Dyfed) provide magnificent exposures of the Lower Palaeozoic rocks that comprise the southern part of the Welsh Basin. Volcanic activity was widespread in this area, particularly during early Ordovician times and volcanic and volcaniclastic rocks, along with contemporaneous high-level intrusions, form an important component of the sequence. The volcanic episode was characterized by contemporaneous subaqueous eruption of basic and acid magmas. Graptolite- and trilobite-bearing black muds reflect the background sediments, which probably accumulated in a low energy outer shelf-like environment.

In this area, three important volcanic centres have been identified, namely, i) Ramsey Island, in the extreme west; ii) the Abereiddi district; and iii) to the northwest of Fishguard.

This itinerary describes the various volcanic products and associated sediments exposed in the Abereiddi district, along with an account of the lower part of the volcanic sequence in the Fishguard area. In addition, an indication is given of the major structures of Caledonian age seen in this area.

Exposures at Abereiddi Bay (Figs 2.7, 2.8) reveal a number of volcanic horizons interbedded with strongly cleaved black pelagic mudstones containing graptolite faunas of Llanvirn age. The headland of Trwyncastell on the north side of the bay (beyond the Porth Gain slate quarry) is composed largely of tuffs and tuffites belonging to the Llanrian Volcanic Group of Cox (1915). This lower Llanvirn volcanic sequence is composed of relatively distal tuffaceous volcaniclastic deposits derived from explosive rhyolitic activity. The sequence is dominated by crystal-vitric tuffs and volcaniclastic sandstones and siltstones, thought to have been deposited from sediment gravity flows of both debris and turbidite flow type. Ash fall units are of minor importance.

On the south side of Abereiddi Bay there is another major volcanic horizon, the <u>Didymograptus</u> <u>murchisoni</u> Ash of Cox (1915). In contrast to the Llanrian Volcanic Group the <u>D. murchisoni</u> Ash was derived from a centre erupting basic magma in a moderately explosive manner. The products of this eruption can be examined in the coastal sections west of Abereiddi for a distance of 2 km where there are exposed tuff units derived from both debris and turbidite flows originating from this volcanic centre.

In the area to the northwest of Fishguard, incorporating Strumble Head, excellent coastal exposures provide a magnificent cross-section through the Fishguard Volcanic Group of lower Llanvirn age. The Group, approximately 1800 m thick at its maximum development, shows a wide variety of submarine volcanic products. Three formations have been recognised on the basis of the nature of the volcanic deposits. The lowermost formation (the Porth Maen Melyn Volcanic Formation) is composed principally of rhyolitic lavas, autobreccias, debris flow breccias and bedded and massive tuffs, although rhyodacitic massive and pillowed lavas

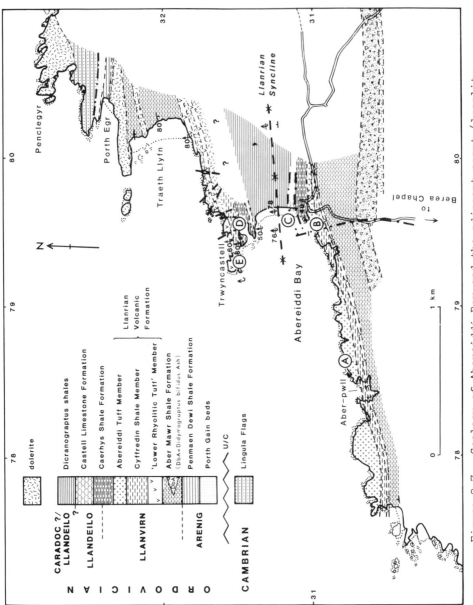

Figure 2.7. Geology of Abereiddi Bay and the adjacent coast (locality 2.1).

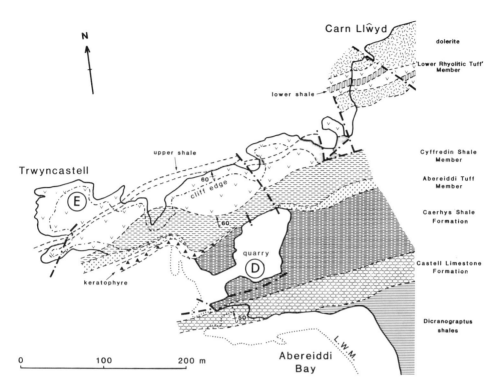

Figure 2.8. Geology of the area around Porth Gain quarry, north side of Abereiddi Bay (localities 2.1D, 2.1E).

also occur. The overlying Strumble Head Volcanic Formation is composed almost entirely of basaltic pillowed and massive lavas with only thin hyaloclastites and hyalotuffs, and very rare rhyolitic tuffaceous horizons. The uppermost formation is described in itinerary 3.

The maps and text for the description of the volcanic rocks of this itinerary and for itinerary 3 are largely adopted from Kokelaar et al. (1984a).

Locality 2.1. Abereiddi Bay and adjacent coast (Figs 2.7, 2.8)

Ample parking space for cars and minibuses is available at the car park (SM 7968 3120) adjacent to the pebble beach at Abereiddi. Coaches, however, should be left at Berea Capel (SM 795 289). Access to the localities both north and southwest of the car park is gained by use of the coastal path.

2.1A. Northeast of Aber Creigwyr (Fig.2.7). The D. murchisoni Ash here comprises two fining-upward sequences of basaltic lapilli tuffs. The dip is c.45° N and the whole 95 m thick lower sequence crops out. The lowermost c.5 m of the upper sequence is preserved as an outlier on the cliff top. The base of each sequence is sharp and planar and that of the upper sequence rests conformably on the finer top of the lower.

The tuffs are composed mostly of angular basaltic clasts, originally glassy and porphyritic but now deformed and highly altered. Sorting is poor with scoriaceous lapilli and sparsely vesicular blocks and rare bombs, up to 40 cm in diameter, supported by a matrix of fine to coarse tuff. The lower sequence is poorly to moderately bedded, with successive massive beds being of finer grade and with more clearly defined coarse-tail grading. The upper sequence is also graded but not obviously bedded. The tuffs are interpreted as deposits from debris-flows and high-density turbidity currents.

2.1B. Melin Abereiddi (Fig.2.7). Here the D. murchisoni Ash is c.60 m thick; the lower sequence is 52 m and the upper 8 m. The base of the lower sequence (exposed in the valley side just above and south of the waterfall) rests on muddy tuffites. The lapilli tuffs near the base, like those at locality 2.1A, are poorly sorted, show crude coarse-tail grading, and bedding is ill-defined. Upwards the tuffs are finer and become thinly bedded, as seen in the outcrops on the foreshore. The base of the upper sequence is marked by a bed of lapilli tuff and the sequence above is of well-bedded fine to coarse tuffs and lapilli tuffs. Bed thickness varies from c.3 to 25 cm, with the thicker beds generally the coarser. Most bed bases are planar although in places erosive contacts can be determined. Some beds show normal grading, but most lack regular variation in grain size. Parallel lamination and ripple drift cross-lamination also occur and the tuffs are interpreted as deposits from both high- and low-density turbidity currents.

The tuffs are overlain by an alternation of thinly-bedded turbiditic tuffs with mudstone intraclasts and muddy debris-flow deposits with tuff clasts. These are succeeded by 2 m of mudstone with horizons of sparse tuff clasts, and then typical black, richly graptolitic (D. murchisoni) mudstones.

The tuffs become thicker, coarser and less well bedded towards the west, and the section is interpreted as the thin distal edge of a submarine volcanic pile. The tuffs were deposited by various sediment-gravity flows from tuffs that had previously been deposited close to the vent. Instability in the accumulating tuffs caused periodic slumping and transport of the tuffs to greater water depths. The improvement in bedding and general fining upwards in the lower unit probably reflects a decreasing rate of effusion, such that slumps became progressively smaller and dispersal of coarse material more limited. A temporary return to a relatively high effusion rate is reflected in the coarse base of the upper unit. Muds subsequently mantling the pile periodically slumped and were redeposited with, and eventually without, further turbiditic tuffs.

2.1C. South-centre of Abereiddi Bay (Fig.2.7). The centre of Abereiddi Bay is occupied by dark shales of Llandeilo-Caradoc age; the dip of the beds is consistently at high angles (60°-85°) to the north, but bedding-cleavage relationships indicate that the northern half of the bay is on the inverted limb of a syncline - the Llanrian Syncline (Fig.2.7). Cleavage is approximately axial planar throughout, dipping northwards at some 5° less than bedding to the north of the axial trace of the main fold (Fig.2.7), and northwards at some 12°-18° more than bedding to the south of this trace (Waltham, 1971; Black et al., 1971; Rickards, 1973). The whole of the sequence is closely faulted. South of the axial trace the sequence descends into 9 m of calcareous and decalcified beds considered to equate with the Castell Limestone Formation north of the bay (Hughes et al., 1982). Rare trilobites occur in these beds. The underlying Caerhys Shale Formation in the southernmost corner of the bay (SM 7965 3104) yields abundant graptolites, particularly Didymograptus murchisoni; thin tuffs occur throughout the succession.

2.1D. Porth Gain, north side of Abereiddi (Fig.2.8). The path and steep faces around the south side of Porth Gain quarry expose a richly fossiliferous, inverted sequence through the Caerhys Shale Formation and Castell Limestone Formation. The boundary between the Llanvirn D. murchisoni Biozone and the lower Llandeilo G. teretiusculus Biozone is within the Caerhys Shale Formation (Hughes et al., 1982); these beds are dark tuffaceous shales and siltstones formed in low energy environments. A prominent graded tuff occurs some 9 m above the base of the shales, probably an air-fall deposit. Where the path narrows at the entrance to the quarry (SM 7952 3142) trinucleid trilobites and graptolites are fairly common near the base of the Castell Limestone.

Exposed in the steep northern face of Porth Gain Quarry is a thick intrusive sheet (termed 'keratophyre' by Cox (1915)). The sheet invaded the sediments while they were still water-logged and unconsolidated, resulting in the production of a series of large, irregular, pillow-like bodies. Thin sections reveal that the rock is extensively altered, being composed predominantly of albitised plagioclase feldspar, along with abundant chlorite.

At the very foot of the cliff, a sequence of basic tuffs is exposed, considered to be a much thinned, lateral equivalence of the D. murchisoni Ash as exposed on the south side of the bay.

2.1E. Trwyncastell (Fig.2.8). On the headland of Trwyncastell, to either side of the tower, various rhyolitic tuff units (of the 'Lower Rhyolitic Tuff' Member) are exposed, again which can be correlated with equivalent strata on the south side of the bay. Immediately below the tower a thin, rhyolitic tuff crops out which is interbedded with sediments. It is composed predominantly of undeformed glass shards, along with rarer quartz and feldspar crystals. Exposed in the steep gulley to the west of the tower, occurring stratigraphically below the above-mentioned tuff, and showing clear evidence of inversion is a thin, graded volcaniclastic sandstone unit, composed predominantly of quartz and feldspar crystals set in a fine, siliceous matrix, along with mud clasts incorporated in the coarse basal part. On the steep, north-facing coast of Trwyncastell are further exposures of graded, rhyolitic tuffs, composed chiefly of shards, along with minor quartz and feldspar crystals, and again providing clear evidence of the inverted nature of the strata on the northern, overturned limb of the Llanrian syncline.

Locality 2.2. Porth Maen Melyn (Fig.2.9)

The starting point for this locality is a small car park above Pwllderi (SM 8936 3862). However, access to this car park is suitable for cars only and coaches must be left at the lay-by at Tref-Asser Cross (SM 8968 3765). Access to the coastal outcrop of this itinerary is gained by following the coastal footpath, past the youth hostel.

2.2A. Lower part of the Porth Maen Melyn Volcanic Formation at Porth Maen Melyn (Fig.2.9). Porth Maen Melyn is eroded into soft, cleaved mudstones, probably of the lower Llanvirn D. bifidus Shales (Cox, 1930). No diagnostic fossils have been found. An intrusive dolerite sheet forms the promontory at the northern end of the bay and can be examined in the crags at the top of the cliff.

The lowermost unit of the Porth Maen Melyn Volcanic Formation crops out in prominent crags adjacent to the coastal path on the north side of a small valley. It comprises 10 m of bedded rhyolitic tuffs. Extensive recrystallization has resulted in the development of a peculiar spherulitic texture which is only readily observable on weathered surfaces. The spherules, up to 2 cm in diameter, and 'cigar-shaped' nodules up to 3 cm in length, have obliterated the original vitroclastic fabric.

Overlying these tuffs is a 35 m thick sequence of white-weathered lithic-crystal-vitric breccias and buff crystal-vitric tuffs (SM 8882 3935). Two units can be identified, both generally fining upwards although the lowermost 50 cm of the lower unit is inversely graded. The breccia clasts are predominantly of angular to sub-rounded rhyolitic lava, commonly with a perlitic texture, although in the lower part of the lower unit basalt and dolerite clasts also occur. Crystals are of plagioclase and quartz (typically bi-pyramidal) and increase in proportion in the fine tail. Shards and shard fragments are abundant and weathered-out streaky clasts probably represent pumices. This sequence of tuffs and breccias reflects sediment-gravity flow reworking of primary products of both explosive and quiet effusion of rhyolitic magmas.

2.2B. Upper part of Porth Maen Melyn Volcanic Formation and lower part of Strumble Head Volcanic Formation (Fig.2.9). Overlying, and partly loaded into, the clastic deposits of locality 2.2A is a massive rhyodacitic lava flow, 40 m thick (SM 8876 3936). The lava is grey to green and cryptocrystalline or spherulitic with a well-developed perlitic texture. Individual perlites reach 1 cm in diameter and in the central parts of the flow flow-banding is locally developed. Rare plagioclase micro-phenocrysts are present, but otherwise the flow is thought to have been originally glassy. The top of the flow is autobrecciated, with clasts up to 5 cm, and is overlain by a thin horizon of basic hyaloclastite of the Strumble Head Volcanic Formation (SM 8872 3938). This is overlain in turn by epidotized, pillowed basaltic lava. Well developed radial joints can be observed in a number of pillows which

A GEOTRAVERSE THROUGH THE CALEDONIDES OF WALES

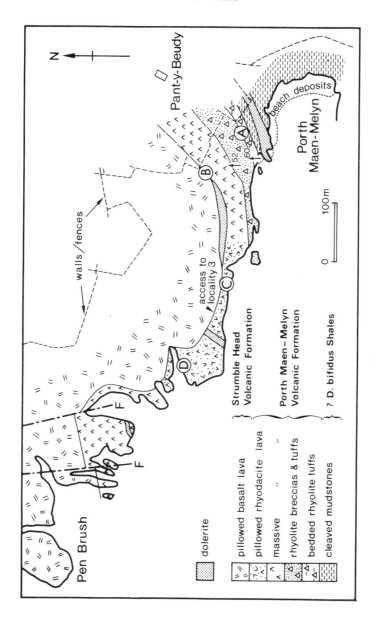

Figure 2.9. Geology of the coast north of Porth Maen Melyn (locality 2.2).

generally also possess highly vesicular margins. Pumpellyite, epidote, albite and chlorite of metamorphic origin are abundant, although augites are pristine and pseudomorphs of igneous textures are generally preserved.

2.2C. Upper part of the Porth Maen Melyn Volcanic Formation (Fig.2.9).
The massive rhyodacite lava of locality 2.2B passes laterally into pillows and elongate tubes and then into isolated-pillow breccias. Excellent exposures occur in the S-facing cliff section above a narrow wave-cut platform, some 300 m west of Porth Maen Melyn. Access to this platform is possible at only one place, and even here the descent is hazardous and the greatest care is necessary. The pillows, 1-3 m in diameter, are separated from each other by cleaved inter-pillow breccia generated by desquamation of the pillow margins. The rhyodacite is purple to green in colour and is petrographically similar to that at locality 2.2B, although here distinctive spherulites are present at pillow margins and within the inter-pillow breccias. These spherulites are particularly rich in K-feldspar and are thought to have been produced during initial crystallization of the lava, in contact with seawater. The tubes and pillows result from the rapid effusion of hot magma at the steep front of the lava flow. Photographs and a full description of the rhyodacite pillows and pillow breccias are given in Bevins and Roach (1979).

2.2D. 500 m WNW of Porth Maen Melyn (Fig.2.9). From the cliff top, the contact between the pillowed rhyodacitic lavas and the basaltic lavas of the Strumble Head Volcanic Formation can be recognized along the northern side of the E-W trending inlet. The lighter-coloured rhyodacite pillows are noticeably larger than the more basic ones.

ITINERARY 3

Ordovician (Llanvirn) volcanic processes of the southwest margin of the basin; and Silurian (Llandovery) trough sedimentation.

The first part of this itinerary represents a continuation of itinerary 2, examining the upper part of the volcanic sequence of the Fishguard Volcanic Group exposed in the northeast of Pencaer. This section illustrates a complex interdigitation of pillowed and massive basaltic and dacitic lavas within the upper part of the Strumble Head Volcanic Formation and the predominantly rhyolitic volcanics of the overlying Goodwick Volcanic Formation. The latter formation here includes rhyolitic lavas and associated autobreccias, rhyolitic ash-flow tuffs, rhyolitic tuffs and tuffites, as well as a thick, bedded, basic hyalotuff horizon and high-level contemporaneous basic intrusions.

Locality 3.1. Pencaer (Figs 2.10, 2.11)

Access to the localities described here is gained by following the coast path from the northwestern end of the road (SM 9491 3917) at Harbour

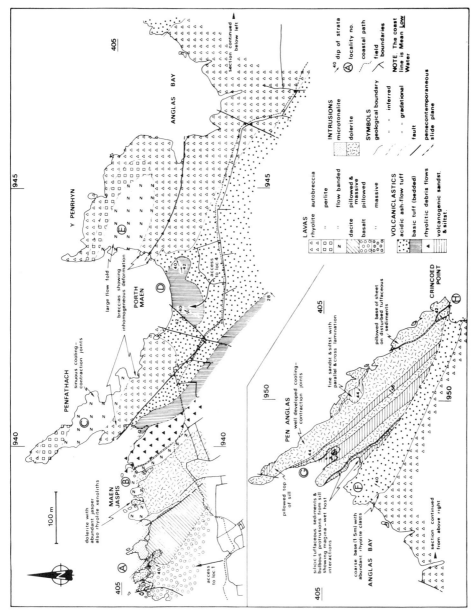

Figure 2.10. Geology of northeast Pen Caer (locality 3.1).

Village, Goodwick. Parking for two or three cars is generally available at the end of the road; coaches can be parked on the road and there is adequate turning space.

3.1A. 100 m west of Maen Jaspis (SM 9375 4047) (Figs 2.10, 2.11).
Basaltic pillow lavas, dipping 40° NNE, form dip-slopes down to a few metres above the high water mark. A thin veneer of fine volcanogenic sediments is locally preserved on this pillowed surface. In the west the sediments are overlain and locally disturbed by a 1.7 m sheet of basaltic lava, with pipe vesicles at its upper and lower margins. The generally smooth top of this lava is mostly overlain by further thin, fine sediments although hollows are infilled with coarse grit and sand. Towards the east this lava sheet passes laterally into vesicular pillow lava which wedges-out up-dip, and further east the pillows become smaller and intensely cleaved. Jasper is abundant, in places forming a geopetal partial fill in drained pillows. The irregular surface at the top of the lava is partly subdued by an infill of fine silicic siltstone, which in turn is overlain by up to 1 m of interbedded acidic and basic turbiditic sandstones and further extremely fine silicic siltstones. Locally these sediments are intruded by a thin, pillowed basaltic sill.

This sequence is overlain and locally disturbed by large-pillowed plagioclase-phyric dacite lava, up to 12 m thick, which is well exposed in the cliffs. To the east the pillows are less well developed, and in the west the lava wedges-out on top of a wedge of intrusive dolerite. The dolerite contains well-formed, curved columnar joints and has been intruded by a small body of dacite. The pillowed dacite lava is overlain by thin porcellanous sediments and the strongly draping base of an indistinctly banded sill, 12 m thick, of porphyritic dacite. Towards its top, on a small promontory, the sill contains prehnitic spherules. To the east, across a small inlet, the dacite sill is overlain by thin sediments and a rhyolite lava with pronounced flow-folded banding. The latter is cut and partly underlain by another dolerite sill.

This locality illustrates the extremely complex interdigitation of lavas and associated intrusions at the junction between the basaltic Strumble Head Formation and the predominantly rhyolitic Goodwick Formation. The abundance of intercalated sediments at this horizon, including some which may in part be chemically precipitated cherts, shows that basaltic activity waned considerably before the onset of rhyolitic activity. The presence of silicic turbidites, with no known local source at this level, indicates that the basaltic Strumble Head Formation did not form a marked positive topographic feature, but rather filled a depression. Fissure eruptions along faults defining a possible graben are envisaged. Weak horizons of intercalated sediment facilitated penecontemporaneous intrusions.

3.1B. Maen Jaspis (SM 9388 4048) (Fig.2.10). The flow-banded rhyolite, in places spherulitic, is brecciated near its base where it has locally disturbed underlying laminated and cross-laminated fine to medium silicic volcaniclastics. Beds of pumice with normal grading also occur. To the east the thin autobrecciated top of the rhyolite is overlain by a veneer of silicic volcanogenic sediment and several massive debris-flows of

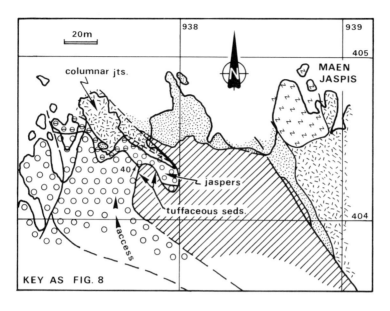

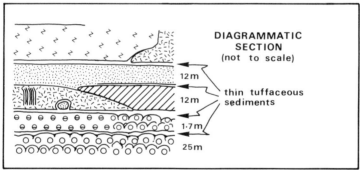

Figure 2.11. Detailed geology 100 m west of Maen Jaspis (locality 3.1A). Key as Figure 2.10.

rhyolitic breccia, which in turn are overlain by a sequence of basaltic tuffs and acidic ash-flow tuffs, although these are best examined at locality 3.1D.

3.1C. Penfathach (Fig.2.10). On the Penfathach headland (SM 9406 4053) various facies of a thick rhyolite lava flow (or small lava dome) are well exposed. The lowest facies comprises autoclastic breccias, with angular, flow-banded clasts averaging 4-5 cm and showing minimal post-brecciation movement. The passage into the overlying massive flow-folded banded lava is sharply gradational. Large- and small-scale

flow folds occur and columnar cooling joints are well developed. The
white-weathered rhyolite contains sparse plagioclase phenocrysts and
spherulitic quartzo-feldspathic intergrowths in a dark-grey to green
groundmass of similar mineralogy, including accessory sphene, zircon and
apatite. Perlitic rhyolite, with fractures up to 10 cm in diameter,
overlies the massive facies and reflects the originally glassy nature of
the flow. The contact with the overlying carapace of autobreccia is
gradational.

During late Caledonian tectonism, strain developed inhomogeneously
in the rhyolite so that intensely cleaved zones alternate with virtually
undeformed zones.

3.1D. Porth Maen (SM 9435 4036) (Fig.2.10). At the foot of the cliffs
on the southeast side of Porth Maen is a sequence of generally
parallel-bedded basaltic tuffs showing both normal and inverse grading,
in beds up to 2 m thick. Average clast size in the coarser beds is 2-3
mm. Maximum basaltic clast size in coarser beds is c.1.5 cm, although
rarer rounded to subangular rhyolite clasts up to 6 cm also occur. There
are no bomb sags and the deposits are interpreted as those of
sediment-gravity flows which were contemporaneous with a nearby eruption.
Soft-tuff faulting during accumulation is evident. In thin section the
basic clasts are seen to be vesicular and composed of plagioclase
microlites in chlorite with sphene, presumably after glass.

At the east end of the bay the basaltic tuffs are succeeded by
rhyolitic ash-flow tuffs containing clasts of pumice and lava. The
latter are well exposed in the cliffs on the faulted west side of the
bay, and in fallen blocks.

3.1E. Y Penrhyn (Fig.2.10). This is the eastward continuation of the
rhyolitic lava described at locality 3.1C. On the west of the headland a
large-scale flow-fold is exposed (SM 9440 4052) and towards the southeast
the 'core facies', of massive and perlitic rhyolites, passes into
autobrecciated rhyolite which constitutes the entire thickness for <700
m along strike.

3.1F. East side of Anglas Bay (Fig.2.10). Approximately 1.5 m of
coarse, lithic-rich, ash-flow tuffs rest on the rhyolite autobreccia.
The lithic blocks, up to 30 cm, are of the underlying rhyolitic lava
proving its extrusive (subaqueous) emplacement. The coarse, basal unit
is overlain by crystal-, lithic- and fiamme-rich tuffs, with fiamme
mostly 10 cm across, but up to 50 cm, commonly moulded around lithic
clasts of rhyolite. The tuffs fine through the succeeding 50 m and are
crystal-rich at the top of the sequence. These ash-flow tuffs are
succeeded by silty mudstones, fine sandstones and fine silicic tuffs
which have been intruded by a series of basic sills at a shallow level.
Close examination of locality 3.1H shows that the contact is a
penecontemporaneous wet slide plane.

3.1G. Pen Anglas (Fig.2.10). Bulbous protrusions and pillows along the
contacts of the dolerite sills, and convolution, homogenization and
vesiculation of the host sediments and tuffs indicate shallow-level

A GEOTRAVERSE THROUGH THE CALEDONIDES OF WALES 53

intrusion into unlithified wet deposits. Polygonal jointing is well developed locally.

3.1H. Crincoed Point (Fig.2.10). Here the basic sheets are pillowed with intervening fine sediments. Along a ledge just above high-water mark the sediments underlying a pillowed body are in contact with the subjacent ash-flow tuffs. The sediments and tuffs are locally interleaved and contorted and at one point silts have been injected into the tuffs suggesting sliding of wet, unlithified materials. Clearly sedimentation, intrusion and sliding were penecontemporaneous.

Route to locality 3.2

Northeastwards from Fishguard, the Ordovician volcanic rocks are overlain by Caradoc - Ashgill sediments that then pass up into Silurian rocks near New Quay. From there the whole of the coastline of Cardigan Bay northwards to Aberystwyth is occupied by thick turbiditic sandstones with interbedded mudstones that make up the well known Aberystwyth Grits of Llandovery age (Wood and Smith, 1959). Together with the underlying Ordovician sediments, the Aberystwyth Grits provide evidence of repeated emplacement of submarine fans across the low energy muds of the Welsh trough. Changes in bed thicknesses and internal structure, grain size, grading, sole structures and arenite/lutite ratios all point to increasing distality of environment from south to north along the axis of the trough.

Locality 3.2. Aberarth (SN 480 642 to 498 654)

Large vehicles should be left on the main road close to the centre of Aberarth; the beach is reached by the minor road and footpath north of the stream. The section begins some 150 m to the north along the beach. Turbidites here are in mid fan facies, in beds of about 30-50 cm thickness but with some extensive units reaching up to 1 m. Some of the thick beds show spectacular intraclasts and large 'rafts' of included sediment, with numerous examples of slurrying and slump folding. Sole structures are common throughout, particularly well seen in a shallow cave along this section.

Some 360 m to the north the thick turbidites are faulted against thinner, finer beds, but in turn these are succeeded northwards by thick units that preserve large crescentic and horse-shoe shaped flute casts, and climbing ripples.

Locality 3.3. Aberystwyth, north end of promenade (SN 583 825) (Fig.2.12)

The shore platform and cliffs running northwards from the end of the promenade at Aberystwyth display well exposed tectonic and sedimentary structures in the Aberystwyth Grits. Most of the information here is based on Bates (1982).

Features of turbidites indicate distal environments throughout this section, with individual beds relatively thin and finer grained than at

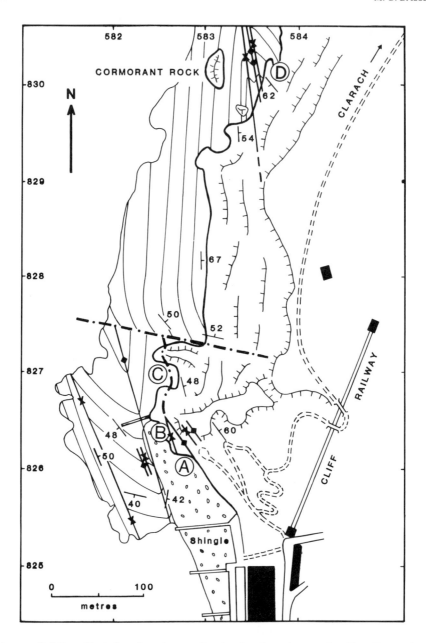

Figure 2.12. Sketch of main geological structures in the Aberystwyth Grits exposed in coastal platform and cliffs between the Promenade and Cormorant Rock, Aberystwyth (locality 3.3).

Aberarth; slurrying and intraclasts are rare, but bottom structures are common together with convolute, parallel and ripple drift lamination. The first exposures to the north of the promenade, before the breakwater, contain well preserved trace fossil associations on the undersides of beds, particularly the network feeding burrows assigned to Paleodictyon which is considered to indicate a distal, deep water environment (Nereites facies of Crimes, 1970b). Between locality 3.3A and locality 3.3B (Fig.2.12) the easterly dip is interrupted by a fold pair that passes along strike into an E-dipping thrust, exposed in the cove at locality 3.3C. Sole structures are well exposed in the roof of a small cave south of the breakwater at locality 3.3B.

Access to the section at locality 3.3C and further to the north is dependent on the tide. The above mentioned thrust strikes N-S across the cove at locality 3.3C, running through the cave on the north side; sole structures are well preserved here. Further north, the cliffs are set back to the east by a prominent E-W tear fault with a sinistral movement of some 75 m; towards locality 3.3D the beds occupy a complex, faulted synclinal zone with ripples whose origin has been suggested to be tectonic (eg. Davies and Cave, 1976). The latter authors have suggested that much of the faulting and cleavage in the Aberystwyth Grits is a result of soft sediment sliding and décollement, although others have argued against this view in favour of a 'conventional' tectonic origin (eg. Lisle, 1977; Howells et al., 1977).

ITINERARY 4. CENTRAL SNOWDONIA

Ordovician (Caradoc) volcanic processes and environments

Route to locality 4.1

The route northwards from Aberystwyth to Machynlleth (Fig.2.1) mainly traverses Silurian (Llandovery) strata, comprising distal turbiditic sandstones and siltstones of the Aberystwyth Grits, and pelagic graptolitic mudstones. North of Machynlleth, there is a conformable Ordovician - Silurian transition just south of Corris, and from there to the Tal y llyn valley, which lies along the Bala Fault, the route cuts down through Ashgill sediments to near the base of the Caradoc.

The S-facing flank of Cader Idris, on the north side of the Tal y llyn valley, is composed of the Aran Volcanic Group, of pre-Caradoc Ordovician age. This group crops out on the south, east and north sides of the Harlech Dome, in southern Snowdonia. Approaching Dolgellau, the disposition of the Cambrian rocks in the core of the Harlech Dome and the Aran Volcanic Group in the scarp feature of Cader Idris should be visible.

Northwards from Dolgellau to Trawsfynydd the route crosses Cambrian strata on the east limb of the N-S anticline. The prominent hills to the west lie in massive turbiditic sandstones of the Rhinog Formation (Lower Cambrian).

From Trawsfynydd the route ascends through the Upper Cambrian sequence. To the north, the Moelwyn Mountains form a prominent scarp in lower Ordovician rocks on the north side of the Vale of Ffestiniog. The

scarp includes the feather edges of the Aran Volcanic Group. At Blaenau
Ffestiniog the slate waste is from excavations in the lower Ordovician.
From here the route runs along the southeast side of central Snowdonia to
Dolwyddelan.

Central Snowdonia

The rocks of central Snowdonia are of Ordovician age. The pre-Caradoc
sequence is composed dominantly of siltstones, with impersistent
sandstones. There is no indication of the Aran Volcanic Group of
southern Snowdonia. In the Caradoc two volcanic groups have been defined
(Howells et al., 1983), a lower, Llewelyn Volcanic Group and an upper,
Snowdon Volcanic Group (after Williams, 1927). The former crops out
mainly in north Snowdonia, whilst central Snowdonia is dominated by the
Snowdon Volcanic Group.

The Snowdon Volcanic Group has figured largely in the description
and interpretation of Ordovician volcanism in North Wales (see Fitch,
1967 for a review). The association of the group with marine sediments
led workers to conclude that the extrusive volcanic rocks had been
emplaced in a marine environment. Rast et al. (1958) recognised that the
dominant acidic lithology of the group was of ash-flow tuff and, as it
was then considered that such eruptions and emplacement could only occur
subaerially, the Caradoc palaeogeography was reappraised. A subaerial
volcano complete with caldera, caldera fault and rim syncline was
proposed by Rast (1969), Bromley (1969) and further refined by Beavon
(1980). However, Howells et al. (1973) and Francis and Howells (1973)
proposed that acidic ash-flow tuffs could be emplaced subaqueously and
still retain sufficient heat to weld. Subsequently this proposal was
supported by Sparks et al. (1980a, 1980b).

The Lower Rhyolitic Tuff Formation (LRTF) is the lowest formation of
the Snowdon Volcanic Group. It comprises mainly acidic ash-flow tuff and
its thickness varies markedly. This excursion is to examine the
formation and its environmental context at two localities. The
variations in the ash-flow tuff are related to different positions with
respect to the eruptive centre and the depositional setting.

Locality 4.1. Capel Curig (SH 7270 5760) (Fig.2.13)

The Cwm Eigiau Formation, the strata subjacent to the Snowdon Volcanic
Group, is traversed following a route to the northeast side of Capel
Curig village (British Geological Survey 1:25 000 Sheet SH75; Howells et
al., 1978). The sequence includes marine siltstones and sandstones with
basaltic tuffs (exposed at locality 4.1A), acidic airfall tuffs
(localities 4.1B and 4.1D) and debris flow breccias (locality 4.1C). The
base of the Lower Crafnant Volcanic Formation conformably overlies
cleaved fossiliferous siltstones.

The Lower Crafnant Volcanic Formation here comprises acidic ash-flow
tuff, c.50 m thick (Howells et al., 1973). The basal zone (locality
4.1E), c.2 m thick, is well cleaved, feldspar crystal-rich with
occasional small clasts, including fragments of brachiopods and
trilobites. The tuff of the massive central zone (locality 4.1F)
comprises devitrified, recrystallized, non-welded shards, and feldspar

crystals in a matrix of devitrified volcanic dust. Clasts are few and mainly of ragged tubular pumice. In the lower part, weathered-out carbonate nodules are common. Within this central zone a crude internal foliation is accentuated slightly by chloritic laminae. The top of the tuff is marked by <1 m of fine grained acidic dust tuff and is overlain by marine siltstones.

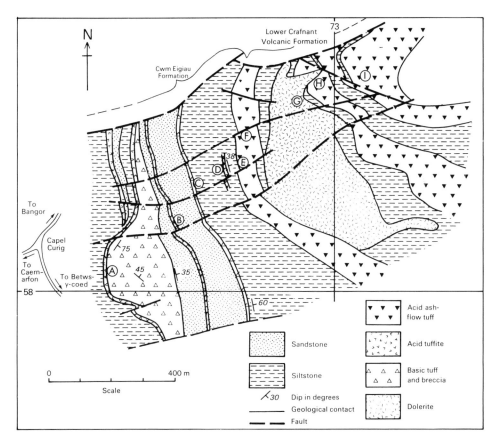

Figure 2.13. East of Capel Curig (locality 4.1). Part of British Geological Survey 1:25 000 Sheet SH 75.

The tuff is the deposit of a single discrete ash flow. The included fossil clasts in its base indicate that it was transported in a marine environment. The crude internal foliation is interpreted as reflecting the development of sub-flow units during transport which became progressively more pronounced distally. The fine grained dust tuff at the top of the flow represents the elutriation of fine dust into the water column during flow transport and subsequent settling on emplacement. The similar siltstones above and below the ash-flow tuff indicate that the emplacement of the tuff had little effect on the depositional environment.

It is proposed (Howells et al., in press) that this ash-flow tuff escaped from an eruptive centre in central Snowdonia. From this locality, on the northeast side of Capel Curig, the thick (>500 m) proximal accumulation at the eruptive centre can be seen in the scarp of Lliwedd on the south side of the Snowdon massif, 11 km to the west. Here the thick ash-flow tuff shows no internal bedding and is interpreted as having accumulated in a NW-SE aligned rifted depression which developed synchronously with the eruption.

The overlying siltstones are intruded by a dolerite sheet, exposed at locality 4.1G. To the northeast, at localities 4.1H and 4.1I, two further ash-flow tuffs crop out, which are similar in character to that seen at localities 4.1E and 4.1F.

Route to locality 4.2

The route from Capel Curig to Ogwen Cottage crosses the axial trace of the main Snowdon syncline, here trending NE-SW. On its NW side the complementary Tryfan anticline and Cwm Idwal syncline are defined clearly by the outcrops of Capel Curig Volcanic Formation ash-flow tuffs and subjacent sandstones to the south of the road.

Locality 4.2. Cwm Idwal (SH 645 590) (Fig.2.14)

The Cwm Eigiau Formation of this area comprises massive and flaggy bedded, coarse to fine grained, cross and parallel laminated sandstones, which indicate a high energy tidal and shallow subtidal environment (Howells et al., in press). The sandstones include thin flaggy beds of acidic, water-settled, dust tuffs (exposed at locality 4.2B), tuff-turbidites and, locally, basalts, which are commonly pillowed, with associated hyaloclastites. Also at this locality the most northerly outcrop of the Pitt's Head ash-flow tuff member, which was erupted from a centre at Llwyd Mawr (Roberts, 1969) c.20 km to the southwest, crops out (locality 4.2A). Further to the southwest, at locality 4.2C, the Pitt's Head Tuff is exposed in roches moutonnées and is intensely silicified, with accentuation of the eutaxitic planes and well developed quartz nodules.

About the syncline in the cwm the LRTF lies conformably and disconformably on the underlying strata. To the south, in central Snowdonia, this contact is locally markedly unconformable. The LRTF is well exposed on the southeast limb of the syncline in the back wall of the cwm. When viewed from a distance the dominant feature is its well bedded character. Closer examination shows that it comprises pyroclastic breccia (locality 4.2D), primary acidic ash-flow tuffs, reworked acidic ash-flow tuffs, sloughed debris-flows of acidic ash-flow tuffs with incorporated sediments (locality 4.2G), a thick rhyolite flow (locality 4.2F), tuffaceous mudstones (locality 4.2E) and intercalated marine siltstones with thin sandstone beds. The pyroclastic breccias and primary ash-flow tuffs are concentrated mainly near the base of the formation and upwards there is a progressive increase in the influence of the depositional environment, with the reworked and remobilized tuffs and

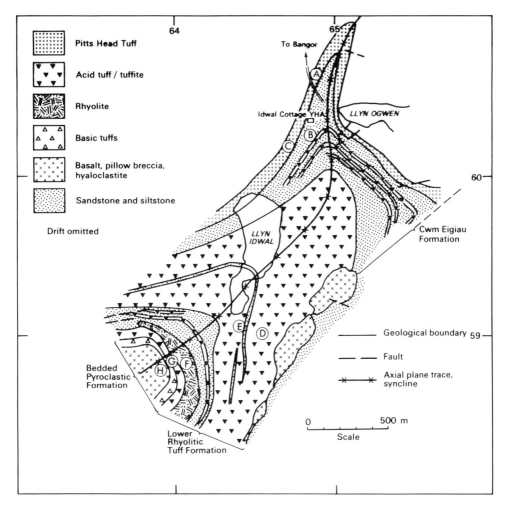

Figure 2.14. Cwm Idwal (locality 4.2). Part of British Geological Survey 1:25 000 Sheet SH 65 and 66 - The Passes of Nant Ffrancon and Llanberis.

interbedded sediments.

The contact between the LRTF and the overlying Bedded Pryoclastic Formation (BPF) is well exposed at the core of the syncline in the back wall of the cwm (locality 4.2H). Here the BPF comprises basaltic lapilli tuffs, containing marine fossils and a thick pillowed basalt, clearly indicating subaqueous emplacement.

At Cwm Idwal all the strata below, within and above the LRTF show evidence of accumulation in a marine environment. It is considered

(Howells et al., 1984) that the LRTF here is equivalent to the single
acidic ash-flow tuff at Capel Curig and to the thicker unbedded,
non-welded ash-flow tuff accumulation about the Snowdon massif to the
south. The latter area is interpreted (Howells et al., in press) as the
proximal accumulation about the eruptive centre which lies in a NW-SE
rifted depression. The eruption of the distal ash-flow tuff in eastern
Snowdonia was coincident with the subsidence in this depression. The
sequence at Cwm Idwal reflects early escape of primary ash flows from
this centre, and subsequent reworking of ash-flow tuffs following uplift
of the centre related to late stage intrusion and extrusion of rhyolite.
Additional localities where the sequence of Ordovician volcanic and
sedimentary rocks can be examined in this area are described by Howells
et al. (1978, 1981).

ITINERARY 5. ANGLESEY

An eastward traverse across Anglesey: Precambrian basement and Lower
Palaeozoic cover

Introduction

Anglesey is best known for extensive exposures of 'Monian' Precambrian
rocks, although important outcrops of Ordovician marine sediments and
volcanics, lower Silurian sediments, Devonian red beds and Carboniferous
rocks also occur. Anglesey geology is dominated by major faults and
thrusts, many of which run NE-SW and divide the island into a series of
narrow, linear belts (Fig.2.15). These faults, some of which were
initiated in the late Precambrian, influenced sedimentation and tectonics
throughout the Lower Palaeozoic.

Monian rocks form the largest Precambrian outcrop area in southern
Britain, but the geology is complicated, hampered by faulting and poor
inland exposure, and is still not well understood. The main contribution
to our knowledge was provided by Edward Greenly who published a complete
map and two-volume memoir on the geology of Anglesey in 1919 after many
years of detailed mapping (Greenly, 1919, 1938). Later publications
include Greenly (1923, 1940, 1946), Shackleton (1954, 1956, 1969, 1975),
Wood (1974), Barber and Max (1979), Barber et al. (1981), Cosgrove
(1980), Gibbons (1981, 1983a, 1983b, 1984) and Gibbons and Mann (1983).
Although Greenly coined the term 'Mona Complex', and this has been used
in all publications since 1919, the term includes diverse and sometimes
unrelated rock units and is avoided in this guide. The more general
adjective 'Monian' (Blake, 1888) is preferred, after the Welsh name for
Anglesey.

The Monian exposures can be subdivided broadly into three units:
  (i)   Monian granites and gneisses. This unit includes the Sarn
Complex of Llŷn, named after the Sarn Granite (Gibbons, 1983b;
Beckinsale et al., 1984), the Coedana Complex of Central Anglesey, named
after the Coedana granite, the Nebo Complex of northeast Anglesey, the
Gader Complex of northwest Anglesey, as well as several slivers of
sheared rocks found along major faults (eg. the Holland Arms Complex
along the Berw Fault in southeast Anglesey).

A GEOTRAVERSE THROUGH THE CALEDONIDES OF WALES

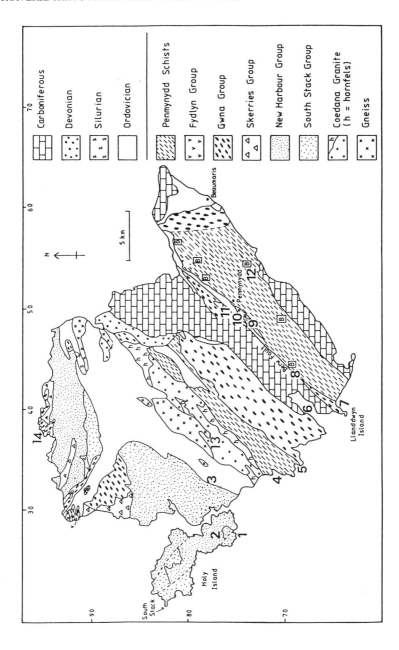

Figure 2.15. Geological map of Anglesey (localities 5.1 - 5.14)

(ii) Monian schists. These belong to Greenly's 'Penmynydd Zone' and occur in three linear NE-SW oriented belts in central and southeast Anglesey and Llŷn. They include various metasediments and metabasic volcanics, many of which are of upper greenschist grade although amphibolite facies rocks and, in southeast Anglesey, blueschists also occur. They are typically highly strained, intensely foliated, and sometimes strongly lineated, and show several deformation phases.

(iii) Monian Supergroup. A succession of low grade, variably deformed, metasediments and volcanic rocks. The lower part of the succession is dominated by turbidites and mudstones (South Stack and New Harbour groups), whereas the upper part includes coarser clastics, acid volcanics (Skerries Group) and large areas of mélange (Gwna Group).

The boundaries between these units are always steep faults, thrusts or shear zones, and their relative ages are therefore not well defined. The gneisses and granitic rocks have been dated as latest Precambrian (c.610-560 Ma). The Monian schists remain undated but some of them include mylonitised granitic rocks and are therefore interpreted as younger than c.610 Ma. A minimum age for the schists is suggested by very similar lithologies occurring as clasts in Middle Cambrian sediments on the Welsh mainland. The Supergroup is also undated, with the only palaeontological evidence available (worm burrows, stromatolites, and acritarchs) suggesting a late Precambrian to possibly early Cambrian age (Wood and Nicholls, 1973; Muir et al., 1979; Barber and Max, 1979; Gibbons, 1984). Outliers of ignimbritic rocks in Anglesey are post-Monian but older than the late Lower Cambrian sediments on the Welsh mainland (Reedman et al., 1984). Overall, therefore, the evidence indicates a latest Precambrian to earliest Cambrian age for the majority of Monian rocks (for further review and references see Gibbons, 1983b, 1984).

Anglesey may be viewed as part of a narrow terrane lying immediately northwest of the late Precambrian, calc-alkaline basement to southern Britain, an area often correlated with the Avalonian terrane of Newfoundland (eg. Gibbons and Gayer, 1985). The Monian and southern Britain terranes exhibit completely different geological histories and, where the contact between them is not covered by younger rocks, are separated by a steep, major, late fault. There has probably been considerable transcurrent movement between these two terranes prior to the late Lower Cambrian, although as yet there is no proof of this. The Monian terrane can be traced into southeast Ireland where the Cullenstown Group (Max, 1975) is lithologically similar to parts of the Monian Supergroup. Several broad correlations between Monian rocks and various terranes along the eastern Appalachian orogen have also been made (Neale and Kennedy, 1975; Rast et al., 1976a, 1976b; Kennedy, 1976; Rast and Skehan, 1981).

To the northwest of the Monian terrane are the Lower Palaeozoic rocks of the Lake District, Isle of Man, and Leinster Massif (Iapetus 'Axial Zone' of Kennedy, 1979; Lakesman terrane of Gibbons and Gayer, 1985). The contact between these two terranes is unexposed. It is clear, however, that the Monian terrane lay on the southeast side of Iapetus during the Lower Palaeozoic, possibly after having been accreted during orogenic collision following a period of late Precambrian subduction beneath the margin of an 'Avalonian' plate. Monian rocks were again deformed and mildly metamorphosed during the late Silurian 'end-Caledonian' (= Taconic) closure of Iapetus.

The object of this one day excursion is to make a traverse across Anglesey from west to east, visiting most of the major Monian rock units and concentrating on the nature of boundaries between these units. The first part of the traverse (locality 5.1) provides an opportunity to examine the structural relationships between the polydeformed metasediments of the South Stack and New Harbour groups. Locality 5.2 is in the serpentinite and gabbro within the New Harbour Group, interpreted by Thorpe (1978, 1979) as part of a dismembered ophiolite. Locality 5.3 allows an examination of the contact between the Supergroup metasediments and the Coedana Complex with its cover of Arenig sediments, which bound the Supergroup to the southeast. Locality 5.4 introduces the Monian schists of central Anglesey, here re-interpreted as a major metasedimentary nappe, thrust over severely mylonitised Coedana Complex. The 33 km traverse across southeast Anglesey between here and the famous blueschists at Llanfairpwllgwyngyll includes optional visits to the Gwna Group - Monian schist contact (Porth Cwfan), the Carboniferous-Gwna Group unconformity at Bodorgan, the Gwna volcanics of Llanddwyn Island, and sheared Monian and Carboniferous rocks along the important Berw Fault system. An alternative route allows examination of granites and gneisses of the Coedana Complex and the famous exposures of Gwna mélange around Llanbadrig in northern Anglesey.

Locality 5.1. Rhoscolyn, Holy Island (Fig.2.16)

Vehicles should be parked at Rhoscolyn Church (SH 268 757) from where first walk 800 m southwest to the coastguard lookout, then 250 m northwest along the clifftop before descending to just above high water mark on the west side (locality 5.1A) of a precipitous fault gully (SH 2606 7510). A spectacular exposure of strongly folded, low grade metasediments of the South Stack Group in the core of a major NE-plunging fold (the Rhoscolyn Anticline - Shackleton, 1969) is seen in the cliff face to the east. These rocks belong to the South Stack Formation, the oldest formation of the Monian Supergroup exposed on Anglesey. The rocks show an alternation of psammitic and pelitic beds, with the pelites exhibiting a more complex deformation history than the psammites. Associated with the asymmetric, SE-verging Rhoscolyn Anticline are abundant minor folds and a steep NW-dipping foliation which is much more pronounced in the pelites. Cosgrove (1980) interpreted this fold as an F^2 structure, although evidence for this is not obvious at this locality. In places, usually within the pelitic beds, the steep foliation is folded by minor folds which possess an axial surface dipping more gently NW. Quartz veins, originally intruded along the steep foliation and subsequently folded by these flat-lying structures, are particularly prominent in the pelitic bands. Superb examples of both the (early) steep and (later) shallow minor structures may be examined in the exposures underfoot. About 80 m along the cliff to the west, following the gently NW-dipping bedding on the shallow limb of the Rhoscolyn Anticline, the upward-facing sediments locally exhibit convoluted dewatering structures (locality 5.1B).

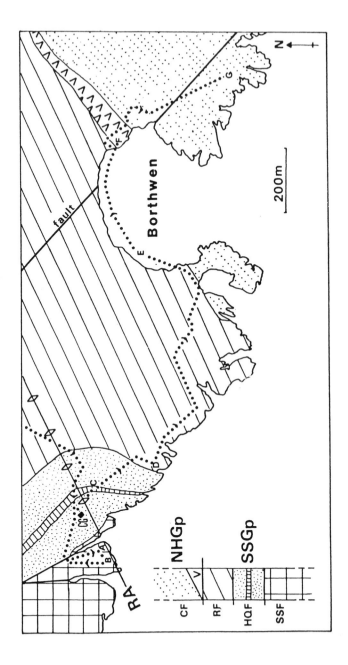

Figure 2.16. Geological sketch map of Rhoscolyn area, southeast Holy Island (locality 5.1). CG = Coastguard lookout; RA = Rhoscolyn Anticline (SE vergence, NE plunge); NHGp = New Harbour Group; CF = Celyn Formation; v = Basic metavolcanics; SSGp = South Stack Group; SSF = South Stack Formation; HQF = Holyhead Quartzite Formation; RF = Rhoscolyn Formation.

A GEOTRAVERSE THROUGH THE CALEDONIDES OF WALES 65

To traverse the steep SE limb of the anticline (Fig.2.16), return to the Coastguard lookout, built upon a prominent ridge of psammite (the Holyhead Quartzite Formation of Shackleton, 1969). 100 m southeast of the lookout, a narrow band of pelite is sandwiched between two thick psammites (locality 5.1C) and forms a prominent, topographic depression (SH 2636 7509). Continue southeast for 150 m over the upper part of this massive Holyhead Quartzite Formation (HQF) to where it is overlain by the more pelitic Rhoscolyn Formation (RF). The latter is lithologically similar to the South Stack Formation (SSF). The contact between the HQF and the RF is marked by a sudden drop from bare rock to grass. There are excellent exposures (locality 5.1D) of pelite with a strongly folded foliation along the clifftop by a rubbish tip (SH 2646 7495). Moving southeast along the coast, alternating pelites and psammites of the Rhoscolyn Formation show sub-vertical or steeply NW-dipping (overturned) bedding cut by a strong steep fabric, itself cut by a NW-dipping flat foliation (with many associated folds in the pelites). Late, upright, conjugate kinks, folding both the earlier steep, and later shallow foliation add to the complexity of polyphase structures.

The overlying New Harbour Group is best examined on the east side of Borth Wen. Follow the footpath to Borth Wen and examine exposures of steeply dipping, cross bedded, SE-facing quartzites at the top of the Rhoscolyn Formation (locality 5.1E) on the west side of the bay (SH 2715 7498). Walk to the northeast side of the beach (locality 5.1F) to where a greenschist facies foliated metabasite marks the boundary between the South Stack and the New Harbour groups (SH 2749 7506). Follow the footpath south-east from just north of this metabasite exposure for 200 m to the rough land east of Borth Wen and head south (locality 5.1G) for the New Harbour Group exposed at the cliffline (eg. SH 2755 7476). The dominantly pelitic Celyn Formation, which forms the lower part of the New Harbour Group (Greenly, 1919), appears more strongly deformed than the South Stack Group. A steep NW-dipping bedding is only rarely discernible, usually having been obliterated by an intense NW-dipping foliation. This early foliation is intensely folded by numerous minor NE-plunging 'S' folds with a shallower, NW-dipping axial surface. Late, upright kinks are associated with broad upright folds, both of which complicate the structural geometry. However, the similarities between the more highly strained New Harbour Group pelites and the pelitic beds in the South Stack Group indicate that both groups have responded to a similar deformational history (see discussion to Barber and Max, 1979). The lithological contrasts between the two groups have governed the differences in structural style (Cosgrove, 1980).

Locality 5.2. Monian gabbros and serpentinites

Park at a public footpath sign (left) by the road 2 km north of Rhoscolyn Church (SH 2685 7720). The New Harbour Group here contains an horizon of deformed serpentinites and metagabbros. The serpentinites are mostly altered dunites and hartzburgites, and the gabbroic rocks belong to the picrite-metagabbro range (Maltman, 1975, 1977). Although Maltman (1977, 1979) interpreted these rocks as a deformed magmatic complex intruded into the New Harbour Group prior to deformation, Thorpe (1978) regards them as part of a dismembered, tectonically-emplaced ophiolite complex (see Gibbons, 1983b, p.156). Exposures of the serpentinite occur just west of the road at SH 2685 7720. The gabbro is well exposed 50 m west of here where it forms a prominent knoll just north of the footpath (SH 2680 7720). If more time is available, a visit to the magnificent exposures of South Stack Formation at the type locality, South Stack (SH 205 823), should be made. Unlike at Rhoscolyn, the folding at South Stack is upright, and the imprint of the later, flatter-lying minor folds is much less pronounced.

Locality 5.3. Llyn Traffwll (Fig.2.17)

The east margin of the New Harbour Group outcrop is exposed on the west side of Llyn Traffwll. The strongly deformed New Harbour Group lies adjacent to intensely cataclastic granitic and amphibolitic gneisses belonging to the Coedana Complex, upon which Arenig sandstones rest

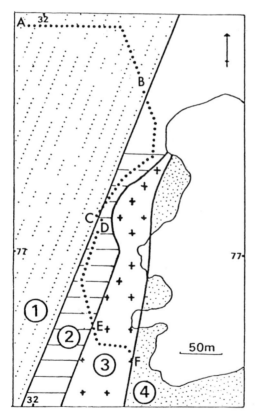

Figure 2.17. Geological sketch map of west side of Llyn Traffwll (locality 5.3). (1) = New Harbour Group; (2) = Dolerite dyke; (3) = cataclastic amphibolite and granite; (4) = Arenig sandstones.

unconformably. A late dolerite dyke has intruded up the sub-vertical shear zone that exists between the New Harbour Group and Coedana Complex. Park vehicles at SH 3213 7749, walk 250 m southwest along the road, turn left (locality 5.3A on Fig.2.17) and walk 125 m east through a housing estate and down a lane to a stile (right). Cross the field to a ladder stile, noting poor exposures of New Harbour Group pelites (locality 5.3B). Cross the stile and walk south-west for 120 m, skirting the west side of a rocky ridge to reach a low knoll of highly deformed New Harbour Group pelites (locality 5.3C) (SH 3206 7703). Similar structures to those examined east of Borth Wen may be seen. Note a prominent SW-plunging lineation.

The ridge immediately east of here exposes a dolerite dyke (locality 5.3D). Walk 100 m south then southeast around this ridge to observe the contact (locality 5.3E) between this dyke and cataclastic gneiss (please do not hammer) (SH 3207 7691). Follow a small path southwest past exposures of sheared gneiss for another 100 m or so, over an old concrete sluice, to reach exposures of coarse Arenig sandstones (Treiorwerth Formation of Bates, 1972). These immature, conglomeratic sandstones, composed almost entirely of derived Monian fragments, show a crude bedding which dips 25°-35° ENE (locality 5.3F). The sediments lie above the Coedana Complex basement, which forms a NE-SW trending, 5 km wide strip of gneiss and granite separating the Monian Supergroup of western Anglesey from the schists and deformed mélanges of central and southeast Anglesey (Fig.2.15).

Locality 5.4. Porth Trecastell

Drive east (A5) then south (A4080) across the Ordovician and Coedana Complex to the beach car park at SH 333 707. The Coedana Complex is bounded on its southeast side by a major, SSE-dipping mylonite zone interpreted as having been produced by the obduction of a metasedimentary nappe, the schists of which form part of Greenly's 'Central Penmynydd Zone'. The best locality for viewing the contact between mylonitised granites and the overlying schists is here at Porth Trecastell. From the beach car park walk beneath the low cliffs on the south side of the bay to examine metasedimentary quartz, mica and graphite schists (upper greenschist - lower amphibolite). The rocks show evidence for a high strain early deformation (intense fabric folded by tight to isoclinal folds) with at least two later phases of folding.

At the northeast corner of the bay, beyond a break in exposure, intensely foliated mylonitic schists derived from a granitic rock dip gently S or SSE. A stretching lineation, interpreted as defining the direction of nappe transport, trends E-W or ESE-WNW. Most of the headland north of here is composed of the same mylonitic rocks, interpreted as derived from the intense shearing of the Coedana granite.

Localities 5.5 - 5.14. Porth Trecastell to
Llanfairpwllgwyngyllgogerychwyrndrobwllllantysiliogogogoch

From Trecastell, one of two alternative routes may be taken; the first of these (A) continues examination of the south coast as far as the Berw

Fault escarpment, whereas the second (B) demonstrates the Gwna mélange on the north coast. Ideally, the north coast should be visited as part of a separate excursion, combining the exposures east of Cemaes Bay with those at Porth Wen, Parys Mountain and Traeth Lligwy (see Bates and Davies, 1981).

Route A. Localities 5.5-5-11. From Trecastell drive southeast through Aberffraw where the metasedimentary 'Penmynydd' schists give way to lower grade Gwna Group mélange. The relationship between the Gwna mélange and the schists is greatly complicated by faulting and is interpreted as a wide, sub-vertical shear zone. It can be examined along the coast around Braich-Llwyd (SH 338 676) (locality 5.5; Fig.2.15 - leave vehicle at Porth Cwfan), although it should be emphasised that these exposures are difficult to interpret and may not repay a short visit.

Beyond Aberffraw the A4080 continues over blown sands that cover the Gwna Group. At Malltraeth the road descends almost to sea level on to marshes which overlie a narrow strip of Carboniferous rocks resting unconformably upon the Gwna Group. Coal was mined in the 19th Century just northeast of here (near Glantraeth, Ty'n Fflat and Pencrug) and at Holland Arms (7 km northeast). Exposures of the Carboniferous rocks are few, but Millstone Grit (Namurian) sediments resting unconformably on Gwna Group are visible at low tide on the shore at SH 393 674 (locality 5.6; Fig.2.15). (Leave cars at SH 399 683 and walk south along beach at low or middle tide for 1 km. Not suitable for coaches).

The southeast side of the Carboniferous outcrop is bounded by the Berw Fault (Fig.2.15), a major crustal lineament which forms a subdued but distinct feature right across Anglesey. The fault juxtaposes more 'Penmynydd' schists which are also strongly affected by shearing close to the Berw Fault. The schists form a NE-SW striking belt up to 7 km wide in southeast Anglesey (the 'Aethwy' unit of Greenly, 1919). Unlike the central Anglesey metasedimentary schist nappe the schist belt in southeast Anglesey includes blueschists (Gibbons and Mann, 1983).

The A4080 climbs across the Berw Fault system just northwest of Newborough. To visit the famous exposures of Gwna Group basic volcanics at Newborough Warren, caught within a splay from the main Berw Fault (obtain permission from the Nature Conservancy Council for party visits; no hammers, collecting, or coaches), turn right at Newborough to reach the NCC car park 6 km to the southwest. Walk across the beach for 1 km to exposures of SE-facing pillow lavas at the north end of Llanddwyn Island (SH 391 634 - locality 5.7; Fig.2.15). Fine exposures of basic lava and limestone melange cut by dykes occur on the southwest tip of the island (SH 3845 6255) (Bates and Davies, 1981, p.12).

From Newborough follow the B4421 northeastwards along the top of the Berw Fault escarpment to the A5 at Holland Arms. Sheared slices of several Monian lithologies occur along the fault, including the Gwna Group (eg. locality 5.8 in quarry at SH 440 687), granite (eg. locality 5.9 mylonites at SH 4670 7175), and gneisses (eg. locality 5.10 in quarry at SH 480 728 - see Beckinsale and Thorpe, 1979 for radiometric dates). Continue to follow the Berw Fault escarpment northeast from Holland Arms for 3 km to the Llangefai road at SH 4895 7485. Turn right, park on the right, and walk for a short distance to roadside exposures (left) of crinoidal limestones and shales of early Carboniferous age (<u>locality 5.11</u>). Proximity to the long-active Berw Fault system has resulted in these rocks having been recumbently folded and thrust – elsewhere on Anglesey rocks of this age are mostly undeformed.

Route B. Localities 5.13-5.14. From Trecastell drive north to the A5. Localities in this area display lithologies typical of the Coedana Complex (<u>locality 5.13</u>; Fig.2.15). Banded amphibolites cut by granite veins are well exposed in an old quarry at SH 374 777 (not suitable for coaches), and the Coedana granite itself can be examined in Gwalchmai Quarry on the A5 at SH 381 765 (permission needed from the Quarry Manager, Kingston Minerals Ltd., Gwalchmai Quarry). The coarse muscovite granite, often badly affected by cataclasis, is cut by breccia zones and basic dykes. Drive north on the B5112, west on the B5109, north on the A5025, by-pass Cemaes Bay and turn north (left) up a narrow lane at SH 379 936 (difficult for coaches). After 600 m turn left to reach the beach car park at SH 375 938, and walk north along the beach (<u>locality 5.14</u>; Fig.2.15).

Greenly described the Gwna mélange exposed along this coast as 'absolutely indescribable'. The mélange, now interpreted as a deformed olistostrome (Shackleton, 1954, 1969), displays a much faulted, chaotic mixture of limestones, sandstones and black phyllite. Follow the clifftop northwards to reach a large olistolith of stromatolitic limestone (Wood and Nicholls, 1973). The stromatolites are best seen just north of the quarry entrance on the shore. If time allows, continue along the clifftop to the lane leading to Llanbadrig Church where there are good exposures of the mélange. Continue along the clifftop (crossing Ordovician sediments) to Ogof Gynfor (SH 378 949). At this famous locality Arenig conglomerate rests unconformably on the Gwna mélange. Above this, cherty mudstones of Caradoc age rest disconformably upon the Arenig sediments (Greenly, 1919; Bates, 1972; Barber and Max, 1979).

To rejoin route A at <u>locality 5.11</u> (Fig.2.15), head east to Amlwch, south on the B5111 and B5110 to Llangefni, then east on the B5420 for 3 km to SH 491 749.

<u>Locality 5.12</u>. Continue east along the road, across the schist belt, through Penmynydd, turn right at a road junction at SH 527 733 and left after 300 m. Follow the lane (Ffordd Penmynydd) for 1 km to a bridge over the A5 Llanfairpwllgwyngyll by-pass (SH 530 723). Fresh roadside exposures of basic crossite-epidote schists interfolded with phengitic schists are typical of Greenly's Aethwy Penmynydd Zone (<u>locality 5.12</u>). The blue amphiboles contain barroisitic and actinolitic cores recording an earlier greenschist facies metamorphism (Gibbons and Gyopari, in press). The blueschists display a gently dipping, intense foliation with a gently N-plunging mineral lineation, interpreted as defining the trend of nappe transport over deformed Gwna mélange now exposed 6 km to the northeast. There are further exposures of ancient crossite-epidote blueschist (no hammers or collecting) in the woods beneath the Marquis of Anglesey Column (SH 5353 7156). The blueschists display an intense, highly folded, flat-lying fabric and a strong N-S lineation. These rocks

provide the best evidence in the Monian terrane for the former existence of an active plate margin prior to the development of the Welsh Basin further southeast. On a clear day it is worth climbing the column for wide views across Anglesey, Snowdonia and the Llŷn Peninsula.

REFERENCES

BAKER, J.W. 1973. 'A marginal late Proterozoic ocean basin in the Welsh region'. Geol. Mag., 110, 447-455.
——— 1982. 'The Precambrian of south-west Dyfed'. In: BASSETT, M.G. (ed.), Geological excursions in Dyfed, south-west Wales. National Museum of Wales, Cardiff, 15-25.
BARBER, A.J. & MAX, M.D. 1979. 'A new look at the Mona Complex (Anglesey, North Wales. J. geol. Soc. London, 136, 407-432.
———, MAX, M.D. & BRÜCK, P.M. 1981. 'Geologists' Association - Irish Geological Association Field Meeting in Anglesey and southeastern Ireland, 4-12 June 1977'. Proc. Geol. Assoc., 92, 269-292.
BATES, D.E.B. 1972. 'The stratigraphy of the Ordovician rocks of Anglesey'. Geol. J., 8, 29-58.
——— 1982. 'The Aberystwyth Grits'. In: BASSETT, M.G. (ed.), Geological excursions in Dyfed, south-west Wales. National Museum of Wales, Cardiff, 81-90.
——— & DAVIES, J.R. 1981. Anglesey. Geol. Assoc. Guide No.40.
BEAVON, R.V. 1980. 'A resurgent cauldron in the early Palaeozoic of Wales, U.K.'. J. volcan. geoth. Res. Amsterdam, 7, 157-174.
BECKINSALE, R.D. & THORPE, R.S. 1979. 'Rubidium-Strontium whole rock isochron evidence for the age of metamorphism and magmatism in the Mona Complex of Anglesey'. J. geol. Soc. London, 136, 433-439.
———, EVANS, J.A., THORPE, R.S., GIBBONS, W. & HARMON, R.S. 1984. 'Rb-Sr whole rock isochron evidence, $\delta^{18}O$ values and geochemical data for the Sarn Igneous Complex and the Parwyd gneisses of the Mona Complex of Llŷn, N Wales'. J. geol. Soc. London, 141, 701-709.
BEVINS, R.E. & ROACH, R.A. 1979. 'Pillow lava and isolated-pillow breccia of rhyodacitic composition from the Fishguard Volcanic Group, Lower Ordovician, S.W. Wales, United Kingdom'. J. Geol., 87, 193-201.
——— & ROWBOTHAM, G. 1983. 'Low-grade metamorphism within the Welsh sector of the paratectonic Caledonides'. Geol. J., 18, 141-167.
———, KOKELAAR, B.P. & DUNKLEY, P.N. 1984. 'Petrology and geochemistry of lower to middle Ordovician igneous rocks in Wales: a volcanic arc to marginal basin transition'. Proc. Geol. Assoc., 95, 337-347.
BLACK, W.W., BULMAN, O.M.B., HEY, R.W. & HUGHES, C.P. 1971. 'Ordovician stratigraphy of Abereiddy Bay, Pembrokeshire'. Geol. Mag., 108, 546-548.
BLAKE, J.F. 1888. 'On the Monian System of rocks'. Q. J. geol. Soc.

London, **44**, 463-546.
BROMLEY, A.V. 1969. 'Acid plutonic igneous activity in the Ordovician of North Wales'. In: WOOD, A. (ed.), The Pre-Cambrian and Lower Palaeozoic rocks of Wales. University of Wales Press, Cardiff, 387-408.
COPE, J.C.W. 1977. 'An Ediacara type fauna from South Wales'. Nature, London, **268**, 624.
COSGROVE, J.W. 1980. 'The tectonic implications of some small scale structures in the Mona Complex of Holy Isle, North Wales'. J. struct. Geol., **2**, 383-396.
COWARD, M.P. & SIDDANS, A.W.B. 1979. 'The tectonic evolution of the Welsh Caledonides'. In: HARRIS, A.L., HOLLAND, C.H. & LEAKE, B.E. (eds), The Caledonides of the British Isles - reviewed. Geol. Soc. London, Spec. Publ., **8**, 187-198.
COX, A.H. 1915. 'The geology of the district between Abereiddy and Abercastle'. Q. J. geol. Soc. London., **71**, 273-340.
—————— 1930. 'Preliminary note on the geological structure of Pen Caer and Strumble Head, Pembrokeshire'. Proc. Geol..Assoc., **41**, 274-289.
CRIMES, T.P. 1970a. 'A facies analysis of the Cambrian of Wales'. Palaeogeogr. Palaeoclimat. Palaeoecol., **7**, 113-170.
—————— 1970b. 'The significance of trace fossils in sedimentology, stratigraphy and palaeoecology with examples from Lower Palaeozoic strata'. In: CRIMES, T.P. & HARPER, J.C. (eds), Trace fossils. Geol. J. Spec. Issue, **3**, 101-126.
DAVIES, W. & CAVE, R. 1976. 'Folding and cleavage determined during sedimentation'. Sedim. Geol., **15**, 89-133.
DEWEY, J.F. 1969. 'Evolution of the Appalachian/Caledonian orogen'. Nature, London, **222**, 124-129.
EDWARDS, D. & DAVIES, E.C.W. 1976. 'Oldest recorded in situ tracheids'. Nature, London, **263**, 494-495.
—————— & ROGERSON, E.C.W. 1979. 'New records of fertile Rhyniophytina from the late Silurian of Wales'. Geol. Mag., **116**, 93-98.
EVANS, W.D. 1948. 'The Cambrian-Ordovician junction, Whitesand Bay, Pembrokeshire'. Geol. Mag., **85**, 110-113.
FITCH, F.J. 1967. 'Ignimbrite volcanism in North Wales'. Bull. volcan., **30**, 199-219.
FRANCIS, E.H. & HOWELLS, M.F. 1973. 'Transgressive welded ash-flow tuffs among the Ordovician sediments of N.E. Snowdonia, N. Wales'. J. geol. Soc. London, **129**, 621-641.
GEORGE, T.N. 1963. 'Palaeozoic growth of the British Caledonides'. In: JOHNSON, M.R. & STEWART, F.H. (eds), The British Caledonides. Oliver & Boyd, Edinburgh, 1-33.
GIBBONS, W. 1981. 'Glaucophanic amphibole in a Monian shear zone on the mainland of North Wales'. J. geol. Soc., London, **138**, 139-143.
—————— 1983a. 'The Monian Penmynydd Zone of Metamorphism in Llŷn, North Wales'. Geol. J., **18**, 1-21.
—————— 1983b. 'Stratigraphy, subduction and strike-slip faulting in the Mona Complex of North Wales - a review'. Proc. Geol. Assoc., **94**, 147-163.

—————— 1984. 'The Precambrian basement of England and Wales'. Proc. Geol. Assoc., **95**, 387-389.
—————— & GAYER, R.A. 1985. 'British Caledonian Terranes'. In: GAYER, R.A. (ed.), The tectonic evolution of the Caledonide-Appalachian Orogen. Earth Evolution Sciences, Monograph 1, Vieweg & Sohn, Braunschweig/Wiesbaden, 3-16.
—————— & GYOPARI, M. in press. 'A greenschist protolith for blueschist in Anglesey, U.K.'. In: EVANS, B.W. & BROWN, E.H. (eds), Blueschists and related eclogites. Geol. Soc. Am., Spec. Paper.
—————— & MANN, A. 1983. 'Pre-Mesozoic lawsonite in Anglesey, North Wales - the preservation of ancient blueschists'. Geology, **11**, 3-6.
GREENLY, E. 1919. 'The Geology of Anglesey'. Mem. Geol. Surv. U.K. (2 vols), London, 980pp.
—————— 1923. 'The succession and metamorphism in the Mona Complex'. Q. J. geol. Soc. London, **79**, 334-351.
—————— 1938. A hand through time: memories, romantic and geological: studies in the arts, religion and the grounds of confidence in immortality. T. Murby & Co. London, 774pp.
—————— 1940. 'Studies in the Mona Complex'. Geol. Mag., **77**, 50-53.
—————— 1946. 'The Monio-Cambrian Interval'. Geol. Mag., **83**, 237-240.
HOWELLS, M.F. & ALLEN, P.M., ADDISON, R., WEBB, B.C., LYNAS, B.D.T. & JACKSON, A. 1977. 'Folding and cleavage determined during sedimentation - comments'. Sedim. Geol., **17**, 333-335.
——————, FRANCIS, E.H., LEVERIDGE, B.E. & EVANS, C.D.R. 1978. Capel Curig and Betwys-y-Coed: Description of 1:25,000 sheet SH 75. Classical areas of British geology, Institute of Geological Sciences, H.M.S.O., London, 73pp.
——————, LEVERIDGE, B.E., ADDISON, R. & REEDMAN, A.J. 1983. 'The lithostratigraphical subdivision of the Ordovician underlying the Snowdon and Crafnant volcanic groups, North Wales'. Rep. Inst. Geol Sci., **83/1**, 11-15.
—————— & EVANS, C.D.R. 1973. 'Ordovician ash-flow tuffs in eastern Snowdonia'. Rep. Inst. Geol. Sci., **73/3**, 1-33.
—————— & REEDMAN, A.J. 1981. Snowdonia. Rocks and fossils series No.1, Unwin Paperbacks, London, 119pp
—————— in press. Outline of the Geology of the Bangor 1:50,000 Geological Sheet (106). British Geological Survey.
HUGHES, C.P., JENKINS, C.J. & RICKARDS, R.B. 1982. 'Abereiddi Bay and the adjacent coast'. In: BASSETT, M.G. (ed.), Geological excursions in Dyfed, south-west Wales. National Museum of Wales, Cardiff, 51-63.
JONES, O.T. 1940. 'Some Lower Palaeozoic contacts in Pembrokeshire'. Geol. Mag., **77**, 405-409.
KENNEDY, M.J. 1976. 'Southeastern margin of the northeastern Appalachians: late Precambrian orogeny on a continental margin'. Geol. Soc. Am. Bull., **87**, 1317-1325.

────────── 1979. 'The continuation of the Canadian Appalachians into the Caledonides of Britain and Ireland'. In: HARRIS, A.L., HOLLAND, C.H. & LEAKE, B.E. (eds), The Caledonides of the British Isles - reviewed. Geol. Soc. London, Spec. Publ., 8, 33-64.

KOKELAAR, B.P., HOWELLS, M.F., BEVINS, R.E. & ROACH, R.A. 1984a. 'Volcanic and associated sedimentary tectonic processes in the Ordovician marginal basin of Wales: A field guide'. In: KOKELAAR, B.P. & HOWELLS, M.F. (eds), Marginal Basin Geology: Volcanic and associated sedimentary and tectonic processes in modern and ancient marginal basins. Geol. Soc. London, Spec. Publ., 16, 291-322.

────────── ────────── & DUNKLEY, P.N. 1984b. 'The Ordovician marginal basin in Wales'. In: KOKELAAR, B.P. & HOWELLS, M.F. (eds), Marginal Basin Geology: Volcanic and associated sedimentary and tectonic processes in modern and ancient and marginal basins. Geol. Soc. London, Spec. Publ., 16, 245-269.

LISLE, R.J. 1977. 'Folding and cleavage determined during sedimentation: Comments'. Sedim. Geol., 19, 69-72.

MALTMAN, A.J. 1975. 'Ultramafic rocks in Anglesey - their non-tectonic emplacement'. J. geol. Soc. London, 131, 593-605.

────────── 1977. 'Serpentinites and related rocks of Anglesey'. Geol. J., 12, 113-128.

────────── 1979. 'Tectonic emplacement of ophiolitic rocks in the Precambrian Mona complex of Anglesey'. Nature, London, 277, 327.

MAX, M.D. 1975. 'Precambrian rocks of south-east Ireland'. In: HARRIS, A.L. et al. (eds), A correlation of the Precambrian rocks in the British Isles. Geol. Soc. London, Spec. Rep., 6, 97-101.

McKERROW, W.S., LAMBERT, R.St.J. & COCKS, L.R.M. in press. 'The Ordovician, Silurian and Devonian periods'. In: SNELLING, N.J. (ed.), Geol. Soc. London, Spec. Publ.

MUIR, M., BLISS, G.M., GRANT, P.R. & FISCHER, M. 1979. 'Palaeontological evidence for the age of some supposedly Precambrian rocks in Anglesey, North Wales'. J. geol. Soc. London, 136, 61-64.

NEALE, E.R.W. & KENNEDY, M.J. 1975. 'Basement and cover rocks at Cape North, Cape Breton Island, Nova Scotia'. Maritime Sediments, 11, 1-4.

PATCHETT, P.J. & JOCELYN, J. 1979. 'U-Pb zircon ages for late Precambrian igneous rocks in South Wales'. J. geol. Soc. London, 136, 13-19.

RAST, N. 1969. 'The relationship between Ordovician structure and volcanicity in Wales'. In: WOOD, A. (ed.), The Pre-Cambrian and Lower Palaeozoic rocks of Wales. University of Wales Press, Cardiff, 305-335.

────────── & SKEHAN, J.W. 1981. 'Possible correlation of Precambrian rocks of Newport, Rhode Island, with those of Anglesey, Wales'. Geology, 9, 596-601.

──────────, BEAVON, R.V. & FITCH, F.J. 1958. 'Sub-aerial volcanicity

in Snowdonia'. Nature, London, 181, 508.
———, KENNEDY, M.J. & BLACKWOOD, R.F. 1976a. 'Comparison of some tectonostratigraphic zones in the Appalachians of Newfoundland and New Brunswick'. Can. J. Earth Sci., 13, 868-875
———, O'BRIEN, B.H. & WARDLE, R.J. 1976b. 'Relationships between Precambrian and Lower Palaeozoic rocks of the 'Avalon Platform' in New Brunswick, the north-east Appalachians and the British Isles'. Tectonophysics, 30, 315-338.
REEDMAN, A.J., LEVERIDGE, B.E. & EVANS, R.B. 1984. 'The Arfon Group ('Arvonian') of North Wales'. Proc. Geol. Assoc., 95, 313-321.
RICKARDS, R.B. 1973. 'The structure of Abereiddy Bay, Pembrokeshire'. Geol. Mag., 110, 185-187.
ROBERTS, B. 1969. 'The Llwyd Mawr ignimbrite and its associated volcanic rocks'. In: WOOD, A. (ed.), The Pre-Cambrian and Lower Palaeozoic rocks of Wales. University of Wales Press, Cardiff, 337-356.
SHACKLETON, R.M. 1954. 'The structure and succession of Anglesey and the Lleyn Peninsula'. Advmt. Sci. London, 11, 106-108.
——— 1956. 'Notes on the structure and relations of the Precambrian and Ordovician rocks of south-western Lleyn (Carnarvonshire)'. Geol. J., 1, 400-409.
——— 1969. 'The Pre-Cambrian of North Wales'. In: WOOD, A. (ed.), The Pre-Cambrian and Lower Palaeozoic rocks of Wales. University of Wales Press, Cardiff, 1-22.
——— 1975. 'Precambrian rocks of North Wales'. In: HARRIS, A.L. et al. (eds), A correlation of the Precambrian rocks in the British Isles. Geol. Soc. London, Spec. Rep., 6, 76-82.
SNELLING, N.J. in press. 'An interim time scale'. In: SNELLING, N.J. (ed.), Geol. Soc. London, Spec. Publ.
SPARKS, R.S.J., SIGURDSSON, H. & CAREY, S.N. 1980a. 'The entrance of pyroclastic flows into the sea. I. Oceanographic and geologic evidence from Dominica, Lesser Antilles'. J. volcan. geoth. Res. Amsterdam, 7, 87-96.
——— ——— ——— 1980b. 'The entrance of pyroclastic flows into the sea. II. Theoretical considerations on subaqueous emplacement and welding'. J. volcan. geoth. Res. Amsterdam, 7, 97-105.
THORPE, R.S. 1974. 'Aspects of magmatism and plate tectonics in the Precambrian of England and Wales'. Geol. J., 9, 115-136.
——— 1978. 'Tectonic emplacement of ophiolitic rocks in the Precambrian Mona Complex of Anglesey'. Nature, London, 276, 57.
——— 1979. 'Late Precambrian igneous activity in S. Britain'. In: HARRIS, A.L., HOLLAND, C.H. & LEAKE, B.E. (eds), The Caledonides of the British Isles - reviewed. Geol. Soc. London, 8, 579-584.
TURNER, P. 1977. 'Notes on the depositional environment of the Lingula Flags in Dyfed, South Wales'. Proc. Yorks. geol. Soc., 41, 199-202.
WALMSLEY, V.G. & BASSETT, M.G. 1976. 'Biostratigraphy and correlation of the Coralliferous Group and Gray Sandstone Group (Silurian) of Pembrokeshire, Wales'. Proc. Geol. Assoc., 87, 191-220.

WALTHAM, A.C. 1971. 'A note on the structure and succession at Abereiddy Bay, Pembrokeshire'. Geol. Mag., 108, 49.
WATSON, J. & DUNNING, F.W. 1979. 'Basement-cover relations in the British Caledonides'. In: HARRIS, A.L., HOLLAND, C.H. & LEAKE, B.E. (eds), The Caledonides of the British Isles - reviewed. Geol. Soc. London, Spec. Publ., 8, 67-91.
WILLIAMS, B.P.J. (ed.). 1978. 'The Old Red Sandstone of the Welsh Borderland and South Wales'. Palaeontological Association Devonian Symposium, Excursion B2, Field Guide, 55-106.
────────── & STEAD, J.T.G. 1982. 'The Cambrian rocks of the Newgale - St David's area'. In: BASSETT, M.G. (ed.), Geological excursions in Dyfed, south-west Wales. National Museum of Wales, Caridff, 27-49.
WILLIAMS, H. 1927. 'The geology of Snowdon (North Wales)'. Q. J. geol. Soc. London, 83, 346-431.
WOOD, A. & SMITH, A.J. 1959. 'The sedimentation and sedimentary history of the Aberystwyth Grits (Upper Llandoverian)'. Q. J. geol. Soc. London, 114, 163-195.
WOOD, D.S. 1974. 'Ophiolites, melanges, blueschists and ignimbrites: early Caledonian subduction in Wales'? In: DOTT, R.H. & SHAVER, R.H. (eds), Modern and ancient geosynclinal sedimentation. Spec. Publs Soc. econ. Paleont. Miner., 19, 334-344.
WOOD, M.G. & NICHOLLS, G.D. 1973. 'Pre-Cambrian stromatolitic limestones from northern Anglesey'. Nature (Phys. Sci.), London, 241, 65.

FIELD GUIDE TO THE LAKE DISTRICT AND SOUTHERN UPLANDS

Coordinators: G. Kelling, W.S. McKerrow and P. Stone
Contributors: G. Kelling, W.S. McKerrow and P. Stone

A. Introduction - the Lower Palaeozoic of the Lake District and Southern Uplands

B. Field itineraries

    6. The Lake District and Dobb's Linn (GK)

    7. Central and Northern Belts of the Southern Uplands (WSMc)

    8. Girvan - Ballantrae section (PS)

C. References

G. Kelling, Department of Geology, University, Keele, Staffordshire.
W.S. McKerrow, Department of Earth Sciences, University, Oxford OX3 OBP.
P. Stone, British Geological Survey, Murchison House, West Mains Road, Edinburgh EH9 3LA, Scotland.

INTRODUCTION - THE LOWER PALAEOZOIC OF THE LAKE DISTRICT AND SOUTHERN UPLANDS

In the northern part of the 'paratectonic zone' Lower Palaeozoic rocks are exposed in three areas, namely : 1, Tremadoc to Pridoli rocks in the Lake District, known to be south of the Iapetus suture from the Caradoc and Ashgill benthic faunas; 2, Arenig to Wenlock rocks in the Southern Uplands, interpreted as an accretionary prism north of the Iapetus suture, and 3, Arenig to Wenlock rocks in the Ballantrae-Girvan succession north of the SUF which have been variously interpreted as Early Ordovician oceanic rocks emplaced on the margin of the North American continent, and overlain by a Llanvirn to Wenlock marginal sequence.

The Lake District

The Lower Palaeozoic rocks of the Lake District consist of four main groups; the Skiddaw Slate Group (Tremadoc to Llanvirn), turbidites and slates (1,980m): the Borrowdale Volcanic Group (Llandeilo and Caradoc); tuffs, ignimbrites, andesites and rhyolites (3,300m): the Coniston Limestone Group (Ashgill); shelf mudstones and siltstones with some thin limestones and a rhyolite flow (the Stockdale or Yarlside Rhyolite) in all some 150m: and Silurian (Llandovery to Pridoli) turbidites and graptolite shales (4,800 m).

The Southern Uplands

It is useful to look at the history of geological work in the region. It was at Birkhill Cottage (on the main road, 400m east of Dobb's Linn), that Charles Lapworth spent his summers, mapping the area and collecting graptolites; a monumental task which not only proved the stratigraphic value of graptolites, but outlined the stratigraphy of the Southern Uplands (Lapworth, 1878, 1889). Much of the area was mapped between 1854 and 1898 by the Geological Survey; this work culminated in the Memoir by Peach and Horne (1899), who acknowledge Lapworth's achievements (1899, pp. 23-36). Subsequent investigations by numerous workers in the present century have made clear the meticulous detail with which Peach and Horne recorded the geology across the region. In the light of modern techniques, only their structural conclusions can be faulted. Their work is all the more remarkable when it is realised that their Southern Uplands field seasons were almost entirely in the winter months; during the summers they were occupied with their first love, the Northwest Highlands.

Peach and Horne (1899) recognised 3 belts in the Southern Uplands (McKerrow, this volume): the Northern Belt, where early Ordovician basalts and cherts are normally present below thick greywackes; the Central Belt, where late Ordovician and early Silurian graptolitic shales occur below the greywackes; and the Southern Belt (S of the Riccarton Line), where those graptolitic shales that do occur are interbedded with

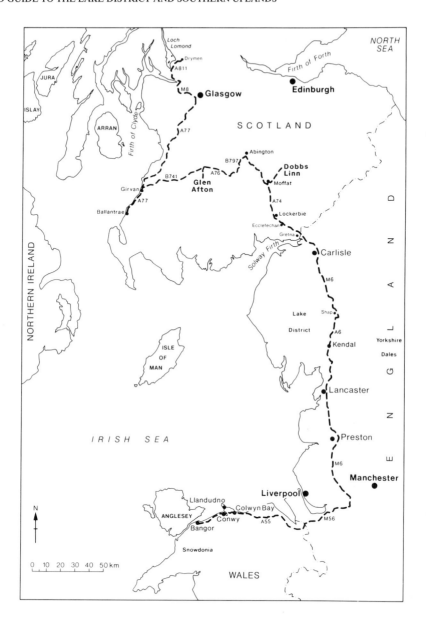

Fig. 3.1. Route through the Lake District and Southern Uplands. Itineraries 6-8.

the thick greywacke sequences. This distinction is now thought to be over-simplified, since there are sequences which differ markedly from each other, especially in the upper ages of the graptolitic shales and in the petrography of the greywackes, within the Northern and Central Belts.

The Southern Uplands contain at least 10 distinct stratigraphic sequences which differ in character and age over distances of a few kilometres across strike, but which can be traced for up to 100 km along strike, and often much further. Some sequences are also repeated across strike; Peach and Horne (1899) considered that this repetition was related to isoclinal folding, but modern work (Craig and Walton, 1959; Walton, 1961, 1963, 1965; Kelling, 1961; Warren, 1964; Rust, 1965; Lumsden et al., 1967; Weir, 1968; Toghill, 1970; McKerrow et al., 1977; Leggett et al., 1979) has demonstrated the importance of reverse faults parallel to the strike, as was first postulated by Harkness (1851, pp 51-3).

Rocks now forming the Southern Uplands were deposited on the NW margin of the Iapetus Ocean during the Lower Palaeozoic. Distinct faunal provinces in the Lake District (where the Ordovician faunas are similar to the rest of England and Wales) and in the Girvan area (where American faunas occur in the Ordovician) show that the ocean was still wide enough during much of the Late Ordovician to separate most benthic faunas with pelagic larval stages (McKerrow and Cocks, 1976). It closed by north-westerly subduction; accretion took place from the Caradoc until the Wenlock, but subduction may have continued through much of the Early Devonian, with underthrusting of English continental crust (Leggett et al., 1983).

It has been suggested (Mitchell and McKerrow, 1975; McKerrow et al., 1977; Leggett et al., 1979) that the Southern Uplands evolved as an accretionary prism along the northern Iapetus margin in Britain during the Ordovician and Silurian. The major fault-bounded tracts in the area are interpreted as individual accreted packets of ocean-floor and trench sediment, added in sequence from NW to SE. This model explains the following facets of the area. 1, The predominant NW (continent-wards) younging of strata. 2, The progressive appearance of younger sequences in the fault blocks towards the SE (ocean-wards).

An important corollary of the model is that the apparent differences in stratigraphy between many of the sequences can be explained by their having been initially deposited on widely separated areas of the ocean floor, and subsequently juxtaposed by accretion.

Ballantrae and Girvan

The Ballantrae Complex (Table 3.1) consists of a tectonic assemblage of pillow basalts, (together with interbedded pyroclastics, olistostromes, shales and cherts) serpentinites, gabbros, trondhjemites and mafic dykes (Bluck, 1978). The volcanic sequences have been variously interpreted as parts of an island arc (Bloxam and Lewis, 1972), as a small ocean basin (Church and Gayer, 1973), as an island-arc-marginal-basin assemblage (Bluck et al., 1980), and as an oceanic island, seamount or plateau (Barrett et al., 1982). Wilkinson and Cann (1974) concluded from trace element (Ti, Zr, Y and Nb) studies that different parts of the sequence may belong to three different genetic groups: hot spot basalts, island arc low-potassium tholeiites, and possible ocean floor basalts. These distinctions fit with the recent finds (Stone and Rushton, 1983) of graptolites north of Bennane Head: the successions south of Balcreuchan Port (NX 099 878) contain fault-repeated sequences of Lower to Middle Arenig sediments and lavas related to hot spot or oceanic island basalts, whilst elsewhere in the Complex, an Upper Arenig sequence has probable island arc affinities. Other proven arc sequences are however much older; 501 ± 12 Ma and 476 ± 14 Ma being reported by Thirlwell and Bluck (1984).

Bluck et al. (1980) suggest that the generation of the basalts, gabbros and trondhjemites occurred at about 483 ± 4 Ma (a U-Pb date from the Byne Hill trondhjemite), and that obduction occurred very shortly afterwards at $478 \pm$ Ma, from a K-Ar date on an amphibole associated with emplacement of the serpentinite (Spray and Williams, 1980). If this interpretation of the data is correct, it follows that Lower and Middle Arenig hot spot and island arc basalts (together with associated serpentinites, gabbros and trondhjemites) were obducted before the Late Arenig. Further arc-related Late Arenig lavas and sediments then covered these allochthonous rocks. These Arenig events are clearly earlier than the accretion of the Southern Uplands, where the oldest and most northerly sequence is Llandeilo/Early Caradoc age. However, the Girvan succession of Caradoc and Ashgill sediments (from the Barr Group upwards) appears to have accumulated prior to and during the early accretion of the Southern Uplands.

Williams (1962) has described the faunas, stratigraphy and structure of the Girvan area. The trilobites and brachiopods of the Ordovician succession all have American affinities, and show that Scotland was on the American side of the Iapetus Ocean. The sequence gets thicker, and its base gets older, to the south of each of a series of ENE-WSW faults. North of Dailly, the Late Caradoc Craighead Limestone rests on Llanvirn basalts and cherts of the Ballantrae Complex, whilst 11 km to the south of this point, the Upper Llanvirn/Llandeilo Stinchar Limestone rests on Lower/Middle Arenig lavas and cherts (Ingham, 1978). Over a distance of around 15 km, the post-Arenig Ordovician sediments increase in thickness from around 800 m to over 4,300 m. Prior to the Caradoc, the Ordovician palaeoslope was mainly down to the south, but after this time current directions are dominantly NE or SW; this fits with the model of accretion in the Southern Uplands commencing during the Caradoc, but direct evidence of an accretionary prism to the south of the Midland Valley is not seen until the appearance of greywacke detritus from the south during the Llandovery (Leggett et al, 1979).

ITINERARY 6.   THE LAKE DISTRICT AND DOBB'S LINN

Route from North Wales to the Lake District

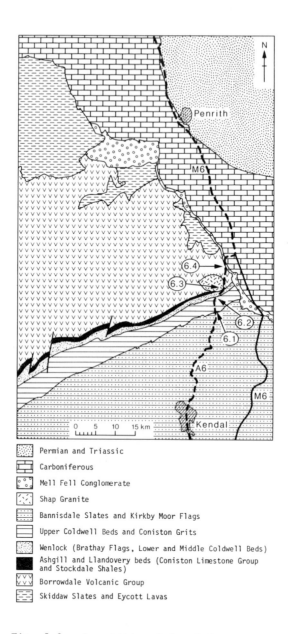

Fig. 3.2. Localities 6.1 - 6.4.

Drive east from Bangor on the A55, along the northern margin of the Snowdon Range, and across the Silurian turbidites of Denbighshire (Fig. 3.1). East of Colwyn Bay, the road passes over Lower Carboniferous (Visean) limestones which are overlain by red beds of Permian or Triassic age, east of Abergele. Further outcrops of Silurian rocks occur east of the Vale of Clwyd; these are again overlain by Carboniferous Limestone, this time followed by deltaic Upper Carboniferous beds, which crop out on the western margin of the Cheshire Basin. This basin contains large thicknesses of Carboniferous rocks, and during the Triassic was part of a graben system extending from Worcestershire to the Irish Sea.

The northern edge of the Cheshire Basin is crossed on the M56; at Junction 9, continue north on the M6. After crossing the high bridge across the Manchester Ship Canal, the motorway traverses Permian sandstones, and then outcrops of Late Carboniferous Coal Measures are present between Junctions 25 and 28. North of Leyland (Junction 28), the motorway crosses the eastern margin of the Irish Sea Basin with thick Triassic sequences. The Pennine Hills standing up to the east of the route are formed predominantly of Middle Carboniferous deltaic sandstones (the Millstone Grit). Around Garstang, the M6 crosses onto these sandstones.

North of Lancaster the Carboniferous Limestone comes to the surface, forming much of the high ground on both sides of the motorway. Between Carnforth and Kendal, the regional dip of these Carboniferous beds is towards the south-east, away from the Lake District. It is thought that the radial drainage in the Lake District is the result of a superimposed system formed during post-Carboniferous doming of this Lower Palaeozoic inlier.

Leave the M6 at Junction 39, and proceed south along the A6 towards Kendal (Fig. 3.2). (Alternatively, if Kendal is not too busy, leave the M6 at Junction 36, drive to Kendal and then north on the A6).

Locality 6.1. Road cuts on the A6 north of the Borrow Bridge (Fig. 3.2)

About ¾ mile north of the Borrow Bridge, exposures of Ludlow turbidites (Coniston Grits and Bannisdale Slates) occur to the east of the road (Fig. 3.3); a stop can be made at a layby on the west of the road (NY 554 058), 200 yards north of the road cuttings (at NY 555 050 to NY 554 056). Here alternating sequences of coarse and fine turbidites are folded with a strike at $055°$ and a steep cleavage at $060°$ (Moseley, 1968).

The Silurian rocks of the Lake District commence with Llandovery and Wenlock mudstones and siltstones; coarse turbidites do not appear until the Late Wenlock and Ludlow. A study of the current directions (Furness et al., 1967) shows a dominant direction from the north-west and a subsidiary direction from the north-east. It has long been assumed that the Iapetus Ocean lay to the north of the Lake District (where the Ordovician shelly faunas are of Scandinavian affinities). How could there be an oceanic source for these sediments? A possible explanation has recently been put forward (Leggett et al., 1983). The calc-alkaline rocks of the Borrowdale Volcanic Group suggest the presence of a Llandeilo island arc lying only some 25 miles south of the Iapetus

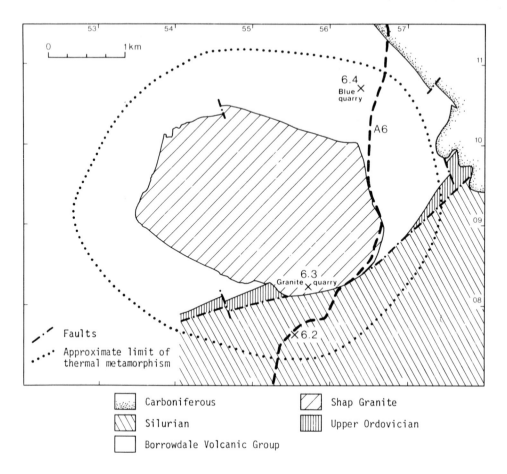

Fig. 3.3. Detail of localities 6.2 - 6.4 at the Shap Granite.

suture, so that clearly some of the Ordovician forearc region is missing. If this missing crust was subducted below the Scottish Southern Uplands between the Late Wenlock and Early Devonian, the Southern Uplands to the north could have been uplifted to form a source for the Late Silurian turbidites in the Lake District. The sedimentary petrography (Furness et al., 1967; Rickards, 1978) is compatible with this hypothesis.

Locality 6.2.   Road cuts on the A6 north of Shap (Fig. 3.2 and 3.3).

Continuing north over Shap summit, it is possible to park on the left shoulder at NX 553 073 and then examine the road cuttings on the east of the road at NX 555 077, where mudstones and fine siltstones of the Wenlock Brathay Flags and Coldwell Beds are exposed (Marr, 1916, Fig.16). The thin partings of silt have been interpreted as turbidites (Furness et al., 1967). The beds are metamorphosed. Looking north one gets a good view of the Shap Granite in the Pink Quarry; from the studies of Harker and Marr (1891, 1893) it is clear that the road cuttings are in the thermal aureole of the granite (Fig. 3.3)

Locality 6.3.   Pink Quarry (Fig. 3.2 and 3.3).

The Pink Quarry (NY 556 083) is owned by Thos. W. Ward (Roadstone) Ltd., Shap. The granite, with its distinctive pink orthoclase phenocrysts has been described by Grantham (1928) and Firman (1978, pp 156-8). It may, with related intrusions, lie under much of the region. Wadge et al., (1978) have obtained an Rb/Sr isochron which indicates an age of 394 ± 3 Ma, which can be interpreted as Early Devonian (McKerrow et al., 1985). As the Shap Granite post-dates the cleavage, but is earlier than the widespread development of kink-bands in the Lake District (Moseley, 1978, pp. 53, 157), it can be concluded that the main Caledonian Orogeny in this region was after the Pridoli (the same structures occur in rocks of this age to the east of Kendal) and before the end of the Early Devonian. The end of subduction in Scotland is marked by the cessation of calc-alkaline lavas in the Siegenian or early Emsian; this is perhaps the time of final collision, and when the main deformation occurred in the Lake District.

Locality 6.4.   Shap Blue Quarry (Fig. 3.2 and 3.3)

The Shap Blue Quarry (NY 564 107) lies $1\frac{1}{2}$ miles north of the Pink Quarry. Here lavas and tuffs of the Borrowdale Volcanic Group are worked by Thos. W. Ward (Roadstone) Ltd. in a quarry to the west of the A6. Thermal metamorphism, epidotisation and formation of garnets have been recorded (Firman, 1954).

Route from the Lake District to the Southern Uplands

The route then continues north for $1\frac{1}{2}$ miles and turns right onto the B6261 to join the M6 motorway at Junction 39. This junction lies on the Carboniferous Limestone, which is dipping gently eastwards, away from the uplifted region of the Lake District. Northwards the motorway passes

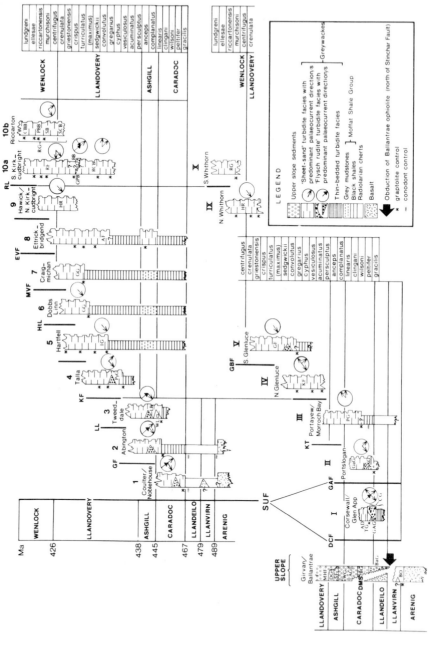

Fig. 3.4. Stratigraphic columns in ten sections in the Southern Uplands (from Leggett, McKerrow and Casey, In: Leggett, 1982, Spec. Publ. Geol. Soc. London, 10).

onto Permian red sandstones around Penrith, and thence around Carlisle where the Permian rocks are overlain by a thick cover of Triassic mudstones and sandstones with some halite. This succession is believed to mask the Iapetus suture (see fig. 1, in McKerrow, this volume).

North of Carlisle, the M6 ends and continues as the A74 road; after 6 miles on the A74, we enter Scotland at the River Sark. In a further 6 miles, between Kirtlebridge and Ecclefechan, the route crosses the outcrop of Lower Carboniferous sandstones and shales. In this area, the basal Carboniferous beds are basalts which form a north-facing escarpment $2\frac{1}{2}$ miles north of Ecclefechan, and also the prominent flat-topped hill to the east, near Burnswark, the site of a Roman camp, 3 miles due north of Ecclefechan (which is clearly visible from the English border). Below the basalt scarp, the A74 crosses the railway to Glasgow on the Upper Old Red Sandstone (fluvial red beds of Late Devonian or Early Carboniferous age), and in less than a mile, the route passes onto Wenlock turbidites of the Southern Uplands, about 3 miles south of Lockerbie.

North of Lockerbie, the A74 runs mainly across outliers of Permian sandstones, which have been more easily eroded than the Lower Palaeozoic rocks and are thus followed by the principal valleys. Many of these Permian sandstones are dune-bedded; they have been extensively quarried for building stone for many of the town between Carlisle and Glasgow.

The route leaves the A74 at Beattock, follows the A701 to Moffat, and then the A708 towards the north-east, up the valley of the Moffat Water.

Locality 6.5.   Dobb's Linn

About 10 miles from Moffat, the A708 passes the car park (on the left) below the Grey Mare's Tail waterfall, and starts ascending towards the head of the valley; about 1 mile further on park at the layby north-west of the road (NT 198 156). Walk down the slope from the road, cross the burn on the right, and ascend the main stream into Dobb's Linn ('linn is a Scots word for waterfall and/or pool or gorge below the waterfall). The Dobb's Linn section (Fig. 3.4, 3.5) is the type locality of the Moffat Shales (Lapworth, 1878; Toghill, 1968; Williams, 1980), a pelagic and hemipelagic succession spanning the interval from Mid-Llandeilo to Mid-Llandovery, and succeeded by a strongly diachronous flysch sequence (Fig. 3.4). The following account is adapted from William (1980), and Weir in Leggett (1980).

(a) <u>Glenkiln Shales</u>. Represented only by the uppermost 2m (<u>Climacogr. peltifer</u> Zone, Harnagian). Poorly fossiliferous, dense black and fissile shales.

(b) <u>Lower Hartfell Shales</u> (? Soudleyian-Pusgillian, 15m). Three zones with contrasting facies - <u>Climacogr. wilsoni</u> Zone, highly pyritous dense black shales with 'platy' fracture, and graptolites preserved in relief as pyrite casts; <u>Dicranogr. clingani</u> Zone, less pyritous 'platy' shales, graptolites preserved as carbonacous compressions; <u>Pleurogr. linearis</u> Zone, more massive and cuboidally-jointed ('blocky') black mudstones, richly fossiliferous. <u>C. wilsoni</u> and <u>D. clingani</u> Zones, climacograptids and dicranograptids; <u>P. linearis</u>

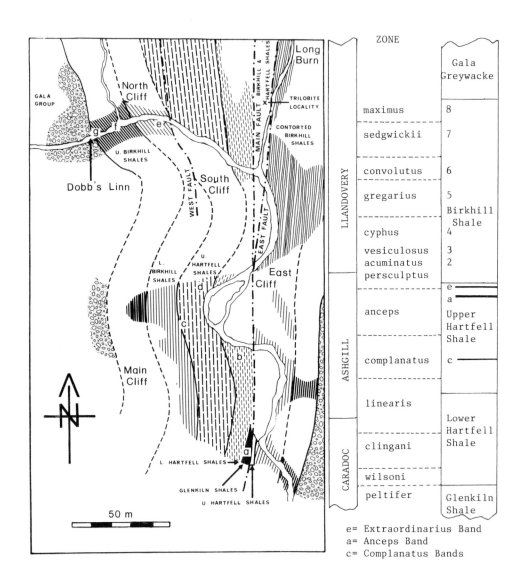

Fig. 3.5. Geological map and succession at Dobb's Linn, locality 6.5. (in part after Williams, 1980).

Zone, large climacograptids, diplograptids, leptograptids.

(c) Upper Hartfell Shales, and contact with Birkhill Shales.
Upper Hartfell Shales (? Cautleyan-Hirnantian, 20m +) - mostly pale grey 'blocky' mudstones (Barren Mudstones). Thin seams and films of fissile black shales occur, and are especially concentrated in the top 2m. The middle of the succession yields the Dicellogr, complanatus fauna (Pusgillian), and the top contains the D. anceps fauna (Hirnantian). A calcareous horizon near the top has yielded a blind dalmanitid (the outcrop in the west bank of the Long Burn, 140 m NNE is more productive). The Barren Mudstones have been correlated with the Ashgill lowering of sea level, when the temporary absence of the greenhouse effect can be correlated with the black shales so characteristic of the Caradoc and Llandovery (Leggett et al., 1979). Barren Mudstones give way suddenly to soft black shales of the Lower Birkhill Shales (Glyptogr. persculptus Zone). These yield an impoverished fauna, mainly of diplograptids and climacograptids.

(d) D. complanatus Band. A thin seam of black shale in the middle of the Barren Mudstone yields this assemblage, mainly of climacograptids, diplograptids and dicranograptids.

(e) Lower Birkhill Shales (Rhuddanian, 14.5m). Again a succession of contrasting facies is displayed - soft black shales succeeded by harder 'platy' shales, G. persculptus Zone; 'platy' shales Akidogr. acuminatus Zone; dark grey 'blocky' mudstones, Cystogr. vesiculous and Coronogr. cyphus Zones. Abundant thin white claystone seams (metabentonites) are present throughout. Richly fossiliferous; climacograptids, diplograptids, monograptids.

(f) Upper Birkhill Shales (Idwian-Fronian, 27m). Alternations of pale grey and dark grey 'blocky' mudstones with fissile black shales and claystones, some thin and nodular. The distal turbidite beds appear; dark horizons die out upwards, and the succession terminates in 10 m of pale grey mudstones. Zone of Corongr. gregarius, Monogr. convolutus (Idwian), M. sedgwickii, Rastrites maximus (Fronian). Fossils confined to darker horizons; monograptids, climacograptids.

(g) Contact of Gala Group and Birkhill Shales. The basal bed of the Gala Group forms the waterfall of Dobb's Linn, and is a thick turbidite with large erosional solemarks.

The Moffat Shale succession comprises two major sedimentary cycles, one Ordovician and one Silurian. Each commences in fissile shales and passes through 'platy' shales and 'blocky' black mudstones into pale grey mudstones, with a sudden reversion to fully euxinic black shales at the top of the Ordovician. The cyclicity has been attributed to modifications in oceanic circulation (Leggett et al., 1981).

ITINERARY 7.   CENTRAL AND NORTHERN BELTS OF THE SOUTHERN UPLANDS

The purpose of this part of the excursion is to provide an overview of the stratigraphic sequence, structural arrangement and sedimentologic attributes of the rocks within the Central and Northern Belts of the Southern Uplands, including some of the Lower Palaeozoic deep marine basin, north of Girvan. South of the Southern Uplands Fault - Stinchar

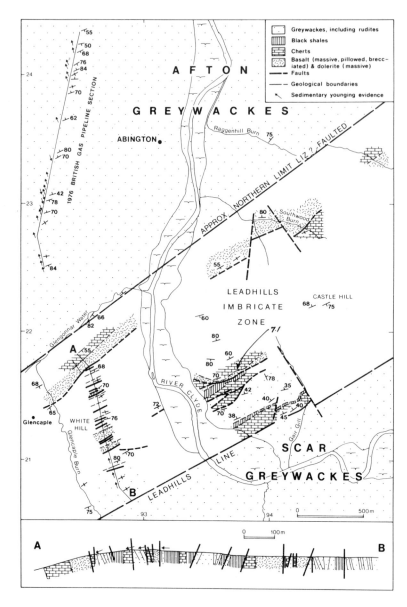

Fig. 3-6. Geological Map of the Abington area (locality 7.1); reproduced from Leggett J.K. and Casey, D.M. 1982, In Watkins, J.S. and Drake C.L. (eds) Memoir American Association of Petroleum Geologists 34, fig. 5. p.385. Reproduced with permission of A.A.P.G.

FIELD GUIDE TO THE LAKE DISTRICT AND SOUTHERN UPLANDS   91

Valley Fault complex the sequence is dominated by greywacke turbidites, disposed in a number of strike-fault bounded slices. Graptolitic shales, cherts and occasional spilites are stratigraphically and structurally interleaved with the greywackes and the faunal evidence indicates that rocks of the Northern Belt (between the Southern Uplands Fault and the Kingledores Fault-Line) (KF, Fig. 2, McKerrow, this volume), are largely of mid- and late Ordovician age (Llandeilo-Ashgill), while the Central Belt sequences are predominantly early Silurian (Llandovery) (Fig. 3.4). Within individual fault-slices the general stratigraphic and structural polarity is towards the northwest, older pelagics and occasional volcanics occurring in disrupted (often imbricated) zones along the southeastern boundaries of the slices.

Route from Moffat

From Moffat the route follows the A74 trunk-road northwards to Abington. A number of quarries and roadcuts in this area reveal thick, poorly graded greywacke-turbidites with the siliceous composition (acid igneous and metamorphic-rock provenance) typical of most Central Belt sandstones. These beds are separated by thin dark shales, occasionally yielding Llandovery graptolies. Small to medium scale folding is locally observed but grading and sole-marks indicate general northerly younging throughout much of this tract. Near Elvanfoot cross the Kingledores Fault onto the Ordovician sequences of the Northern Belt.

Locality 7.1.   Castle Hill, Abington

At Abington leave the A74 and take a minor road that leads east out of the village. After crossing the River Clyde turn south after some 600 m and proceed for about 1.5 km to the foot of Castle Hill (near the railway bridge). Climb to the gully and crags (NS 936 216) (Fig. 3.6).
    The following account is adapted from Leggett (1980).
    Abington is situated in the Leadhills Imbricate Zone, along the southern border of the Afton-Abington tract (Fig. 7.1). The 1976 British Gas pipeline section passed across the hills to the west of the Clyde. In the pipeline section, four basalt strips were laid bare in the Leadhills Imbricate Zone (Fig. 3.6) ranging in thickness from about 20 to 90 m. All are fault-bounded at the base, and overlain (in most cases with an apparently concordant contact) by bedded grey cherts. The basal faults are manifested either as thin (up to 30 cm) zones of crushed shaly rock with irregular quartz veins up to 3 cm across (eg. northernmost body), or as thicker, more diffuse zones of intense shattering in the underlying lithology. Where the attitude of faults in the imbricate zone is clear, they are steeply dipping, frequently sub-parallel to bedding. The distribution of faults across White Hill is suggestive of a crude fan arrangement.
    Details within the basalt strips are usually obscure. Most of these rocks are rotted to varying degrees, and appear to be massive. Exceptions occcur in the southernmost body, where a number of 10-40 cm units are separated by narrow, foliated (almost shaly) zones of weathered basalt, and may represent thin flows. In addition, a 5 m unit of pillow lava

occurs at the top of the body. In the northernmost basalt strip the rock is a rather uniform coarse metadolerite.

Nowhere in the Abington area could the intensity of imbrications in the Leadhills Imbricate Zone, as revealed in the pipeline section, be suspected from contemporary outcrops. However in a small area on the SW flank of Castle Hill a number of strike faults can be mapped.

The prominent gully exposes some 25 metres of pale grey fine-grained lava in crags along its northern face. The lava has a characteristic light reddish-brown colour and rubbly texture on weathered faces, and the rock is permeated by numerous irregular brown planes of weathering. A thick development of basalt breccia occurs towards the top of the gully - a rare feature in the Northern Belt basalts. Most of the clasts are angular fist-sized blocks of the normal pale-grey lava, but some have distinct pillow shapes, with a maximum dimension of 30 cm. The matrix of the breccia is calcareous, with angular fragments of chloritized material, which may originally have been glass shards. Black shales and flaggy mudstones are exposed in the south bank, so the gully floor probably marks a fault. The distribution of small outcrops further downslope suggests that the fault may dip at a moderate to shallow angle to the NW. The basalt is overlain by grey, lenticular-bedded, white-weathering cherts which are exposed for a few hundreds of metres to the NW (highly unlikely to be a true thickness). In the next gully, running up the slope to the NE of an ancient earthwork, screes of black cherty mudstone and shale indicate the possible presence of another fault.

The cherts of the Abington district have yielded conodonts ranging in age from Arenig to Llandeilo, but overlying black shales contain only Caradoc graptolites.

Route to Afton Water

Return to the A74 road at Abington, proceed via the B797 road through the Leadhills-Wanlockhead area. This region has a long history of lead-zinc production from NNW-trending veins transecting Caradoc greywackes. The controlling structural feature for this mineralisation appears to be a major thrust or reverse fault (the Leadhills Line) which superposes cherts, graptolitic shales and occasional spilites of Llandeilo-Caradoc age on the greywackes to the northwest.

Turning right on to the A76 road south of Sanquhar the route passes into the Carboniferous rocks of the Sanquhar Basin. At Bail Hill, on the northern border of this Basin, about 2 km NNE of Kirkconnel Village, a body of late Llandeilo submarine volcanics (mainly basalts, hawaiites and mugearites) with associated cherts and olistostromic sediments recently has been interpreted as a mildly alkaline seamount, tectonically accreted to the northwestern inner wall of the Southern Uplands trench during the Late Ordovician (McMurtry, 1980; Hepworth et al., 1982). About 2.5 km east of New Cumnock the A76 road crosses the trace of the Southern Uplands Fault which here brings Lower Carboniferous sandstones and shales on the north against Ordovician cherts, spilites and black shales on the south. Turn left in New Cumnock, near the junction of the A74 with the B741 road, and follow a narrow road leading to Glen

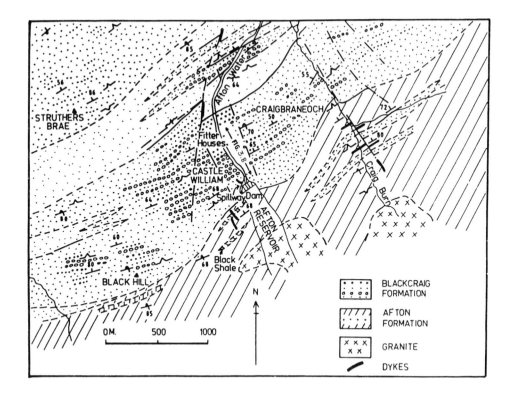

Fig. 3.7. Geology of Afton Water area (locality 7.2), after Holroyd (1978).

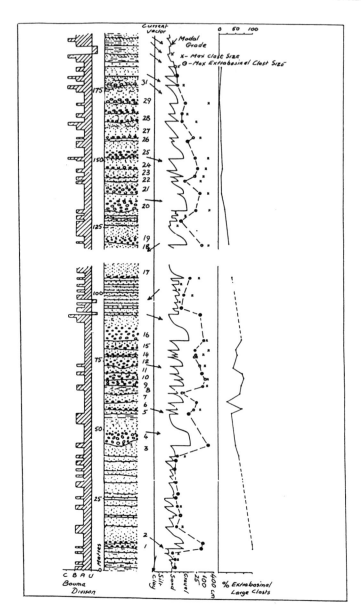

Fig.3.8. Log of the Afton Water section shown on Figure 3.7 (locality 7.2), modified from Kelling and Holroyd (1978).

Afton, made famous by the Scottish Poet, Robert Burns, in his song "Flow gently, Sweet Afton". Proceed south for about 7.5 km to the Afton Reservoir area to examine the succession of Late Ordovician coarse clastics exposed in the bed and on the banks of the Afton Water (ensuring that the controlled outflow from the Reservoir into the stream is at a minimum).

Locality 7.2.   Afton Water

Prominent crags on either side of the Afton Water near the Filter Houses (Craigbraneoch Hill to the east and Castle William to the west) are formed by the coarse clastics of the late Ordovician Blackcraig Formation (Floyd, 1976) that here forms the northward-younging southern limb of a broad, ENE-trending syncline (Fig. 3.7).

The Blackcraig Formation is a 1500 m thick sequence characterised by thick, poorly graded greywacke-turbidites that coarsen upwards rapidly into pebbly and granular greywackes, interbedded with thick units of boulder-conglomerate, most of which are capped by laminated and cross-laminated arenites. Near the top of the Formation the rudites are finer in grade, passing gradually upward into massive, quartz-pebble bearing greywackes.

The lower part of this succession is exposed in the Afton Water just north of the Reservoir. Commencing near the spillway exit (Fig.3.7) steeply dipping, northwards-younging beds forming the sequence depicted in Fig.3.8 may be followed downstream for approximately 600 m to the vicinity of the Filter Houses. Thick, broadly lenticular, indistinctly graded (T_{abe}) coarse greywackes at the base of the sequence contain a few isolated cobbles and boulders of 'extra-basinal' rock-types (mainly acid plutonics) and two thick graded boulder-rudites with strongly channelled bases. About 45 m above the base of this sequence, units of boulder-rudite appear in abundance, heralded by a 6m thick 'mega-graded' unit, strongly channelled into the underlying arenites and siltstones and rich in well-rounded cobbles and boulders of spilite, dolerite, granite and porphyrite.

Downstream, the succeeding 45m sequence of channelled rudites and pebbly arenites includes mainly organised ('mega-graded') types, although a few disorganised (matrix-supported) units also occur. An overall tendency to upwards-thinning in this sequence is accompanied by a less conspicuous decline in the maximum clast size and by a gradual increase in the content of intraclasts. The latter are composed of greywacke and lutite, petrographically identical with the finer-grained units interbedded with the rudites. Greywacke clasts generally display better rounding than the associated lutite cobbles or boulders and a few composite clasts (partly greywacke and partly lutite) also reveal differential rounding. These observations suggest that the intraclasts have been gouged from the walls of the submarine fan channels or lower submarine canyon in a partly lithified condition and have undergone shape modification during transport (presumably relatively short-lived) to their ultimate depositional site   (Kelling and Holroyd, 1978).

A further feature of the rudite units near the top of this sequence is the occurrence of single sets of trough cross-stratification (up to

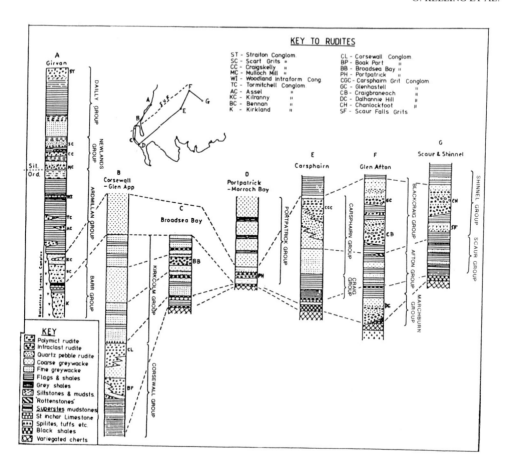

Fig. 3.9. Comparison of the Upper Ordovician and Lower Silurian successions with rudite horizons, in the north-west part of the Southern Uplands (after Holroyd, 1978).

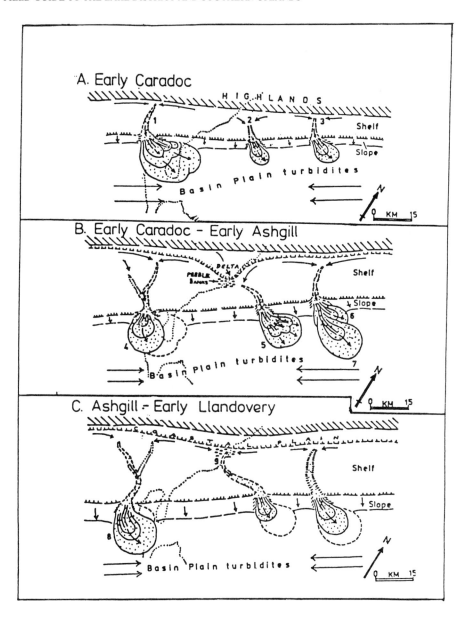

Fig. 3.10. Composite diagram illustrating the general development of the Scottish canyon-fan systems from early Caradoc to early Llandovery times. From left to right these are the Corsewall, Carsphairn and Afton systems (after Holroyd, 1978).
A-1, Corsewall lobes; 2, proto Carsphairn fan; 3, Marchburn lobe.
B-4, Broadsea Bay lobe; 5, Carsphairn lobe; 6, Afton lobe;
    7, Shinnel lobe.
C-8, Portpatrick lobe; 9, Carsphairn canyon head, Girvan area.
Arrows indicate main transport directions.

30 cm thick) as the analogues of the Bouma 'c' interval in the upper part of several mega-graded rudites (see, especially unit 16 in Fig. 3.8). Over-steepened foresets are common in these cross-stratified sets, the trough character of which is best discerned in the crags on Craigbraneoch Rig.

Downstream, beyond the small gap in the exposure, a further upwards-thinning sequence of mega-graded rudites (60 m thick) is conspicuously enriched in large intraclasts and correspondingly impoverished in coarse extrabasinal material, thus continuing the compositional trend apparent throughout the entire rudite succession (Fig. 3.8). The topmost rudites are capped by well-graded (T_{abc}), granule-rich greywackes.

Basal scours, cross-strata and cobble imbrication consistently indicate current transport from the northwest or north - essentially perpendicular to the late Ordovician marginal slope. The Blackcraig rudites attain their maximum thickness and coarseness just west of the Afton Water (Fig. 3.7). Traced along the strike both to the southwest and the northeast the rudites are replaced laterally by quartz-pebble bearing, well-graded greywackes, interbedded with thin graptolite mudstones. The channelled character, internal organisation and lateral/vertical relationships of the rudites indicate that they were formed within the channelled apical portion of a major submarine fan system, prograding from the North into the Southern Uplands trench in late Ordovician times.

The present distribution of the major flysch-rudite bodies in the Southern Uplands (Fig. 3.9) reflects the location of a limited number of coarse gravel input systems (? canyon-fan complex). The best-developed and most persistent of these were the Corsewall (Rhinns of Galloway), Girvan-Carsphairn and Afton-Shinnel systems (Fig.3.10), each of which originated early in the Caradoc and persisted spasmodically into the Llandovery (Holroyd, 1978). No significant polymict rudite bodies are known from areas to the northeast of the Afton-Shinnel complex, apart from the rudites associated with localised reworking of submarine volcanic edifices such as the Bail Hill complex or olistostromes, such as that of the Wrae/Stobo area of Peeblesshire (Leggett, 1980). However, granule- and pebble-grade sediments form proximal turbidite sequences within the Northern Belt (such as the "Haggis Rock" of Ritchie and Eckford, 1935) and presumably identify further lateral input systems.

Route to Girvan

Return to New Cumnock and proceed westwards on the B471 road, running virtually parallel to the scarp of the Southern Uplands Fault for the first 8 km or so before traversing across the southern part of the Ayrshire coalfield between Dalmellington and Straiton. West of Straiton the route crosses a faulted inlier of Silurian, Old Red Sandstone and Carboniferous sediments. These beds lie above a thick sequence of (Llanvirn to Ashgill) Ordovician sediments which occur to the north of the Southern Uplands accretionary prism, and which were affected by large faults (downthrowing to the south) during deposition. Posthumous movements on some of these faults control the present outcrops of the later Palaeozoic.

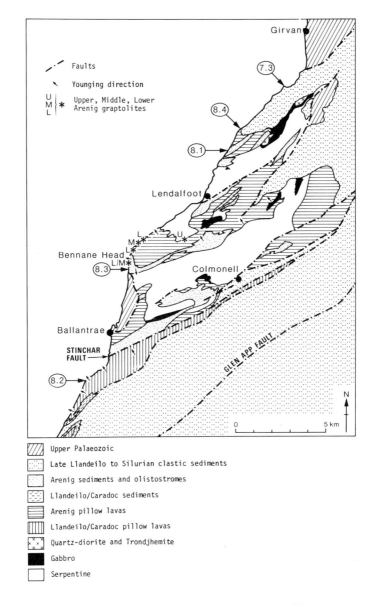

Fig. 3.11. Geological map of the Girvan-Ballantrae area.

Locality 7.3.   Woodland Point

Drive through Girvan, turning right at the roundabout to the south of the town on the A77 coast. In about a mile, park at Woodland Point (marked Black Neuk on the 1:50,000 map).

At Woodland Point (NX 169 953) Lower Llandovery sediments rest with a smaller angular discordance on shales and turbidites of the Early Ashgill Shalloch Formation (Ingham, 1978). Ingham (1978, p.174) points out that the unconformity is not subaerial, but reflects shallowing and down-slope channelling in earliest Silurian times. The basal Silurian formation is the Craigskelly Conglomerate (only in 1 m lenses at Woodland, but up to 40 m thick at the Haven (NX 179 961) 1 km to the north-east). The succeeding Woodland Formation consists of shales and thin sandstones with, successively, a Stricklandia Community, a Clorinda Community and graptolite shales (Cocks and Toghill, 1973). Although most of the brachiopod horizons are death assemblages, both a Stricklandia and a Clorinda Community have been recorded in life position, with umbones facing downwards. The Woodland Formation can thus be interpreted as being deposited on a deep shelf environment, which towards the top becomes too deep for brachiopods. The main mass of Woodland Point (Black Neuk) consists of 45 m of the Scart Grits, massive turbidites with units up to 2 m thick grading from conglomerate to siltstones, and containing igneous and quartzite pebbles. The Scart Grits were deposited before consolidation of the upper shales of the Woodland Formation which are deformed in places (Kuenen, 1953; Cocks and Toghill, 1973).

ITINERARY 8.   GIRVAN - BALLANTRAE SECTION.

The Glen App Fault (Fig 3.12) has traditionally been taken as the westward continuation of the Southern Upland Fault, but to the north of it there is a thick development of the N. gracilis Zone (Glenn App and Tappins Groups) which is rich in basic volcanic debris similar to the Marchburn Group (Floyd, 1982), and the Coulter-Noblehouse sequence (Leggett et al., 1979) which lie south of the Southern Upland Fault along strike to the east. It now seems probable that these groups form the oldest part of the Southern Uplands accretionary prism, and were emplaced in the Early Caradoc (Leggett et al., 1982). In fault contact to the north of these sediments, the Downan Point lavas may represent part of the same sequence (Barrett et al., 1972); these lavas are separated from the main mass of the Ballantrae Complex by the Stinchar Valley Fault, the likely continuation of the Southern Upland Fault in this area.

Locality 8.1.

Drive south from Girvan on the A77 coast road past Woodland Point, (loc. 7.3) to Kennedy's Pass (loc. 8.4) below where the younger Ordovician sediments (Fig. 3.11, 3.12, Table 1) rest unconformably on the Arenig lavas of the Ballantrae Complex, and thence to Pinbain Beach (NX 137 915)

An olistostrome around 200 m thick crops out on the shore. It contains blocks of basalt, limestone, greywacke, amphibolite,

TABLE 3.1   SUCCESSION IN THE GIRVAN - BALLANTRAE COASTAL SECTION

Early Wenlock and Llandovery	Conglomerates, sandstones and shales (Graptolitic to shallow shelf)	80 m
----non-sequence----------------------		
Ashgill	Ardmillan Group: Shalloch Formation (graptolitic sandstones and muds)	330 m
	Upper Whitehouse Formation (green and red deep shelf mudstones)	30 m
Caradoc	Lower Whitehouse Formation (calcareous turbidites, greywackes and shales)	100 m
	Ardwell Formation (greywakces)	1,200 m
	Balclatchie Formation (represented on the coast by greywakces and the deep water Kilranny Conglomerate)	250 m
Llandeilo and Late Llanvirn	Barr Group: Benan Conglomerate (coastal)	180 m
----unconformity----------------------		
Late Arenig	Ballantrae complex lavas at North Baillard and possible Games Loup, with olistostromes, conglomerates, cherts and shales (deep to shallow water). Conglomerates include clasts of amphibolite and trondhjemite.	
478± 8 Ma 483± 4 Ma	Thrusting and imbrication of Early and Middle Arenig lavas and sediment, serpentinite. Intrusion of gabbro and trondhjemites. Development of amphibolites at base of ultramafic thrust sheets.	
Middle and Early Arenig	Ballantrae Complex lavas at Bennane Head, with interbedded conglomerates, cherts and shales.	
505± 11 Ma 576± 32 Ma	Crystallisation of ultramafic segregation within cooling upper mantle or lower oceanic crust (Hamilton et al., 1984).	

NOTES:
1. The Craighead Limestone (equivalent to the Ardwell Formation) rests directly on lavas of the Ballantrae Complex to the north of Girvan.
2. The Stinchar Limestone and the Kirkland Conglomerate (basal units of the Barr Group) rest on the Ballantrae Complex east of Ballantrae. Higher formations of the Ardmillan Group are exposed inland, but not on the coast; rapid increase in thickness develops towards the SE (Fig. 3.11) (Ingham, 1978).
3. The Silurian Sequences reaches a maximum of 1,100 m to the east of Girvan (Cocks and Toghill, 1973).
4. To the south of the Stinchar Fault (south of Ballantrae) the Downan Point lavas (formerly included in the Ballantrae Complex) and greywackes and conglomerates of the Tappins and Glen App Groups are all probably of Llandeilo/Early Caradoc age.

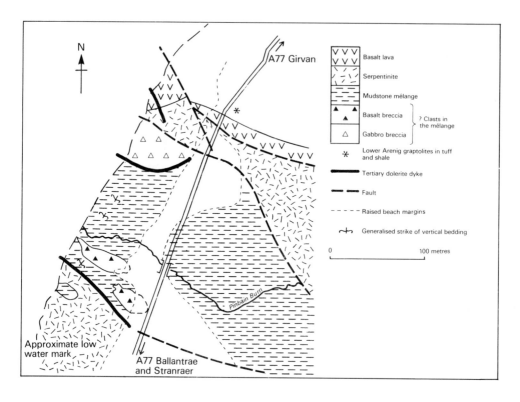

Fig. 3.12. Plan of strata on the shore at Pinbain, north of Lendalfoot (locality 8.1) (after Bailey and McCallien, 1957; J.L.Smellie, pers. comm.).

diorite, dolerite, black shale and, more rarely, blueschist, all contained in a matrix of siliceous mudstone. A large mass of basalt lava and breccia may divide the olistostrome into two separate units or may simply be a very large clast. The black shale clasts contain fragmentary graptolites but these are undiagnostic of age (Rushton et al., in press). However, the correlation of the amphibolite clasts with similar lithologies dated at 478 ± 8 Ma (Bluck et al., 1980) and contained within the metamorphic "sole" of the adjacent ultramafic body (Spray and Williams, 1980) requires a maximum age for the olistostrome of Middle Arenig. The sequence must have been deposited at a late stage in the development of the Ballantrae Complex subsequent to a phase of major thrusting (Table 3.1). A narrow, faulted sliver of serpentinite separates the northern margin of the olistostrome from a sequence of lavas and interbedded volcaniclastic sediment, chert and shale which is steeply inclined and youngs consistently northwards, away from the olistostrome. The basal beds of the volcaniclastic sequence, adjacent to the olistostrome margin, contain an early Arenig graptolite fauna (Rushton et al., in press) which emphasises the structural isolation of the olistostrome. Despite the presence of graptolitic shales at the base of the volcanic sequence there is evidence that its upper part, exposed further north, was laid down in shallow water (Bluck, 1982; Smellie, 1984).

At its southern margin the olistostrome is faulted against serpentinite although the actual plane of contact is obscured by the intrusion along it of a Tertiary dyke. The ultramafic rock is best exposed in this vicinity on a small rocky promontory about 400 m south of the olistostrome in association with a striking pegmatitic gabbro (Bonney's Dyke; Balsillie, 1932). The whole assemblage was probably comagmatic particularly as chilling is evident at the margins of the gabbro. However, three distinct types of contact are displayed: sharp contacts against serpentinised harzburgites, including serpentinite enclaves within the gabbro; less well defined, rather diffuse margins abutting coarse clinopyroxenite (diopside) which in turn forms a vein-like network throughout the harzburgite; and sheared contacts that are flinty, fine-grained and intensely rodingitised. This latter effect is more or less pervasive throughout the gabbro converting the original clinopyroxene and plagioclase to a secondary assemblage of pectolite, prehnite, pumpellyite, actinolite, chlorite and garnet. The alteration was probably the result of Ca-metasomatism contemporaneous with the serpentinisation of the ultramafic rock.

Locality 8.2.

Drive south to Ballantrae. Beyond the village, follow the main road over the River Stinchar, and in 200 m turn right, and in a further 450 m, right again. Take the next turn on the right down to the shore and park (NX 079 813). Walk south along the shore.

The lavas north of Downan Point (NX 068 806) lie south of the Stinchar Fault, dip steeply to the north-west and are probably younger than the greywackes and conglomerates of the Glen App Group which crop out (with a faulted contact) to the south-east. They are probably

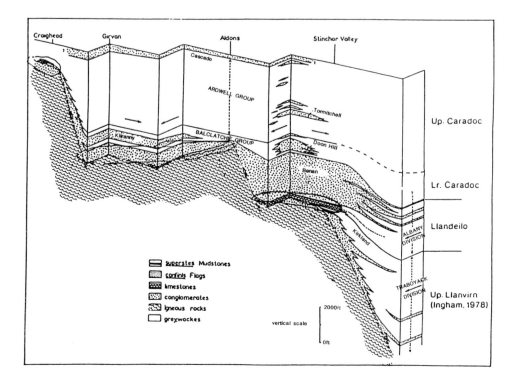

Fig. 3.13. Lateral variations in the Llanvirn to Caradoc sediments at Girvan. After Williams, 1962.

Llandeilo or Early Caradoc in age (Barrett et al., 1982), and thus are now considered distinct from the lavas of the Ballantrae Complex. If the Stinchar Valley Fault is the continuation of the Southern Upland Fault, the Downan Point lavas may be a volcanic edifice in the oldest accreted sequence of the Southern Uplands (Barrett et al., 1982). Individual pillows range from elliptical, through elongate 'bolsters' to sheet flows. Most pillows show concave upper surfaces which are often highly vesicular. Bloxam (1960) noted that "there is a tendency for the larger vesicles to occur in series of concentric bands parallel to the margins of the pillows". Chert and siliceous mudstone is present between some of the pillows. Trace element (Ti, Zr, Y and Nb) abundances suggest they are hot-spot basalts (Wilkinson and Cann, 1974); this interpretation would agree with their assumed position in the Southern Uplands accretionary prism.

Locality 8.3

Return to the A77, and drive north towards Girvan. North of Ballantrae, the road is on a raised beach (the Main Postglacial Shoreline) carved about 6,500 years ago (Sissons, 1976) in red sandstones of Permian age, exposed sporadically on the foreshore. At the north end of the raised beach park near Bennane Lea (NX 092 860) and descend to the shore (Fig. 3.13).

The Permian red sandstones are exposed intertidally and at this point contain isolated angular pebbles of basalt possibly derived from the Ballantrae Complex; the contact between the latter and the Permian sandstones seems unconformable but has been considerably modified by faulting. North of the sandstone sheared serpentinite containing large reddened tectonic inclusions of dolerite is exposed below High Water Mark for a N-S distance of about 10 m. The northern margin of the serpentinite is faulted against doleritic rock, the actual fault zone being marked by a silicified serpentinite breccia a few tens of centimetres thick. The doleritic rock north of the fault has been variously described as extrusive, intrusive and as a block within a melange. It has a curious "autobrecciated" internal texture but does appear to be in conformable contact with bedded red chert at its northern margin. The chert is deformed by extremely well-developed slump folds but has also been tectonically deformed by a series of tight folds which plunge steeply seaward. These are well exposed on the cliffs forming the eastern, inland margin of the raised beach and have the effect of exposing the highest parts of the sequence in synclinal hinge zones on the shore. Conglomerates, interbedded with and stratigraphically overlying the cherts are thus seen close to Low Water Mark; they contain a clast assemblage dominated by basalt with accessory dolerite, greywacke, carbonate (possible altered serpentinite) and chert. Northwards the cherts continue with, overall, an E-W strike and southward younging sense, i.e. by walking north we are moving down-sequence. The appearance of basalt sand and breccia interbedded with the cherts coincides with a swing in strike such that, across a faulted "hinge" zone, strike becomes predominantly N-S with westerly facing. The section is now formed mainly of massive basalt lavas but some breccia and black shale horizons also crop out, the latter containing a sparse

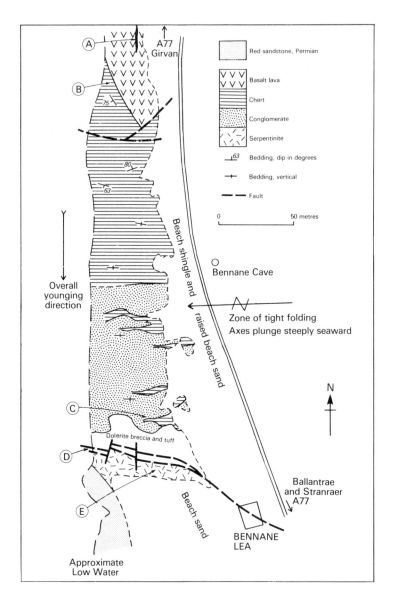

Fig. 3.14. Plan of strata on the shore in the vicinity of Bennane Cave (locality 8.3) (after Bluck, 1978; A.H.F.Robertson, pers.comm.). A, Black siliceous mudstone contains mid-Arenig graptolites. B, Chert and basalt breccia overlies lava. C, Well-developed slump folds in chert. D, Silicified ultramafic breccia. E, Serpentinite contains blocks of rodingitised gabbro.

middle Arenig graptolite fauna.   Down-sequence (north-eastwards),
towards Bennane Head, breccias become the dominant lithology present.
They are oligomict, consisting entirely of basalt debris ranging from
pumice to massive aphyric, and although some of the clasts are rounded
the great majority are angular or sub-angular.   Grain size decreases
locally such that the breccias are interbedded with fine-grained
hyaloclastites.   This part of the succession is best seen on the
roadside sections and upper parts of the sea cliffs at Bennane Head
itself.   It may be preferred to drive from Bennane Lea northwards along
the A77 for about 1 km, park in the layby at Port Vad (NX 093 869) and
then walk south along the road for 200-300 metres.   Stratigraphically
below the breccias a sequence of well-pillowed, aphyric basalt lavas
are well exposed at Port Vad itself and can be best examined at sea level.
This involves a steep scramble down into Port Vad from the layby.   This
pillow basalt sequence lies at the top of the succession from which has
been proved structural imbrication of middle and upper Arenig Strata
(Stone and Rushton, 1983).

Locality 8.4.

From Port Vad drive north past Lendalfoot and Pinbain Beach (Stop 8.1)
to Kennedy's Pass (NX 149 933) which is named after a clan which
occupied this region prior to the 18th Century.
   On the hill to the east, the Llandeilo Benan Conglomerate
unconformably overlies the Ballantrae Complex, but on the shore the
contact is faulted in several places.   Pillow basalts and agglomerates
are in fault contact with both the Benan Conglomerate and the slightly
younger Kilranny Conglomerate.   Large clasts from an inland exposure
of the Benan Conglomerate include some 560 Ma granites which are
unfoliated and thus do not appear to have been affected by the Grampian
Orogeny which affected the adjacent Scottish Highlands around 500 Ma
(Longman et al., 1979).   This evidence suggests that the continental
crust below the southern parts of the Midland Valley (onto which the
Ballantrae Complex was emplaced) has a distinct igneous and metamorphic
history from the Grampian Highlands.   Strike slip faulting along the
Highland Boundary Fault (or below the Midland Valley) may be inferred.
   Above the Benan Conglomerate (but with a faulted contact on the
shore) the Balclatchie Formation is represented by greywackes followed
by the Kilranny Conglomerate, with very varied clasts.   200 m north of
Kennedy's Pass, this conglomerate passes up into greywackes of the
Ardwell Formation with convolute bedding and intraformational flat-
pebble breccias.   This stratigraphic setting of the Kilranny
Conglomerate fits with its interpretation as a deep water slide
conglomerate (Kuenen, 1953; Williams, 1962; Ingham, 1978). In addition
to the soft-sediment deformation, the Ardwell Formation develops small
kink bands near its base;   these increase in size over 200 m into folds
with a wavelength of 50 m.   Which of these folds are tectonic and which
are due to soft sediment deformation remains a matter for argument
(see Ingham, 1978).

Route from Girvan to Drymen

Return to Girvan, thence follow the A77 over the Old Red Sandstone hills around Maybole to the Coal Measure basin north of Ayr. North of Kilmarnock, the A77 rises onto hills formed by thick Early Carboniferous basalts, and then descends into the synclinical basin where Glasgow rests on the Coal Measures. At Thornliebank, turn left onto the A726 to Paisley. Follow the M8 and M898 to Erskine Bridge over the Clyde. Turn right on the A82 to Duntocher, then left on the A810 to Bearsden, and north on the A809, which crosses the basalts on the north limb of the syncline. A stop can be made at Queen's View (NS 511 808), 8 km north of Bearsden, where on clear days there is a panoramic view across the Old Red Sandstone outcrop to the Highland Boundary Fault and the hills beyond; Ben Lomond is particularly prominent to the east of Loch Lomond. Continue north on the A809 to Drymen.

REFERENCES

BAILEY, E.B. & McCALLIEN, W.J. 1957. 'The Ballantrae serpentine, Ayrshire'. Trans. Geol. Soc. Edinburgh, 17, 33-53.

BARRETT, T.J., JENKYNS, H.C., LEGGETT, J.K. & ROBERTSON, A.H.F. 1982. 'Comment on Age and origin of Ballantrae ophiolite and its significance to the Caledonian orogeny and the Ordovician time scale'. Geology, 10, 331.

BALSILLIE, D. 1932. 'The Ballantrae igneous complex, south Ayrshire'. Geol. Mag., 69, 107-31.

BLOXAM, T.W. 1960. 'Pillow structure in spilitic lavas at Downan Point, Ballantrae'. Trans. Geol. Soc. Glasgow, 24, 19-26.

────── & LEWIS, A.D. 1972. 'Ti, Zr and Cr in some British pillow lavas and their petrogenetic affinities'. Nature, Phys. Sci., 237, 134-136.

BLUCK, B.J. 1978. 'Geology of a continental margin 1: the Ballantrae Complex'. In: BOWES, D.R. & LEAKE, B.E. (eds.), Crustal evolution in northwestern Britain and adjacent regions. Geol. J. Spec. Issue, 10, 151-62.

────── 1982. 'Hyalotuff deltaic deposits in the Ballantrae ophiolite of SW Scotland : evidence for crustal position of the lava sequence'. Trans. R. Soc. Edinburgh: Earth Sciences, 72 (for 1981), 217-28.

──────, HALLIDAY, A.N., AFTALION, M. & MACINTYRE, R.M. 1980. 'Age and origin of Ballantrae ophiolite and its significance to the Caledonian orogeny and Ordovician time scale'. Geology, 8, 492-5.

CHURCH, W.R. & GAYER, R.A. 1973. 'The Ballantrae ophiolite'. Geol. Mag., 110, 497-510.

COCKS, L.R.M. & TOGHILL, P. 1973. 'The biostratigraphy of the Silurian rocks of the Girvan district, Scotland'. J. Geol. Soc. London, 129, 209-43.

CRAIG, G.Y. & WALTON, E.K. 1959. 'Sequence and structure in the Silurian rocks of Kirkcudbrightshire'. Geol. Mag., 96, 209-20.

FIRMAN, R.J. 1954. 'Note on metasomatic changes in the rocks adjacent to the Shap Granite'. Proc. Geol. Assoc., 65, 412-4.
─────── 1978. 'Intrusions'. In: MOSELEY, F. (ed.), The Geology of the Lake District. Yorks. Geol. Soc. Occasional Publ., 3, 146-163.
FLOYD, J.D. 1976. 'The Ordovician Rocks of west Nithsdale'. Ph.D. thesis (unpubl.), Univ. of St. Andrews.
─────── 1982. 'Stratigraphy of a flysch succession: the Ordovician of W Nithsdale, SW Scotland'. Trans. Roy. Soc. Edinburgh : Earth Sciences, 73, 1-9.
FURNESS, R.R., LLEWELLYN, P.G., NORMAN, T.N. & RICKARDS, R.B. 1967. 'A review of Wenlock and Ludlow stratigraphy and sedimentation in NW England'. Geol. Mag., 104, 132-47.
GRANTHAM, D.R. 1928. 'The petrology of the Shap Granite'. Proc. Geol. Assoc., 39, 299-311.
HAMILTON, P.J., BLUCK, B.J. & HALLIDAY, A.N. 1984. 'Sm-Nd ages from the Ballantrae Complex, SW Scotland'. Trans. Roy. Soc. Edinburgh: Earth Sciences, 75, 183-187.
HARKER, A. & MARR, J.E. 1891. 'The Shap Granite and Associated rocks'. Q. J. Geol. Soc. London, 47, 266-328.
─────── 1893. 'Supplementary notes on the Metamorphic rocks around the Shap Granite'. Q. J. Geol. Soc. London, 49, 359-71.
HARKNESS, R. 1851. 'On the Silurian rocks of Dumfriesshire and Kirkcudbrightshire'. Q. J. Geol. Soc. London, 7, 46-58.
HEPWORTH, B.C., OLIVER, G.J.H. & McMURTRY, M.J. 1982. 'Sedimentology, volcanism, structure and metamorphism of the northern margin of a Lower Palaeozoic accretionary complex; Bail Hill - Abington area of the Southern Uplands of Scotland. In: LEGGETT, J.K. (ed.), Trench-Forearc Geology. Geol. Soc. London, Spec. Publ., 10, 521-534.
HOLROYD, H. 1978. 'The sedimentology and geotectonic significance of Lower Palaeozoic flysch-rudites'. Ph.D. thesis (unpubl.), Univ. of Wales (Swansea).
INGHAM, J.K. 1978. 'Geology of a continental margin 2: middle and late Ordovician transgression, Girvan'. In: BOWES, D.R. & LEAKE, B.E. (eds.), Crustal Evolution in Northwestern Britain and Adjacent Regions. Geol. J. Spec. Issue, 10, 163-76.
KELLING, G. 1961. 'The stratigraphy and structure of the Ordovician rocks of the Rhinns of Galloway'. Q. J. Geol. Soc. London, 117, 37-75.
─────── & HOLROYD, J. 1978. 'Aspects of clast size, shape and composition in some ancient and modern fan gravels'. In: KELLING, G. & STANLEY, D.J. (eds.), Sedimentation in Submarine Canyons, Fans and Trenches. Dowden, Hutchinson and Ross, Stroudsburg, Penn., 138-159.
KUENEN, P.H. 1953. 'Graded bedding with observations on Lower Palaeozoic rocks of Britain'. K. Ned. Akad. Afd. Nat. Verh., 20, 1-47.
LAPWORTH, C. 1878. 'The Moffat Series'. Q. J. Geol. Soc. London, 34, 240-346.
─────── 1889. 'On the Ballantrae Rocks of the South of Scotland,

and their place in the Upland Sequence'. Geol. Mag., 6, 59-69.
LEGGETT, J.K. (ed.) 1980. A field-guide to the Lower Palaeozoic geology of the Southern Uplands of Scotland. Imperial College of Science and Technology, London. 95pp.
—————— et al., 1981. 'Periodicity in the early Palaeozoic marine realm'. J. Geol. Soc. London, 138, 167-176.
——————, McKERROW, W.S. & EALES, M.H. 1979. 'The Southern Uplands of Scotland: a Lower Palaeozoic accretionary prism'. J. Geol. Soc. London, 136, 755-770.
——————, —————— & SOPER, N.J. 1983. 'A model for the crustal evolution of Southern Scotland'. Tectonics, 2, 187-210.
LONGMAN, C.D., BLUCK, B.J. & van BREEMEN, O. 1979. 'Ordovician conglomerates and evolution of Midland Valley'. Nature, 280, 578-81.
LUMSDEN, G.I., TULLOCH, W., HOWELLS, M.F. & DAVIES, A. 1967. 'The geology of the neighbourhood of Langholm'. Mem. Geol. Surv. Scotland, Sheet 11, 225pp.
MARR, J.E. 1916. The Geology of the Lake District. Cambridge University Press. 220pp.
McKERROW, W.S. & COCKS, L.R.M. 1976. 'Progressive faunal migration across the Iapetus Ocean'. Nature, 263, 304-06.
——————, LEGGETT, J.K. & EALES, M.H. 1977. 'Imbricate thrust model of the Southern Uplands of Scotland'. Nature, 267, 237-239.
——————, LAMBERT, R.St.J. & COCKS, L.R.M. 1985. 'The Ordovician, Silurian and Devonian'. In: SNELLING, N.J.(ed.), Geochronology and the Geological Record. Geol.Soc.London, Spec. Publ., 20, 73-80.
McMURTRY, M.J. 1980. 'The Ordovician rocks of the Bail Hill area, Sanquhar, South Scotland: volcanism and sedimentation in the Iapetus Ocean'. Ph.D. thesis (unpubl.) Univ. of St. Andrews.
MITCHELL, A.H.G. & McKERROW, W.S. 1975. 'Analogous evolution of the Burma orogen and the Scottish Caledonides'. Bull. Geol. Soc. Am., 86, 305-15.
MOSELEY, F. 1968. 'Joints and other structures in the Silurian rocks of the Southern Shap Fells, Westmoreland'. Geol. J., 6, 79-86.
—————— (ed.) 1978. The Geology of the Lake District. Yorks. Geol. Soc. Occasional Publication, 3, 284pp.
PEACH, B.N. & HORNE, J. 1899. 'The Silurian rocks of Britain, 1, Scotland'. Mem. Geol. Surv. Scotland, 749pp.
RICKARDS, R.B. 1978. 'Silurian'. In: MOSELEY, F. (ed.), The Geology of the Lake District. Yorks. Geol. Soc. Occasional Publ., 3, 130-141.
RITCHIE, M. & ECKFORD, R.J.A. 1935. 'The Haggis Rock of the Southern Uplands'. Trans. Geol. Soc. Edinburgh, 13, 371-377.
RUSHTON, A.W.A., STONE, P., SMELLIE, J.L. & TUNNICLIFF, S.P. (in press) 'An early Arenig age for the Pinbain sequence, of the Ballantrae Complex'. Scott. J. Geol.
RUST, B.R. 1965 'The stratigraphy and structure of the Whithorn area of Wigtownshire, Scotland'. Scot. J. Geol., 1, 101-33.
SISSONS, J.B. 1976. The geomorphology of the British Isles: Scotland. Methuen, London. 150pp.
SMELLIE, J.L. 1984. 'Accretionary lapilli and highly vesiculated

pumice in the Ballantrae ophiolite complex; ash-fall products of subaerial eruptions'. Rep. Br. Geol. Surv. 16, No.1, 36-40.
SPRAY, J.G. & WILLIAMS, G.D. 1980. 'The sub-ophiolite metamorphic rocks of the Ballantrae Igneous Complex, SW Scotland'. J. Geol. Soc. London, 137, 359-68.
STONE, P. & RUSHTON, A.W.A. 1983. 'Graptolite faunas from the Ballantrae ophiolite complex and their structural implications'. Scot. J. Geol., 19, 297-310.
THIRLWALL, M.F. & BLUCK, B.J. 1984. 'Sr-Nd isotope and geochemical evidence that the Ballantrae "ophiolite", SW Scotland, is polygenetic'. In: GASS, I.G., LIPPARD, S.J. & SHELTON, A.W., (eds.), Ophiolites and oceanic lithosphere. Geol. Soc. London, Spec. Publ., 13, 215-230.
TOGHILL, P. 1968. 'The graptolite assemblages and zones of the Birkhill Shales (Lower Silurian) at Dobb's Linn'. Palaeontology, 11, 654-68.
─────────── 1970. 'The south-east limit of the Moffat Shales in the Upper Ettrick valley region, Selkirkshire'. Scot. J. Geol., 6, 233-42
WADGE, A.J., GALE, N.H., BECKINSALE, R.D. & RUNDLE, C.C. 1978. 'A Rb-Sr isochron for the Shap Granite'. Proc. Yorks. Geol. Soc., 42, 297-305.
WALTON, E.K. 1961. 'Some aspects of the succession and structure in the Lower Palaeozoic rocks of the Southern Uplands of Scotland'. Geol. Rdsch., 50, 63-77.
─────────── 1963. 'Sedimentation and structure in the Southern Uplands'. In: JOHNSON, M.R.W. and STEWART, F.H. (eds.), The British Caledonides. Oliver and Boyd, Edinburgh, 71-97.
─────────── 1965. 'Lower Palaeozoic rocks: stratigraphy, palaeogeography and structure'. In: CRAIG, G.Y. (ed.), The Geology of Scotland. Oliver and Boyd, Edinburgh, 161-227.
WARREN, P.T. 1964. 'The stratigraphy and structure of the Silurian rocks south-east of Hawick, Roxburghshire'. Q. J. Geol. Soc. London, 120, 192-222.
WEIR, J.A. 1968. 'Structural history of the Silurian rocks of the coast west of Gatehouse, Kirkcudbrightshire'. Scot. J. Geol., 4, 31-52.
WILKINSON, J.M. & CANN, J.R. 1974. 'Trace elements and tectonic relationships of basaltic rocks in the Ballantrae igneous complex, Ayrshire'. Geol. Mag., 111, 35-41.
WILLIAMS, A. 1962. 'The Barr and Lower Ardmillan Series (Caradoc) of the Girvan district, south-west Ayrshire'. Mem. Geol. Soc. London, 3, 267pp.
WILLIAMS, S.H. 1980. An excursion guide to Dobb's Linn. Proc. Geol. Soc. Glasgow,

THE CALEDONIAN GEOLOGY OF THE SCOTTISH HIGHLANDS

Coordinators: A.L. Harris and D.J. Fettes
Contributors: A.L. Harris, D.J. Fettes, W.G. Henderson, J.L. Roberts, J.E. Treagus, A.J. Barber, M.P. Coward, R. Strachan

A. Introduction - the Caledonian Geology of the Scottish Highlands

B. Field itineraries

  9. The Highland Border and Southern Dalradian (WGH)

  10. (Part 1) Rannock Moor and Glencoe Cauldron Subsidence (JRL)
      (Part 2) The Ballachulish Dalradian (JRL & JET)

  11. Morar (ALH)

  12. A traverse from the Great Glen Fault to the Foreland (ALH & AJB)

  13. (Part 1) The Moine Thrust Zone (MPC)

  14. (Part 2) The Moine Thrust Zone (MPC)

C. References

A.L. Harris, Jane Herdman Laboratories of Geology, University, Liverpool L69 3BX.
D.J. Fettes, British Geological Survey, Murchison House, West Mains Road, Edinburgh EH9 3LA, Scotland.
W.G. Henderson, British Geological Survey, Murchison House, West Mains Road, Edinburgh EH9 3LA, Scotland.
J.L. Roberts, Department of Geology, University, Newcastle.
J.E. Treagus, Department of Geology, University, Manchester.
A.J. Barber, Department of Geology, Royal Holloway and Bedford New College, University of London, Egham, Surrey.
M.P. Coward, Department of Geology, Imperial College of Science and Technology, Prince Consort Road, London SW7.
R. Strachan, Department of Geology, Polytechnic, Oxford.

INTRODUCTION - THE CALEDONIAN GEOLOGY OF THE SCOTTISH HIGHLANDS

Introduction

Late Middle Proterozoic to Cambrian metasedimentary and meta-igneous rocks (Moine and Dalradian) comprise much of the Scottish Highlands. These are limited to the SE by the Highland Boundary fault and to the NW by the Moine Thrust zone. Many geologists would now divide the Moine into two (the 'Old' and the 'Young') on the basis of radiometric age and structural and metamorphic history. 'Old' Moine rocks have suffered the tectonothermal effects of the Grenville at 1000 - 1100 Ma as well as later orogenesis. 'Young' Moine rocks are part of a cover sequence which passes conformably upwards into the Riphean-middle Cambrian Dalradian rocks and is included by some workers in the Dalradian Supergroup. This cover sequence underwent only the Grampian (530 - 490 Ma) event. An unconformity between 'Old' and 'Young' Moine is inferred but has not yet been demonstrated. 'Young' Moine rocks have not been proved to crop out to the NW of the Great Glen.

The Moine and Dalradian rocks have been intruded by major plutons varying in composition from acid to ultrabasic and ranging in age from 1000 Ma to 400 Ma. The youngest of these which are volumetrically the most plentiful were generally contemporaneous with molasse sediments (Old Red Sandstone) and intermediate-to-acid lavas of late Silurian to Lower Devonian in age which are strongly unconformable on the metamorphic rocks. Abundant vein complexes and stockworks of minor intrusions of intermediate-to-acid composition appear slightly to predate many of the younger granites. It is likely, on geophysical and structural grounds, that the metamorphic Caledonides are underlain by Lewisian gneisses.

For a fuller discussion of the geology of the Highlands the reader is referred to Fettes and Harris, this volume).

'Old' Moine

These wholly metasedimentary rocks occupy much of the Northern Highlands and a small area SW of Inverness in the Grampian Highlands. Three tectono-metamorphic divisions of 'Old' Moine rocks are distinguished in the N Highlands: the Morar, Glenfinnan and Locheil divisions. These divisions were originally believed to have mutual tectonic contacts across which stratigraphic units could not be traced or correlated. It has been confirmed that the Morar and Glenfinnan divisions are separated by a major ductile thrust (the Sgurr Beag slide) and that the relative age of the two divisions cannot be determined. Since the three-fold sub-division of the 'Old' Moines was set up in 1969, however, it has been discovered that the Loch Quoich line separating Glenfinnan and Locheil divisions is not a major tectonic break and that there is a stratigraphic upward passage from the Glenfinnan metasediments into the Locheil metasediments.

The stratigraphic order for the rocks within the Morar Division

THE CALEDONIAN GEOLOGY OF THE SCOTTISH HIGHLANDS

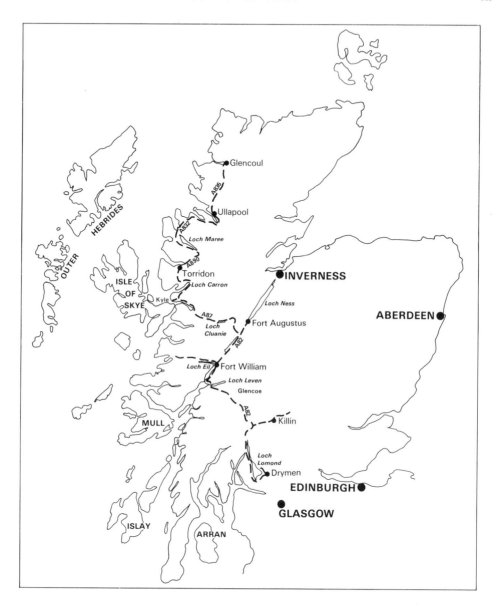

Fig. 4 1. Generalised route map through the Scottish Highlands Itineraries 9 to 14.

is well established on the basis of sedimentary structures. The oldest rocks rest on a basement of Lewisian gneisses and a number of formations of generally psammitic and pelitic nature has been worked out and refined by the vertical distribution of minor lithologies such as calc-silicate bands and heavy mineral laminae. Sedimentary structures are sparse in the Glenfinnan metasediments although sufficient occur to establish that they are older than the largely psammitic Locheil rocks. The ductile thrust between the Morar and Glenfinnan divisions is almost certainly Caledonian, probably being the ductile, in depth extension of a thrust higher than but comparable with the Moine thrust. Many tens or even hundred of kilometres of displacement across the Sgurr Beag thrust has been accomplished and the Glenfinnan rocks are normally at a higher metamorphic grade than the Morar. The Sgurr Beag thrust is folded by NNE-trending upright, strongly curvilinear folds which deform the 456 ± 5 Ma Glen Dessary syenite and are cut by only weakly deformed 440 Ma granite pegmatites; these folds die out abruptly eastwards and in doing so mark the NNE-trending junction between flat-lying Moine (largely Locheil) and steep Moine (largely Glenfinnan) which is strongly folded by the upright post 456 Ma folds. This junction was termed the Loch Quoich line. The Sgurr Beag ductile thrust is not only marked by a major change in lithology and contrast in metamorphic grade but, north of Loch Hourn its course is marked by a series of slices of basement-derived Lewisian rocks. The status of the Lewisian 'inliers' is not yet properly established; they may be slices of basement carrying Glenfinnan rocks as a cover sequence or they may be basement slices thrust into place between two major units.

Radiometric evidence suggests that the two or more episodes of deformation which preceded the thrusting and the upright folds were probably Precambrian - c. 1000 Ma and further evidence for orogenesis at this time is offered by the 1030 Ma (Rb-Sr) adamellitic Ardgour orthogneiss. A strongly folded pegmatite dated at 760 Ma (Rb-Sr) which postdates two earlier fabrics (has recently been reported). This is valuable evidence that the episode of pegmatite emplacement referred to as <u>Knoydartian</u> or <u>Morarian</u> postdated deformation inferred to be late Middle Proterozoic and often somewhat loosely called <u>Grenville</u>.

'Young' Moine and Dalradian

In line with increasing usage the 'Young' Moine is here termed the Grampian Group and included as the lowest group of the Dalradian Supergroup. The other groups in order of ascending age are the Appin, Argyll and Southern Highland group (see Fettes & Harris, this volume). In general the Dalradian youngs from north to south and shows a transition from stable esturine or near shore facies to turbiditic greywackes characteristic of unstable, tectonically active environments. Within the Dalradian a major tillite, the Port Askaig Tillite, has been identified, and this formation is inferred to be c.668 Ma old by somewhat tenuous correlation with sediments associated with the Varanger tillite of Finnmark. Several kilometres higher in the stratigraphic sequence is the Tayvallich Limestone - a thin but persistent marble - with which are associated the oldest widespread basic volcanic rocks of the

TABLE 4.1 SEQUENCE OF EVENTS IN THE SCOTTISH HIGHLANDS

NW OF THE GREAT GLEN			SE OF THE GREAT GLEN		
STRATIGRAPHIC UNITS	TECTONISM	PLUTONISM	STRATIGRAPHIC UNITS	TECTONISM	PLUTONISM
	FAULTING			FAULTING	
OLD RED SANDSTONE			OLD RED SANDSTONE		
	CALEDONIAN (c. 440 Ma)	NEWER GRANITES (c. 420 Ma)			NEWER GRANITES (c. 420 Ma)
					NE BASIC COMPLEXES (c. 490 Ma)
			HIGHLAND BOUNDARY COMPLEX —— Tectonic Contact ——	GRAMPIAN (c. 510 Ma)	BEN VURICH GRANITE (c. 515 Ma)
		CARN CHUINNEAG GRANITE (c. 550 Ma)	DALRADIAN { SOUTHERN HIGHLAND; ARGYLL [TAYVALLICH, CRINAN, EASDALE, ISLAY]; APPIN [BLAIR ATHOLL, BALLACHULISH, LOCHABER]; GRAMPIAN }		
		STRATH HALLADALE GRANITE (c. 650 Ma)			
	MORARIAN? (c. 750 Ma) 'GRENVILLE' (c. 1000 Ma)	ARDGOUR GRANITIC GNEISS (c. 1030 Ma)	MOINE { CENTRAL HIGHLAND DIVISION }	MORARIAN? 'GRENVILLE'?	
MOINE { LOCH EIL DIVISION/GROUP; GLENFINNAN " "; MORAR " " }					

Dalradian succession. The Tayvallich Limestone is the oldest formation in the Dalradian to yield unequivocal Cambrian acritarchs although several formations below the Tayvallich contain acritarchs which have a wide range from Precambrian to Cambrian. It is clear that the Dalradian contains rocks which are transitional from Precambrian to Cambrian. The Leny Limestone some 4-5 km above the Tayvallich Limestone contains Pagetia assigned to the late Lower Cambrian.

The Dalradian Supergroup, cumulatively some 25-30 km thick, is now disposed in a major nappe complex. Within the nappe complex it is possible to distinguish a northwestern zone in which early folds are overturned to face northwestwards and a southeastern zone in which early folds are almost recumbent and face southeastwards. A zone of upright, upward facing early folds intervenes between the two regimes of opposing facing and the recognition of this divergence has suggested a nappe-root zone to many workers. Most localities where Dalradian rocks are exposed offer evidence of severe polyphase deformation and low-to-moderate regional metamorphism. Peripheral areas and areas at a structurally high level in the nappe complex are at greenschist facies while elsewhere grade is as high as the upper amphibolite facies. Pressure conditions vary such that Buchan (intermediate low-pressure) metamorphism characterised by andalusite and cordierite bearing assemblages occurs in NE Scotland, while the rest of the Dalradian tract suffered Barrovian metamorphism with kyanite and almandine garnet bearing assemblages. It is believed that although Buchan metamorphism was already taking place in NE Scotland the emplacement of major gabbroic bodies, the Aberdeenshire (Younger) gabbros, at 489 ± 17 Ma (Rb-Sr) contributed to the crystallisation of regional sillimanite and the development of migmatites in that area. In the central part of the Southern Highlands at Pitlochry the Ben Vuirich granite emplaced at $514 +6/-7$ Ma (U-Pb) calibrates the emplacement of the Tay Nappe. The granite cuts major folds formed during the local second episode of deformation, but was itself deformed during the third.

The demonstration that Caledonides deformation occurred at 456 Ma - 440 Ma (or shortly after) in the N Highlands but was largely pre-514 Ma in the Central Highlands poses a major problem of timing in Highland orogenesis. A model involving deformation becoming progressively later towards the NW over a period of 60-70 Ma may be less valid than one which brings together two terrains which had acquired their separate deformation chronology far apart.

Highland Boundary Complex

The Highland Boundary Complex which outcrops sporadically along the Highland Border, consists of sandstones, shales, cherts, spilites and serpentinite lying largely in a series of tectonically bounded slivers. Arenig fossils have been found in the sediments. It is generally believed that the Complex represents part of an oceanic sequence brought into contact with the Dalradian during major strike slip movements on the Highland Boundary fault.

Igneous activity

The earliest recorded igneous rocks in the Scottish Highlands is the Ardgour orthogneiss referred to above. This body is cut by metabasic rocks, now in the form of garnetiferous hornblende schists which have a wide distribution not only in the granitic orthogneiss but in the Locheil and youngest Glenfinnan metasediments. These amphibolites, like the granitic gneiss, seem to have suffered all the deformation and metamorphism imprinted on the metasediments.

Igneous bodies useful in calibrating the deformation and metamorphic sequence of the Moine include the Carn Chuinneag (560 Ma; Rb-Sr) and Strath Halladale (649 Ma; Rb-Sr) granitic bodies. The Carn Chuinneag granite cuts isoclinal folds in the Moine, but is itself deformed and augened into an orthogneiss whose fabric has the characteristic ESE plunge of the Caledonian mylonites. Undeformed Strath Halladale granite cuts Moine-like rocks which had already suffered polyphase deformation and intense migmitization. This suggests that parts of the N Highlands largely escaped Caledonian reworking and the Strath Halladale situation is reminiscent of the Locheil Division in this respect. As mentioned above the Ben Vuirich granite (514 Ma) and the 'Younger' gabbros of the NE Highlands are valuable in calibrating Grampian deformation and metamorphism in the Dalradian.

Later granites are largely Caledonian I-type granites which have an abundant distribution in the Grampian Highlands but which are scarce in the N Highlands. They range in age from about 440-400 Ma, but the large majority are 415-400 Ma. It was probably the buoyancy lent to the crust of the orthotectonic zone by abundant rising granitic plutons that caused the orogen to rise at the close of the Silurian. As it rose it shed the conglomeratic molasse deposits of the Old Red Sandstone, while some granitic bodies breaching the surface yielded volcanics which are now interbedded with Old Red Sandstone molasse.

Faults

The major thrusts which separate the Moine rocks from the NW foreland are part of the Caledonian reworking of the N Highlands. A series of successively younger and lower thrusts is envisaged as part of a foreland - propogating sequence. Not only the Lewisian basement and Moine rocks are involved but in the Moine thrust zone Torridonian clastic sediments and Cambro-Ordovician quartzites and carbonates are thrust, imbricated and mylonitised. The relationship in time of this thrusting to activity in the orogenic zone is not firmly established. Nevertheless it is established that the thrusting must be post Lower Ordovician while, in the orogenic zone, the set of folds which fold the Sgurr Beag thrust and the Glen Dessary Syenite are Upper Ordovician-Lower Silurian.

At the SE margin of the Highlands the Highland Border fault may have had a prolonged history the earliest part of which may have contained extensive strike slip movements. Lower Devonian rocks are unconformably flat-lying on downward facing Dalradian to the NW of the fault, but becomes steep or vertical southeastwards across the fault. The fault has been shown to have been active as a major basin margin during Lower

Devonian times and much of the displacement across it was probably
accomplished during the early and middle Devonian.
    The Scottish Highlands are transected NE-SW by the Great Glen
fault - a major fault with a large component of lateral slip, and a
small component of displacement down to the SE. Conventionally the
fault is believed to have had a sinistral lateral slip of about 100 km
and a subsequent dextral slip since the Eocene. The very large
displacements required by some palaeomagnetic evidence are thought to
be unlikely because of the general similarity of igneous and
metamorphic rocks on either side of the fault. Rather superficial or
even spurious correlations of igneous bodies and stratigraphic horizons
across the fault are now generally recognised as such and with
increasing knowledge of the geology on either side of the fault is is
increasingly apparent that no confident correlation of any feature,
igneous or metamorphic, can be made within Scotland. The left-lateral
slip on the fault is inferred only because the Great Glen appears to be
one of a number of Upper Palaeozoic faults to affect the Highlands;
the others having smaller displacements across them, can in some cases
be demonstrated to be sinistral.

ITINERARY 9.   THE HIGHLAND BORDER AND SOUTHERN DALRADIAN

Introduction

The Highland Boundary fracture-zone, a major Caledonian lineament,
divides the late Silurian and younger sedimentary and volcanic rocks of
the Scottish Midland Valley from the Cambrian and Precambrian
metamorphic rocks of the Scottish Highlands. The fracture-zone has
been traced across Ireland, but debate surrounds its precise route. It
is a complex zone which has probably involved substantial (?sinistral)
strike-slip and normal faulting in the Silurian and Devonian.
Intermittently along its length there are very narrow outcrops of
altered ultrabasic rocks, mafic extrusives, cherts, limestones, black
pelites, arenites and rudites containing mafic and ultramafic detritus,
collectively called the Highland Boundary Complex (HBC) (Henderson and
Robertson, 1982; Henderson and Fortey, 1982). Sediments within the HBC
have recently revealed fossils ranging in age from Arenig to Llandeilo
and, possibly, to Ashgill (Curry et al., 1984).
    The exposures at Balmaha on Loch Lomondside are very accessible
and show a wide range of the sedimentary and altered ultramafic igneous
components of the HBC. Other components of the HBC are better
represented elsewhere. Mafic igneous units (Ikin, 1983) are well
exposed on the Isle of Arran, at Stonehaven and in the section in the
River North Esk. There are garnetiferous hornblende-schists more than
70 m thick on the Island of Bute and in the forests SW of Aberfoyle.
    For up to 10 km north of the Highland Boundary fracture-zone,
quartz-mica-schists of the Southern Highland Group (Dalradian)
generally have a steep attitude. This 'Highland Border steep belt' is
attributed to (? late Ordovician) downbending of formerly gently
inclined rocks about a lineament, the Highland Border downbend, which

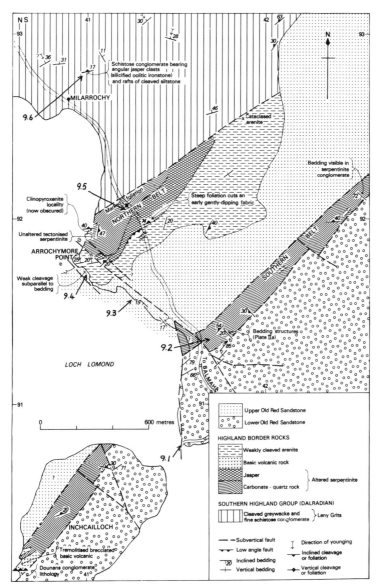

Fig. 4.2. Geology of the Highland Border at Balmaha, Loch Lomond, showing localities 9.1 to 9.6 (after Henderson and Fortey with the permission of the Scottish Academic Press).

runs north of the Highland Boundary fracture-zone and nearly parallel
to it (Fig. 4.2). The Highland Border may represent a long lived
fundamental crustal feature (Harte et al., 1984).

North of the Downbend the regional foliation remains generally
flat for some 25 km, with the rocks arching gently over the Cowal
antiform. North of Loch Lomond the Southern Highland Group rocks dip
northwards under the stratigraphically older Argyll Group formations.
This demonstrates that the 'flat belt' is generally inverted and forms
the lower limb of the major southward facing Tay Nappe complex, the
nose of which is downbent in the Highland Border steep belt. The
Argyll Group rocks maintain their generally inverted nature around Loch
Tay where they are disposed in the late Caledonian Ben Lawers synform.

The section north from the Highland Border to Loch Tay illustrates
a story of increasing structural complexity (see Harte et al., 1984).
Within the southern part of the 'steep zone' the rocks generally
exhibit only the effects of the first deformational phase (D1).
Northwards across the 'steep belt' the successive phases D2, D3 and D4
appear. D2 folds and the associated axial plane fabrics (S2)
characteristically produce a coarse crenulation cleavage or
'herringbone' structure in the more quartzose lithologies and a
composite S1/S2 fabric in the more pelitic rocks. D3 forms SE-facing
open to tight folds with near horizontal axes and axial planes dipping
at c.$45°$ to the NW. D4 folds are generally upright open folds commonly
with subvertical crenulation cleavages developed in pelitic rocks. D4
is responsible for at least the late phase in the Downbend formation
and also for the Ben Lawers synform.

Around the north end of Loch Lomond there are a number of late
Caledonian diorite bodies the most notable of which is the one above
Arrochar (Anderson, 1935).

Note on localities 9.1 to 9.9

Coaches are not permitted on the narrow road north of Balmaha village.
Although localities 9.1 to 9.5 can be conveniently reached on foot from
Balmaha, alternative transport is required to reach the other
localities.

Locality 9.1.   Balmaha Pier, 550 m WSW of large car park at Balmaha
                (NS 4153 9080).

Drive from Drymen to Balmaha and park in the large car park. Access to
the locality is unrestricted along a public footpath. At the lochside
30 m N of the pier, coarse Lower Old Red Sandstone conglomerates dip at
$45°$ to the SW. Conglomeratic units several metres thick are
intercalated with sheets and lenses of pebbly sandstone up to 500 mm
thick. The rounded clasts, up to 400 mm in diameter, were probably
redeposited from an earlier sediment and are mainly of red-stained vein
quartz and a psammite which cannot be matched with lithologies
presently exposed north of the Highland Boundary fracture-zone.
Pervasive closely-spaced fractures cutting the clasts relate to tear-
faults normal to the Highland Boundary fracture-zone. The route to

THE CALEDONIAN GEOLOGY OF THE SCOTTISH HIGHLANDS 123

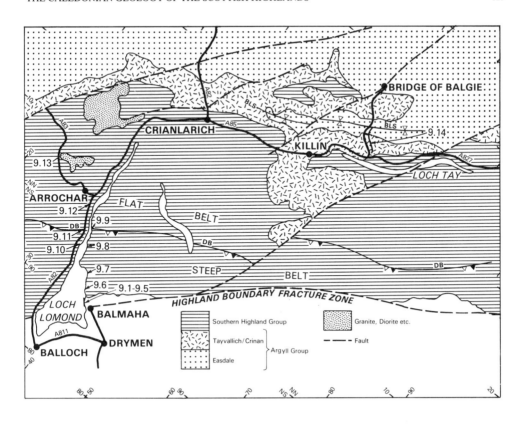

Fig. 4.3. Simplified geological map of the Loch Lomond – Loch Tay area, showing the localities of itinerary 9.

locality 9.2 is via Craigie Fort, an excellent view point situated on a ridge of conglomerate which runs south-west parallel to the fracture-zone from the heights of Conic Hill across to the islands of Inchcailloch, Creinch and Inchmurrin in Loch Lomond. The dip of the Lower Old Red Sandstone increases towards the north, finally becoming vertical or overturned next to the serpentinitic rocks of the next locality.

Locality 9.2.   Druim nam Buraich, 550 m NW of large car park at Balmaha (NS 4178 9138).

Access is by means of a footpath which starts at a stile at (NS 4163 9138) on the B837 Balmaha-Rowardennan road. No dogs are permitted. The outcrops are approximately 350 m from the road. At this locality, serpentinite-clast conglomerates and redeposited serpentinitic breccias of the HBC are strongly altered to carbonate and silica minerals. The HBC includes a number of other isolated units of sediments containing mafic and ultramafic detritus, informally called the 'Loch Lomond clastic unit'. The serpentinite-clast conglomerates here at Balmaha are essentiallty uncleaved, unlike the Green conglomerate in the River North Esk section and the Loch Fad conglomerate on the Isle of Bute which are strongly cleaved and flattened.

Locality 9.3.   Lochside exposures, 1000 m NW of Balmaha (NS 4131 9154)

Access via the West Highland Way is unrestricted. Brick-red sandstones and breccias contain fragments mainly of cleaved greywacke and vein quartz which can be matched closely with local Dalradian Supergroup lithologies. The Upper Old Red Sandstone is considered to have been deposited by braided rivers draining the Highlands to the north. The outcrop is traversed by NNW-trending close-set joints which may relate to the tear-faults which affect the Lower Old Red Sandstone conglomerates south of the Highland Boundary fracture-zone. These sediments lie unconformably on HBC sandstones, the contact in Loch Lomond being exposed only at low water to the south of Arrochymore Point (locality 9.4).

Locality 9.4.   Arrochymore Point, 1400 m NW of Balmaha (NS 4092 9177).

Access via the West Highland Way is unrestricted. Another component of the Loch Lomond clastic unit is exposed here in the form of weakly cleaved immature sandstones which contain grains of chert, altered spilite and chromite. Repeated cross-lamination at this locality shows that the gently inclined bedding is inverted, which is similar to the attitude of local Upper Dalradian metasediments (locality 6). North-trending faults separate these rocks from a strip of foliated antigorite-lizardite-serpentinites which are converted, to varying degrees, to carbonate and silica minerals. The route to the next locality passes a small outcrop (in the woods) of antigorite-serpentinite and (at the lochside) others of brecciated carbonated serpentinite and foliated talc-dolomite-serpentinite.

Locality 9.5. Roadcut beside the B837 Balmaha-Rowardennan road, 1350 m NW of Balmaha (NS 4127 9200).

At the N end of the roadcut, reddened rocks composed of ferroan dolomite and quartz contain streaks of chromite. In the central part of the roadcut, reddened jaspers derived from serpentinite preserve ghosted pyroxene grains, fine serpentinous fabrics and chromite grains rimmed by chromian micas and clay minerals. At the south end of the exposure, a gently inclined fault separates the jaspers from weakly cleaved quartzose sandstone.

Locality 9.6. Hillslope, 170 m NE of Milarrochy Cottage, 2 km NW of Balmaha (NS 4088 9265).

Access to the hill is from the stile on the B837 Balmaha-Rowardennan road beside the Blair Burn. No dogs are permitted. The outcrops are approximately 400 m from the road. Cleaved sandstones and purple siltstones are correlated with the Leny Grits (Dalradian Supergroup) of Cambrian age are inter-stratified with irregular beds, up to 4 m thick, of cleaved conglomerate and pebbly sandstone. Clasts in these conglomertes are mostly subangular and between 5 mm and 30 mm long, and include vein quartz, quartzite, plagioclase, microcline and jasper. The jasper clasts probably originated from an iron-formation; internal oolitic and polygonal features are remarkably well preserved under the microscope. The conglomerates may represent mass-flow deposits which incorporated rafts of finer sediment during downslope movement in a submarine fan setting. For rocks within the Highland Border steep belt, the dip of bedding and cleavage here is anomalously gentle. This is probably due to a late (? F4) overfolding, since fold zones comparable to F4 can be traced on the hill slopes between this locality and the next.

Locality 9.7. Foreshore of Loch Lomond, 300 m N of mouth of Cashell Burn, 8 km NW of Balmaha (NS 3931 9439).

The section starts at the northern limit of continuous exposure. The rocks are medium-grained cleaved impure turbiditic sandstones with one bed of matrix-supported conglomerate 300 mm thick. The bedding and the slaty-looking cleavage show the characteristic attitude of rocks within the Highland Border steep belt. Fine ripple-drift cross-lamination and grading indicate that the rocks young to the south. Over the next 50 m to the south, two downward-facing penetrative cleavages dip SE more gently than the bedding. The steeper cleavage shows as a fine alignment of grains and is traversed by a more gently dipping spaced cleavage, strongly oblique to bedding. Towards the top of graded units, the spaced cleavage ('S2') appears to refract into the fine cleavage ('S1'). Note that two cleavages are not apparent in all exposures; in the roadside quarry 160 m N of Cashell Burn, the one dominant cleavage is presumably composite S1/S2 fabric.

Locality 9.8. Foreshore of Loch Lomond, 700 m S of Rowardennan Hotel on B837 Balmaha-Rowardennan road (NS 3607 9765).

Coarse sandstones with granules of quartz and feldspar show clear graded bedding which is inverted and youngs northwards overall. The beds are tightly folded by small folds plunging steeply SW. A strongly-developed fine scale grain flattening fabric runs approximately parallel to bedding. It is folded with the bedding and can be seen in fold hinges to be transected by the dominant foliation of the outcrop, a pressure-solution cleavage spaced at approximately 10 mm intervals. In fold limbs both the grain flattening (S1) cleavage and the spaced (S2) cleavage appear to form a composite 'stripping' fabric. Outside F2 fold hinges, therefore, the dominant fabric of these rocks appears to be composite in nature.

Locality 9.9. Road cuts on forestry road above Ptarmigan Lodge (Section from (NN 3522 0089) to (NN 3509 0104)).

Excellent examples can be examined of D_3 and D_4 structures. The D_3 folds are subhorizontal with ENE-WSW trends, D_4 folds have subvertical axial planes. Examples of D_2 'herringbone' structures folded by D_3 and D_4 can also be seen.

Route to Locality 9.10.

Return to Drymen and follow the A811 around the end of Loch Lomond to join the A82 west of Balloch. Drive northwards up the west side of Loch Lomond on the A82.

Locality 9.10. Roadcuts next to old quarry (NS 3494 9660).

The quarry contains a worked lamprophyric dyke. The series of crags north of the quarry exhibit steeply dipping rocks with excellent examples of D_2 'herringbone' structures.

Locality 9.11. Rubha Mhor (NN 346 001).

Park in lay-by to south of Rubha Mhor headland and examine the series of crags on the west side of the road. Please note that these localities are close to the road and could be dangerous for large parties. The crags contain excellent examples of the four deformation phases, and interference structures can be readily examined. Rubha Mhor is close to the Downbend axis.

Locality 9.12. Stuckivoulich (NN 325 320)

A large lay-by allows easy parking. The crags on the west side of the lay-by characterise the generally flat rolling foliation of the area. Examples of D_3 and D_4 structures can be readily identified. The latter with subvertical axial planes.

Locality 9.13.   Arrochar Diorite

A detour of 25 km is necessary to examine the Arrochar diorite (Anderson, 1935). Leave the A82 at Tarbet and follow the A83 around the head of Loch Long and up the hill to the small quarry at (NN 2406 0648) where altered quartz-diorite and tonalites, commonly with large knobs of pyrite, can be examined. Return down the road a short distance to the Croe water section at (NN 242 060). Start in the stream below the road-bridge and work upwards to examine the various igneous lithologies, starting with a contact altered igneous breccia cut by granite veins and a sheet of microdiorite, followed, further upstream, by pyroxene-mica-diorites.

Route to Locality 9.14.

Rejoin the A82 at Tarbet and drive north to Crianlarich. Drive east along Glen Dochart to Killin on the A85 and A827. Follow around the north side of Loch Tay for c. 7 km to the small road running north over the shoulder of Ben Lawers to Bridge of Balgie.

Locality 9.14.   Lochan na Lairige section.

On the road to Lochan na Lairige park at a small lay-by close to a right hand bend and proceed to the small quarry at (NN 621 364) to examine outcrops of the Loch Tay Limestone. Proceed up the road and park at the National Trust Visitor Centre to examine outcrops of Ben Lui Schist in the Burn of Edramucky. The rock is a characteristic garnetiferous mica-schist. Continue northwards and park at the Lochan na Lairige dam. In the nearby roadcuts and exposures at (NN 600 390) calcareous and locally pyritous Ben Lawers Schist may be seen. This point is close to the axis of the Ben Lawers synform and associated minor structures can be seen (Treagus, 1964). Return to Killin.

ITINERARY 10.   PART 1.   RANNOCH MOOR AND GLENCOE CAULDRON SUBSIDENCE.

Route to Locality 10.1

Leave Killin on A827 to join the A85 and proceed west to meet the A82 at Crianlarich. Follow the A82 north through Tyndrum past the foot of Ben Dorain to Bridge of Orchy. The contact of the Moor of Rannoch Granite is crossed between Loch Tulla and Loch Ba. Good views can be obtained in fine weather westwards to the Black Mount, formed by the Cruachan Granite of the Etive Complex, and eastwards across the Moor of Rannoch. Beyond Loch Ba, the road turns northwest towards the Devonian volcanics of the Glencoe Cauldron-Subsidence. Meall a Bhuiridh and Sron na Creise form prominent hills southwest of the road, while Buchaille Etive Mhor is seen as the very precipitous mountain farther to the west beyond the entrance to Glen Etive.
    Between Kingshouse and Altnafeadh, the road passes from the Moor of Rannoch Granite, across the Glencoe Fault-Intrusion, and into the cauldron-subsidence itself. The low ground immediately below Sron na

Creise and Buchaille Etive Mhor is floored by psammitic rocks of the Eilde Flags and Quartzite lying within the cauldron, beyond which volcanic rocks make up the high ground to the southwest. The actual contact between the metamorphic rocks and the volcanic rocks, which elsewhere within the cauldron is marked by an angular unconformity is now thought to represent the inner wall of the cauldron-subsidence, where it is exposed on the lower slopes of the Buchaille Etive Mhor. The line of the Main Ring-Fault can be seen where it crosses the skyline ahead of Altnafeadh in the small but prominent craig of Stob Mhic Mhartuin.

Beyond Altnafeadh, the road crosses on to the volcanic rocks of the cauldron-subsidence. South of the road, Buchaille Etive Beag, Beinn Fhada, Gearr Aonach and Aonach Dubh form precipitous hills, all capped by a thick ignimbrite sheet, lying above rhyolite and andesite lavas. North of the road, the ridge of Aonach Eagach is formed by andesite lavas turned up against the Main Ring-Fault on the other side of the mountain. At the white cottage halfway down the Pass of Glen Coe, a good view to the south, can be obtained of the large rock-fall blocking Coire Gabhail, known popularly as the Lost Valley. Passing the farm of Achtriochtan, the dark slit of Ossian's Cave, formed by a porphyrite dyke, can be seen to the south of the road on the steep slopes of Aonach Dubh. Turn left off the road at the western end of Loch Achtriochtan along a short track leading to Achnambeithach, and park opposite the bridge.

Locality 10.1. Small quarry near Achnambeithach (NN 138 568) (Figs. 4.6 and 4.5)

Cross the main road to the small quarry. This quarry exposes the porphyrite of the Glencoe Fault-Intrusion, together with calc-silicate rocks representing impure Ballachulish Limestone where it has been thermally metamorphosed by the fault-intrusion. Good exposures of the xenolithic facies of the fault-intrusion are seen at the western end of the quarry, close to a small stream.

The view south towards Coire nam Beith (Fig. 4.6) shows the salient features of the geology. An't Sron is built by granitic rocks of the Glencoe Fault-Intrusion, while the metamorphic rocks (Leven Schists and Ballachulish Limestone) outside the cauldron outcrop farther down the glen to the right. The inner contact of the fault-intrusion follows the line of the prominent gully shown on the diagram, along which the fault-intrusion chills against the volcanic rocks downfaulted to form the cauldron subsidence. The lowest part of the volcanic sequence is formed by relatively thin flows of basic andesites, which form the precipitous lower slopes of Aonach Dubh to the left. These lavas rest unconformably on phyllites, representing the low-grade equivalent of the Leven Schists within the cauldron, and exposed on the lowermost slopes of the glen. A basal conglomerate is present locally along this contact, north of the road. The andesite lavas pass upwards into two thick flows of rhyolite lava, above which a thick ignimbrite horizon is developed, capping the upper slopes of Aonach Dubh. The volcanic rocks lying above this horizon can be seen

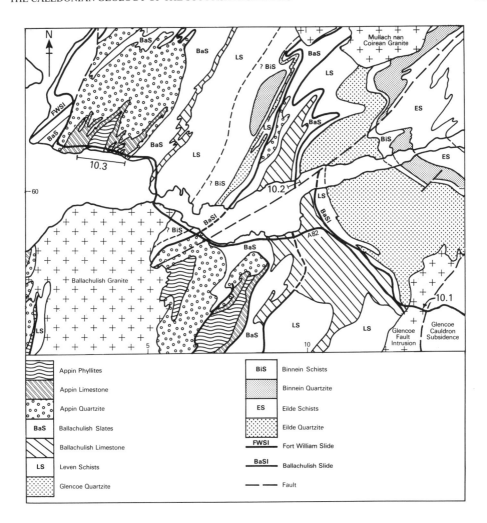

Fig. 4.4 Geology of the area around Loch Leven after Roberts and Treagus (1977) showing the localities of itinerary 10.

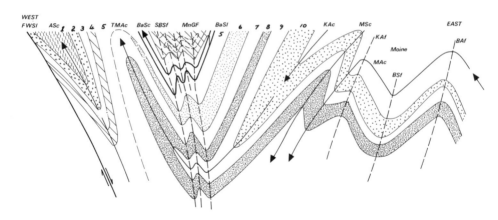

Fig. 4.5 Diagrammatic cross-section across Figure 4.4. FWSl= Fort William slide; ASc= Appin syncline; TMAc= Tom Meadhoin anticline; BaSc= Ballachulish syncline; SBSf= Stob Ban synform; MnGF= Mam na Gualainn folds; BaSl= Ballachulish slide; KAc= Kinlochleven anticline; MSc= Mamore syncline; KAf= Kinlochleven antiform; MAc= Mamore anticline; BSf= Blackwater synform; BAf= Blackwater antiform. After Roberts and Treagus, 1977, fig.2, with the permission of the Scottish Academic Press. 1= Appin Phyllites and Limestone; 2= Appin Quartzite; 3= Ballachulish Slates; 4= Ballachulish Limestone; 5= Leven Schist; 6= Glencoe Quartzite; 7= Binnein Schists; 8= Binnein Quartzite; 9= Eilde Schist; 10= Eilde Quartzite.

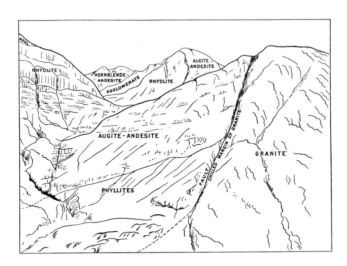

Fig 4.6. View of the boundary fault of the Glencoe Cauldron Subsidence as seen from locality 10.1. From Bailey and Maufe, 1960, with the permission of the Director BGS.

on the upper slopes of Stob Coire nam Beith where they are upturned along the line of the Main Ring-Fault.

ITINERARY 10.   PART 2.   THE BALLACHULISH DALRADIAN

Introduction

The distribution of the major lithostratigraphic formations is shown in Figure 4.4 locally modified after Sheet 53 of the Geological Survey of Scotland. Bailey's (1934, 1960) view of the structure is exemplified by section AA of Sheet 53. In essence, Bailey envisaged three recumbent folds, the Appin Syncline (lowest), the Kinlochleven Anticline and the Ballachulish Syncline (highest) facing towards the north-west. The lower limbs of the two synclines have been considerably affected by tectonic sliding (the Fort William and Ballachulish Slides respectively) and the pile of recumbent folds have been deformed by a series of secondary antiforms and synforms (Bailey, 1960, pp 110-15).

Figure 4.5 shows the present author's modified version of the structure (Roberts, 1976; Treagus 1974). The north-east trending recumbent folds are assigned to the F_1 deformation, associated with an axial plane penetrative cleavage, S_1. These folds face up to the north-west in the west and in the east of the area. Between these there is a zone where the folds face down to the north-west occupying the common limb of a pair of secondary folds, the Stob Ban synform in the north-west and the Kinlochleven antiform in the south-east. The former fold opens to the south-west into the complex around Ballachulish (see Bailey 1960, Chapter VI). The F_2 deformation is responsible for the major secondary folds mentioned above; these folds trend NE, and are associated with an intense strain-slip cleavage (S_2). F_3 is responsible for the major deflection in strike of all previous structures across Loch Leven (the Loch Leven antiform of Bailey, 1960, p.23); F_4 is only locally associated with significant folding. F_3 and F_4 are both related to strain-slip cleavages, the former trending ENE, and the latter N-S.

Such an interpretation may be contrasted with that of Hickman (1978) following the earlier work of Bowes and Wright (1973). These authors identify small-scale structures which pre-date our penetrative cleavage, S_1. In particular, Hickman (1978) considers that the Ballachulish syncline does not exist, while he identifies the Appin syncline, the Stob Ban synform and the intervening Tom Meadhoin anticline, which we consider to be the upward-facing nose of the Kinlochleven anticline to the east, as all belonging to the F_2 deformation.

Route to Locality 10.2.

Follow the A82 from locality 10.1 towards Glencoe Village along the outcrop of the Leven Schists and the Ballachulish Limestone. Passing the Glencoe Visitor Centre, close to the scene of the Glencoe Massacre

(1692), a good view is obtained north toward Sgor nam Fiannaidh. The prominent gully follows the crush of a minor fault, while the shoulder to the right shows Glencoe Fault-Intrusion (pinkish exposures cutting) Glencoe Quartzite (bedded appearance) and Ballachulish Limestone (grey exposures) in a complex fashion.

Fork left at Glencoe Village, following the A82 past roadside exposures of the Ballachulish Limestone and through the village of Ballachulish with its quarries in the Ballachulish Slates. After crossing the road-bridge across Loch Leven, turn right onto the B863 towards Kinlochleven. With a large party set them down on the north side of the road just beyond the cottage to the east of Callert House where the road leaves the shore and turns a sharp corner (in Appin Quartzite). The coach can turn in a roadside quarry 2 miles up the road towards Kinlochleven. Walk back down the road to the first exposures, just east of the burn where it enters Loch Leven.

Locality 10.2.   Ballachulish slide at Callert, Loch Leven

The first exposures east of the burn (NN 096 604) show pale quartzose limestone representing part of the Ballachulish Limestone on the upper limb of the Ballachulish syncline. Minor F_2 folds plunge SW at a shallow angle and verge SE thereby supporting the Stob Ban synform. These rocks are succeeded to the east by a narrow outcrop of Ballachulish Slates beyond which there is a transitional contact with a thin band of gritty quartzite, representing the Appin Quartzite in the core of the Ballachulish syncline. This is flanked to the SE by a narrow band of impure quartzites and black slates, representing the Appin Transition Group. These rocks are in tectonic contact with greenish Leven Schists across the line of the Ballachulish slide. The exposures of Appin Quartzite can easily be located as they form a slight headland marked by the foundations of a summer house.

Continuing along the shore, there is a short gap in exposure beyond which banded Leven Schists are exposed at (NN 097 605). These rocks are affected not only by F_2 folds, plunging moderately SW and verging SE, but also by F_1 folds, plunging SW at a shallow angle and verging NW. Both sets of folds are well seen in an exposure formed by a low craig on the shore. The F_1 folds face downwards to the NW on the stratigraphy. The contact of these rocks with the Glencoe Quartzite can be examined some 200 m farther to the east along the shore. Cross-bedding in the quartzite indicates that the beds young to the NW.

Route to Locality 10.3.

Return to the A82 at North Ballachulish, and take this road towards Fort William. A large party should be dropped by the coach opposite the Creag Dhu Hotel, just west of a conspicuous quarry in the Appin Quartzite, slightly over a kilometre from the road junction. The coach should then proceed to the Onich Hotel where it can be parked to pick up the party.

Locality 10.3.   Appin syncline, Onich shore section.

The Appin Quartzite is well-exposed on the slight headland immediately to the east. Festoon cross-bedding shows the beds young to the NW across the whole width of the outcrop. A slaty cleavage S_1 dips SE at a lower angle than bedding, and faces upwards to the NW.

Appin Phyllites are exposed immediately to the west, alongwith bands of Appin Limestone. The F_1 structures become progressively more open and less intense away from the contact with the Appin Quartzite. It is possible to show from bedding-cleavage relations and the vergence of minor folds that a syncline within the Appin Phyllites exists with its axial trace within 10 m of this contact. Quartzite beds exposed on the shore in front of the Creag Dhu Hotel show cross-bedding younging to the SE of the NW limb of this syncline. Although bands of dolomitic limestone are interbedded with the Appin Phyllites, the structure is such that a considerable part of the Appin Phyllites, and all of the Appin Limestone, must be cut out by a slide at the contact of the Appin Quartzite.

Continue west along the shore, noting that the slaty cleavage S_1 dips more steeply than bedding on the NW limb of the syncline. The axial trace of the anticline complementary to this fold is crossed near the outcrop of a pink acid dyke, exposed on the shore near the old school (NN 038 613). Farther west, sedimentary structures, bedding-cleavage relationships and the vergence of minor folds can be used to show that bands of dolomitic limestone are interbedded in the Appin Phyllites on what can be identified as the SE limb of the major fold forming the Appin Syncline in this region. Grading with silty layers showing sharp bases, channelling and ripple-drift bedding alongwith sedimentary dykes, are well-exposed at low tide particularly on the shore below a bungalow (NN 036 613). The core of the Appin syncline is crossed at the back of a small bay (NN 033 613), beyond which the opposite relationships are seen on the NW limb of the Appin syncline. Return to the coach at the Onich Hotel and follow the A82 to Fort William.

ITINERARY 11.   MORAR

A traverse to examine Moine rocks and igneous intrusions between Fort William and Mallaig.

Introduction

About two miles NE of Fort William the road to Mallaig (A830) turn off the main road (A82) which links Fort William and Inverness. The road crosses the major open valley which marks the course of the Great Glen fault zone before turning west Banavie. As far as Lochailort the road has been modernised and is two-lane, with nevertheless bad bends but fast traffic. Beyond Lochailort the road has not been modernised and is single-lane with passing places. Although lay-bys exist which help with parking of vehicles many of the road cuts are dangerous for

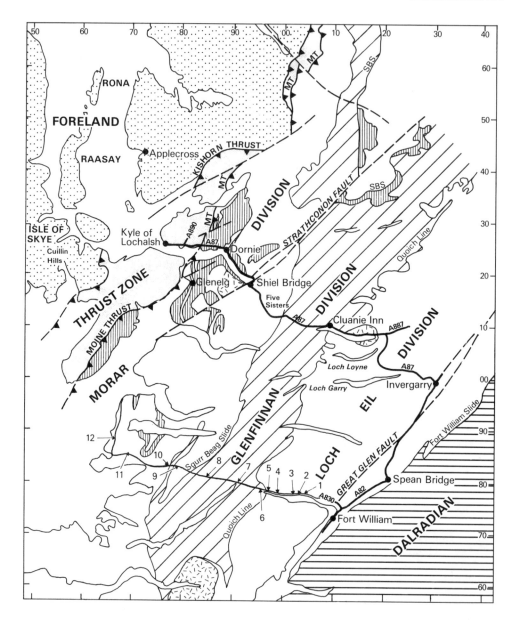

Fig. 4.7. Map of the main Moine Divisions in W Inverness-shire, showing the routes of itineraries 11 and 12. Numbers indicate the localities on itinerary 11.
Close spaced vertical ruling = Lewisianoid rocks within the Caledonian Mountain Belt; cross-hatching = Caledonian (c. 400 Ma) granites; toothed lines = major thrusts limiting the Moine Thrust Zone.

large parties.

The road from Fort William to the west cost of Inverness-shire at Morar traverses the Locheil, Glenfinnan and Morar division of the western Moine distinguished by Johnstone et al. (1969). The area has been subjected to a long and complex igneous history, and igneous rocks ranging in composition from acid to basic and in age from late Proterozoic to Tertiary are of common occurrence.

The nature of the Moine divisions is shown in Table 4.2 and their general outcrop in Figure 4.7. The Morar and Glenfinnan divisions are separated (locality 11.10) by a major ductile thrust, the Sgurr Beag slide (Tanner et al., 1970); Rathbone and Harris, 1979), the course of which is marked by about 100m of blastomylonite. Across this thrust there have been tens or maybe hundreds of kilometres of displacement. Consequently, the relative age of the Morar and Glenfinnan divisions is uncertain. The blastomylonites comprise thoroughly recrystallised quartzofeldspathic metasediments of Morar psammite. All fabric elements within these intensely flaggy rocks are parallel and they are totally devoid of sedimentary bedforms and almost devoid of minor folds. The full range of phenomena relating to this ductile thrust zone has been described by Rathbone et al. (1983).

It has been concluded (Kelley and Powell, 1985) that the Sgurr Beag slide is a Caledonian structure, lying above and to the east of the Moine thrust. It has been folded (Fig. 4.7) by NNE-trending, upright, tight, strongly curvilinear folds which also deform the Glen Dessary syenite. This syenite has yielded an Rb-Sr isochron which suggests an age emplacement of 465 ± 5 (van Breemen et al., 1979). It seems likely that it is the oldest and highest of the foreland-propogating set of thrusts of which the Moine thrust is the best known. If this is the case the inference of its Caledonian age is strong.

The boundary between the Glenfinnan and Locheil division rocks is by contrast a normal stratigraphic junction, the Glenfinnan Division being the older. Flat-lying Locheil Division psammites occur to the east and steeply inclined, strongly folded pelitic gneisses, psammites and quartzites of the Glenfinnan Division lie to the west. Roberts and Harris (1983) have shown that the change from steep to flat-lying rocks occurs across a major asymmetric synform the axial trace of which marks the eastern limit of severe, upright reworking of previously flat-lying rocks. The zone marking the change from one division to the other, together with the change from flat to steep domain was termed the Loch Quoich line by Clifford (1957) although at that time the different Moine divisions were not distinguished by the names now in use. The asymmetric synform referred to above is of the same generation as the folds which deform the Glen Dessary syenite and hence the age of the Loch Quoich line is well constrained. There is no evidence for either a tectonic discontinuity or an unconformity at the contact between the two divisions.

Throughout much of its extent, the Ardgour granite gneiss (1030 ± 45 Ma) coincides approximately with the Loch Quoich line. It is an adamellitic orthogneiss which appears everywhere to be concordant, but which regionally cross-cuts the Moine lithostratigraphy. This orthogneiss is transected for about 3 km by the A830 traverse. It can

TABLE 4.2: Summary of litholgical characters of Moine formations of western Inverness-shire (after Johnstone, Smith and Harris, 1969).

Division	Formation	Description
LOCH EIL DIVISION	LOCH EIL PSAMMITE	Variably quartzose psammitic granulite (locally a micaceous 'salt and pepper' type) with very subordinate bands of pelitic and semipelitic schists. Calc-silicate ribs and lenticles present throughout and locally abundant.
GLENFINNAN DIVISION	GLENFINNAN STRIPED	Banded siliceous psammite (locally quartzite) and pelitic gneisses; pods or lenses of calc-silicate granulites.
	LOCHAILORT PELITE	Pelitic gneiss, with subordinate psammitic or semipelitic stripes. Metasedimentary amphibolite and calc-silicate lenses are usually present. Position of Sgurr Beag slide (ductile thrust).
MORAR DIVISION	UPPER MORAR PSAMMITE	Dominantly psammitic granulite, often pebbly, with common semipelitic bands; calc-silicate ribs throughout.
	MORAR (Striped and Pelitic) SCHIST	Dominantly pelitic rocks locally divided into: a. Rhythmically striped and banded pelite, semipelitic schists and micaceous psammitic rocks with abundant calc-silicate ribs. b. Pelitic schists with some subordinate semipelite stripes. c. Laminated grey, semipelitic and micaceous granulites, locally with thin siliceous and calc-silicate ribs rare except towards the top; heavy-mineral bands present, but most common near the base.
	LOWER MORAR PSAMMITE	Micaceous and siliceous psammitic granulites locally pebbly; subordinate semipelitic rocks developed locally and more thickly towards the top; heavy-mineral bands present, but most common near the base.
	BASAL PELITE	Dominantly pelitic and semipelitic schists, thinly banded with psammite.

be examined at locality 11.7.

As mentioned above, the igneous rocks of the Morar area span a wide age range and are also very diverse in composition. The oldest is the Ardgour orthogneiss which is cut by metabasic amphibolites. These are commonly garnetiferous and form part of a widely developed suite of metabasic rocks which occur in the Moine to the east of the Loch Quoich line. Both the granitic gneiss and the metabasic rocks are believed to have suffered the whole of the tectonic and metamorphic history of the Moine metasediments. Pegmatites dated at c. 760 Ma cut fabrics related to two episodes of deformation (Powell et al., 1983) and are themselves deformed during a third episode which is interpreted as Caledonian. Such pegmatites can be examined at locality 11.9, but it is too dangerous for large parties without police controlling the traffic. Late Caledonian igneous activity includes the emplacement of the Glendessary syenite (456 ± 5 Ma) which predates the Caledonian upright folding. Later quartz-feldspar-muscovite pegmatites and granites occur in ramifying veins while mafic-felsic microdiorites, commonly foliated, occur in sheets typically 1m thick. Pegmatites, granites and microdiorites have suffered varying degrees of, but generally slight, deformation and it is concluded that they were emplaced at a late stage in the late Caledonian upright deformation. Minettes (Devonian), E-W-trending camptonites (Permo-Carboniferous) and NW-trending dolerites (Tertiary) are all entirely post-tectonic.

Locality 11.1.

On the north side of the road a large road cutting (NN 043 785) exposes flat-lying Loch Eil Psammite containing thin concordant hornblende-schists, thought to be metabasic intrusions. Large recumbent almost isoclinal folds are cut by thin cross-cutting granitic veins some of which appear to have been weakly folded. There is a weak axial-planar fabric, and a poorly defined lineation parallel to the north-south trending fold hinges. Despite the highly attenuated nature of the folds, sedimentary structures are present and demonstrate that the folds face to the west. A large layby lies about 200m east of the exposure on the south side of the road.

Locality 11.2.

1 km west of locality 11.1 road cuttings both north and south of the road (NN 035 784) expose Loch Eil Psammite containing abundant sedimentary structures; again the psammites are largely flat-lying. Both tabular and trough-bedded cross-stratification and cross-lamination are present, and good examples of 'herring-bone' bedding occur, demonstrating that at least part of the succession was deposited under tidal conditions. Sedimentary structures of this type are commonly found within the Loch Eil Division of this area. Large tight moderately-inclined folds exposed at the east end of the locality north of the road are thought to have originally been recumbent folds, of the same generation as those at 11.1, but downfolded during subsequent

deformation.

Locality 11.3

Exposures directly beneath the small road bridge at Fassfern (NN 021 7895) comprise quartzite and siliceous psammite dipping gently to the east. Bedding surfaces display numerous symmetrical straight-crested ripple forms which are interpreted to represent the effects of wave activity. A traverse upstream reveals a gradation into grey micaceous psammite containing small-scale cross-laminations which young to the east. 300 m north of the bridge (NN 0220 7925) the psammites are cut by a sheet of foliated microdiorite, part of the late Caledonian microdiorite suite (Smith, 1979). Vehicles may be parked 50m west of the bridge in a layby on the north side of the road.

Locality 11.4

On the north side of the road (NM 988 788) a large road cutting exposes dark micaceous psammites containing calc-silicates. At the east end of the locality the psammites are cut by sheets of granite and aplite which belong to one of the late Caledonian granite vein complexes described by Fettes and MacDonald (1978). The calc-silicate mineral assemblages consists of hornblende + andesine + quartz + garnet $\pm$ biotite. The calc-silicates show every morphological variation from large pods to small elongate lenticles and wisps; they are considered to represent concretions formed during the diagenesis of the enclosing metasediments. A large layby lies immediately east of the locality on the south side of the road.

Locality 11.5

The stratigraphically lowest levels of the Loch Eil Division are exposed north of the road on a small tree-covered hillock (NM 963 794) which consists of flat-lying micaceous psammite and concordant hornblende schists. Uppermost units of the Glenfinnan Division are exposed on low ground 100m west of the hillock and beneath the old road bridge over the Allt Fionne Lighe (NM 9605 7940) where they comprise medium to coarse-grained pelitic gneisses with concordant migmatitic quartzo-feldspathic segregations. Vehicles may be parked off the main road east of the old bridge.

Although the actual contact between the two Divisons is unexposed, the absence of extensively developed high strain fabrics at any of these localities precludes the possibility of the contact being marked by any major tectonic break. For a clear demonstration of the conformable nature of the Loch Eil - Glenfinnan Division contact interested parties are referred to the Invergarry - Kinlochourn road section.

Locality 11.6

Glenfinnan Division rocks are exposed north of the road (NM 9515 7915)

500 m west of locality 11.5. The exposure consists of interbanded migmatitic pelitic gneiss and coarse striped psammite with thin calc-silicates which have been affected by three phases of deformation:
(1) Formation of the main gneissose fabric; since this is invariably parallel to all lithological layering, formation of the fabric is likely to have occurred as a result of isoclinal folding.
(2) Isoclinal almost recumbent folds which deform the gneissose fabric and develop a local axial-planar schistosity. These westerly verging folds are thought to be of the same generation as those at locality 11.1.
(3) Crenulations of all pre-existing fabrics with sub-vertical axial planes trending at $180°$.

Cross-cutting sheets of granite and aplite of late Caledonian age are present at the east end of the locality.

Locality 11.7

On the northern side of the A82 to the east of Glenfinnan (NM 915 803) a road cutting contains units of several suites of igneous rocks ranging from pre-Caledonian granite and mafic rock to late-orogenic Caledonian pegmatite and microdiorite. This exposure occurs near the centre of the adamellite Ardgour granitic orthogneiss which consists of quartz, plagioclase (An20), K-feldspar and biotite. Minor garnet is present. The banding which renders the rock gneissose consists of crudely lenticular quarto-feldspathic layers which carry partial selvages of dark mica. Pegmatitic augen and veins carry similar but thicker and much better defined selvages. The Ardgour gneiss has yielded an Rb-Sr age of 1030 ± 45 Ma interpreted as an age of intrusion. At this locality it is cut by thick ramifying granitic pegmatite veins which are typical of the major pegmatitic vein complexes recorded within the Moine rocks by the Geological Survey on the Morar (61) and Quoich (62W) sheets; similar pegmatites have been dated at c. 440 Ma. The pegmatite here cuts not only granitic orthogneiss but garnetiferous metabasite and is itself cut by a sheet of rather mafic foliated microdiorite, a member of the late Caledonian microdiorite suite (Smith, 1979) deformed, foliated and locally folded during the last stages of deformation (Talbot, 1984). By contrast the garnetiferous metabasite has suffered much if not all of the deformation and metamorphism of the Moine. The amphibolite is thought to be part of the same suite as the amphibolites in the Locheil Division at locality 11.1. Attitudes of foliation at this locality are still generally flat-lying, though irregular and undulating. Westwards towards Glenfinnan the foliation becomes steep-to-vertical across the axial trace of the synform which marks the Loch Quoich line.

Locality 11.8.

Glenfinnan Division siliceous psammites and pelites are splendidly exposed on a rocky hill on the northern side of the road (NM 857 815) near Loch Eilt. Parking of vehicles including a coach is easy on the south side of the road side just to the east of the railway tunnel. Road cuttings on this stretch of rock are very dangerous for study by large parties and police assistance to control traffic should be sought. The excellent glaciated slabs at the top of the hill display typically striped and banded coarse siliceous psammite and garnetiferous pelite characteristic of Glenfinnan Division. Garnetiferous calc-silicate lenticles commonly occur, liberally studded with pink garnet, within psammite bands. Books of muscovite possibly after aluminosilicates are abundant in the pelites. Ductility contrasts between psammitic and pelitic beds has resulted in abundant small- and intermediate-scale folds of two generations. The second generation recognised here are interpreted as those which deformed the Glen Dessary syenite and hence are post 456 ± 5 Ma. The first folds are probably the regional second set. Minor folds exhibit markedly curvilinear hinges and commonly are doubly verging, having a sheath-like form. Interference between folds structures of both generations is very well displayed. Coarse, crenulation cleavage in the pelites transects pre-existing penetrative fabrics defined by muscovite, biotite and quartz and feldspathic lenticles. Thick (1m) K-feldspar, pegmatite veins cut both sets of tectonic structures and most of these carry orientated muscovite books and appear to have been mildly deformed by the later folding episode. These pegmatites belong to the c.440 Ma suite and their mild deformation puts an approximate age to the end of the intense Caledonian deformation of the Glenfinnan rocks.

Locality 11.9

At the western end of a road cutting (NM 799 826) on a road bend and very dangerous, psammites on both sides of the road have a pervasive almost vertical fabric and carry tight almost isoclinal upright similar folds. Deformed beryl-bearing pegmatite veins up to 1m thick bear a strong fabric clearly coeval with that in the psammite. Some thin ($\sim$5cm) pegmatite veins are folded with the banding in the psammite. These tectonic fabrics in the psammite are cut by scarcely deformed irregular granitic pegmatite veins. The earlier deformed pegmatites have been dated at c. 730 Ma. The later pegmatites are similar to those dated elsewhere at c. 450 Ma. Cross-cutting mafic microdioritic sheets (c. 20-30 cm) occur, these rocks are similar to those cutting pegmatites and metabasic rocks at the Ardgour orthogneiss locality, in that they are strongly foliated oblique to their margins, and at this locality are also weakly folded.

Locality 11.10

At Lochailort rocks of the Sgurr Beag ductile thrust zone (slide) are exposed both in the road cutting to the east of the village and at a large rocky knoll (NM 770 819) where the contact between pelitic gneisses of the Glenfinnan Division (Lochailort Pelite) to the east and Morar Division psammites (Upper Morar Psammite) to the west is exposed. In the road cutting intense deformation of the psammites has obliterated cross-bedding and has brought almost every element of the rock fabric into strict parallelism; where deformation has declined sufficiently, several hundred metres to the west, cross-bedding youngs towards the pelite gneiss. The intense parallelism of the psammite fabric throughout this thick zone has been interpreted as the result of high tectonic strain marking the course of the Sgurr Beag slide. The displacement across this zone is so large that the contact cannot be interpreted as stratigraphic.

Locality 11.11.   Drumindarroch (NM 690 842).

The pelite at Drumindarroch has yielded an Rb-Sr age of 1024 ± 96 Ma; this has been variously interpreted as dating the peak of regional metamorphism broadly coeval with the second episode of deformation to affect these rocks. There is clear evidence at this locality that the rocks have experienced three deformation episodes.
1. Isoclinal folding – structures modify only bedding or lithological banding. (These may be modified sedimentary folds).
2. Tight, similar, almost recumbent structures around which both bedding planes and the axial planes of the isoclinal folds mentioned above are folded. The strong axial planar crenulation schistosity accompanies these folds. This schistosity dips at $10°$ towards $250°$. The folds on all scales at this locality plunge $10°$ towards $240°$.
3. Sub-vertical crenulations trend $160°$ and locally are accompanied by a rudimentary crenulation cleavage.

Microcline porphyroblasts are strongly flattened in the plane of, and are wrapped by, the second schistosity. It is therefore inferred that they predate most of the strain imposed during the second episode of deformation. Other microcline porphyroblasts show no preferred orientation and are almost certainly entirely post-tectonic.

Locality 11.12.   Back of Keppoch.

Follow the minor road west from the A830 at Back of Keppoch and park cars at (NM 644 881). This road is unsuitable for a coach. Coach parties should proceed on foot from the main Mallaig road.
The psammites which form excellent coastal exposures are referred to the Upper Morar Psammite occurring on the western limit of the Morar Anticline. They are only slightly deformed, although thin garnetiferous pelites carry cleavage of at least two generations one of which at least is a crenulation cleavage. Calcsilicate ribs and lenticles of the type used by Kennedy to determine metamorphic facies in this area are common. Cross-bedding and soft-sediment deformation structures in westward-dipping, greatly inclined commonly gritty psammites indicate conclusively that these rocks are right-way-up. The complexity of the igneous history of the Morar district receives further testimony here. Thin sheets of late-orogenic, strongly deformed and foliated mafic microdioite occur and represent the late Caledonian suite of Smith (1979). Later events are recorded by an E-W trending Permo-Carboniferous camptonite which is cut by a NNW-trending basic dyke probably related to the adjacent Tertiary complex of Skye. Magnificent views of Skye, Rhum and Eigg can be obtained on a clear

day.

ITINERARY 12.   A TRAVERSE FROM THE GREAT GLEN FAULT TO THE FORELAND.

Route from Fort William to Kyle of Lochalsh.

From Fort Willian on Loch Linnhe, the route follows the A82 northeastwards to Invergarry, branching off the A87 through Glen Shiel and along the shores of Loch Duich and Loch Alsh to Kyle of Lochalsh.
   This route crosses the Northern Highlands of Scotland from the Great Glen Fault to the northwestern margin of the British Caledonides in the Moine Thrust Zone. The traverse crosses the Locheil, Glenfinnan and Morar Divisions of the Moinian (Johnstone et al., 1969), including the Sgurr Beag slide, a major ductile shear zone which separates the Glenfinnan and Morar Divisions (Tanner, 1970; Rathbone and Harris, 1979).   The Cluanie (Leedal, 1952) and Ratagan (Nichols, 1950) Caledonian granites are intruded into the Moinian metasediments. Along the shores of Loch Duich and Loch Alsh the traverse passes through the Glenelg Lewisian inlier, an upthrust mass of basement gneiss enclosed by the Moinian metasediments.   The rocks of the inlier have been described by Peach et al. (1910), Ramsay (1957), Sutton and Watson (1958), Barber and May (1976), and Sanders (1979). At the northwestern end of the traverse the Moine Thrust Zone in Lochalsh has been described by Peach et al. (1910), Barber(1965) and Coward and Whalley (1979).   Previous field guides to localities in the Moine Thurst Zone and the Glenelg Lewisian inlier in Lochalsh and Kintail include Barber and Soper (1973), and Barber et al. (1978, 1980).
   NE from Fort William the route (A82) follows the spectacular trench of the Great Glen floored by drift deposits consisting variously of glacial moraines, fluvioglacial sand and gravels (especially around Spean Bridge) and lacustrine and fluvial alluvium.   The hollow forming the Great Glen is followed by the Caledonian Canal built by William Telford in 1822 which links the west and east cost of Scotland and which joins natural fresh-water lochs such as Lochy, Oich and Ness, all of which lie along the floor of he Great Glen.   Northeast of Spean Bridge the road runs along the SE shore of Loch Lochy offering spectacular views of hanging valleys on the NW slopes of the Glen. The country rocks over which the road passes between Fort William and Loch Lochy are largely of 'young' Moine psammites while the rocks on the NW side of the Glen are 'old' Moine - largely Locheil Division.   Ben Nevis, the highest point in the British Isles, which lies on the SE side of the road as it leaves Fort William is made of Devonian granite and andesite emplaced within Dalradian quartzite and limestone.
   Driving NE along Loch Lochy the route crosses fault-bounded, largely conglomeratic, Old Red Sandstone sediments, but at the NE end of the loch it crosses the Caledonian Canal by means of a swing bridge and for the next four kilometres runs along the NW shore of Loch Oich passing over adamellitic granitic gneiss similar to the Ardgour granitic orthogneiss seen on itinerary 11.   In this vicinity the gneiss lies within the Locheil Division psammites.   These are well

exposed in several road cuttings west of Invergarry where the route A87 diverges from the Great Glen into Glen Garry to commence the western traverse across the Moine divisions.

Westwards from Invergarry the granite orthogneiss and Locheil psammites with Caledonian granitic and pegmatitic vein complexes are essentially flat-lying so that the thin (c. 50-100m) granitic gneisses have several wide outcrops. On the south side of the road lies Loch Garry which has been extended and deepened into a hydro-electric reservoir by a dam placed across the narrow gorge at its eastern end. After following the loch for about four kilometres the A87 climbs steeply away from the loch from a point where the minor road to Tomdoun and Kinlochourn turns off. The steep climb from Loch Garry across the watershed between it and Loch Loyne and Glen Moriston is rewarded by one of the most spectacular views of the Highlands. From the viewpoint at the top of the climb a panorama of the Western Highlands can be seen with spectacular views of Lochs Garry and Loyne and many mountains of over 1000m. The occurrence of the high mountains to the west marks the incoming of the Glenfinnan Division across the Loch Quoich line. To the south on a clear day Ben Nevis is visible about 35 km away. The road descends from the viewpoint in a general northerly direction past the dam which has turned Loch Loyne into a hydro-electric reservoir and follows the River Loyne northwards to the point at which it joins the River Moriston. Here the route (A87) is joined from the east by the A887. The lower part of the Loyne valley is remarkable for well defined glacial morainic mounds deposited by valley glaciers during their last advance between 12500 and 10500 years BP. The large well exposed hill to the west is formed of the Cluanie Granite - a Caledonian intrusion (425 Ma) which is a quartz-oligoclase-perthite-hornblende-sphene-granite with zoned plagioclase. The perthitic feldspar largely occurs as small megacrysts. This is a handsome rock in thin section and will be examined and collected in a quarry at locality 12.9 where granite is cut by a mafic microdiorite sheet. The quarry was opened to provide fill for the dam which forms the eastern end of Loch Cluanie; yet another hydro-electric reservoir.

The Cluanie Granite lies on both sides of Loch Cluanie for much of its length but cross-bedded psammites are well exposed in road cuttings towards the western end of Loch Cluanie where examples of deformed cross-bedding can be studied (locality 12.8). Near the western end of Loch Cluanie the road crosses the Loch Quoich line and the Locheil Division rocks, generally flat-lying, give way to steeply inclined striped and banded pelites and psammites of the Glenfinnan Division. These are displayed in many large road-cuttings between Loch Cluanie as far west as the Sgurr Beag slide (ductile thrust) which crosses Glen Shiel (NH 007 136) separating the diverse lithologies of the Glenfinnan Division to the east from the montonous Morar Division psammites to the west. The course of the slide in Glen Shiel is marked by a narrow persistent sliver of hornblende gneiss associated with diopside marble, both lithologies being referred to the Lewisian. The steep-sided lower slopes of Glen Shiel are largely made of Morar Division psammites which form the spectacular mountain known as The Saddle on the south side of the glen and the towering crags of the Five Sisters of Kintail in the north. Legend has it that these five peaks all rising to approximately 1000m OD are five sisters turned to stone as they waited in vain for their lovers to return from war by way of Loch Duich a long arm of the sea penetrating eastwards to the foot of Glen Shiel.

There is a dramatic view northwards up Loch Duich to Eilean Donan Castle at Dornie. Hills covered by the forestry plantation in the southwestern side of the loch form part of the Ratagan Granite (415 ± 5 Ma). Farther northwest, and on the northeastern side of the loch the lower slopes are composed of basement gneisses of the Glenelg Lewisian inlier, overlain by Moinian psammites of the Morar Division forms the summits of the hills.

On reaching the Loch the road turns northeastwards, and beyond the turning to the Kintail Lodge Hotel, passes through a road-cutting in granite, regarded as a sliver of the Ratagan complex caught up in the Strathconon Fault. The road then crosses a causeway across the tidal estuary of the River Croe. The northeasterly trending valley of Strath Croe marks the course of the Strathconon Fault, which is one of a series of NE-SW sinistral transcurrent faults which cut through the heart of the Moinian outcrop. Due east the craggy mountain is Ben Attow (1032 m) composed of Moine psammites with intrusive granitic rocks.

From the head of Loch Duich the A87 follows the northeastern shore of the loch to Dornie. In the hills at the southern end of the loch, Moinian rocks extend right down to the lochside, but near Inverinate Lodge (NG 923 217) the road passes across the contact between the Moinian and the underlying basement gneisses of the Glenelg Lewisian Inlier. The contact is not exposed in the road, but road-cuttings south of the lodge show Moinian siliceous psammites, while to the north they show pink granitic gneiss forming part of the Lewisian.

The Glenelg Lewisian Inlier extends along the western margin of the Caledonian Mountain Belt for a distance of 30 km from Arnisdale in the south to Attadale in the north (Barber and May, 1976). The eastern boundary between the Moinian and the Lewisian rocks of the inlier is highly deformed and folded into large scale folds on easterly plunging axes (Ramsay, 1957; Sanders, 1979). Lithological units and structures such as foliations and lineations within the units on either side of the boundary are entirely concordant, but evidence of an earlier structural and metamorphic history in the Lewisian which is not represented in the adjacent Moinian, and the local occurrence of a basalt conglomerate along the contact (Peach et al., 1910) show that the boundary between the two units is an unconformity.

Geographically the inlier is divided into two parts, the Eastern and Western Lewisian which are also lithologically distinct, Lewisian by a 'slide', a narrow ductile shear zone along which both Lewisian and Moinian rocks have been intensely deformed. The Eastern Lewisian includes a great variety of rock types, some clearly of sedimentary origin; forsterite-diopside marble, garnet-kyanite-biotite gneiss, iron-rich (fayalite-hedenbergite) rocks (eulysites - Tilley, 1936), magnetite and graphite-schists, as well as eclogite, amphibolite, hornblende-schist, hornblende-gneiss and quartzo-feldspathic gneiss, which were derived from igneous parents. Some of the gneisses are mixed acid and basic migmatites and all the rocks of the complex, especially near the margins against the Moinian rocks, may be intensely deformed and recrystallised in the amphibolite facies as blastomylonites.

The Western Lewisian is less variable, being composed predominantly of migmatitic gneisses and more homogeneous granodioritic quartzo-feldspathic biotite-gneisses, interspersed with basic and ultrabasic rocks, including serpentinites, amphibolites, hornblende-gneisses and hornblende-schists. The basic rocks represent layered gabbroic complexes, intrusive dolerites and basaltic dykes (Barber and May, 1976). The complex is largely in the amphibolite facies, but locally granulite facies relics occur and an eclogite has been found in the Western Lewisian at Sandaig (Sanders, 1979). The Western Lewisian has been less intensely deformed by Caledonian (including Grenville and Morarian) events than the Eastern Lewisian and a detailed correlation with the structural and metamorphic history of the Lewisian complex of the Laurentian craton has been established (Barber and May, 1976).

The A87 continues northwestwards along the shores of Loch Duich where road-cuts are in the eastern part of the Glenelg Lewisian Inlier. Equivalent rocks form the lower part of the hillslopes on the southwestern side of the loch by Letterfearn and Totaig, capped by overlying Moine psammite which forms bare crags on the hilltops. Intermittent exposures along the road north of Inverinate show pink acid gneiss. More continuous exposure commences to the north of the bridge over the An Leth Allt (NG 907 232). Outcrops for half a kilometre immediately north of the bridge show high grade pelitic garnet-biotite gneiss with occasional kyanite.

Beyond the landslip retaining wall outcrops show fine-banded hornblende-and biotite-gneiss with more massive amphibolites and occasional eclogites and forsterite-marbles with diopside-nodules. About 2.5 km north of the bridge (NG 890 246) the gneisses become finer grained and are best described as blastomylonites. Northwards at (NG 885 250) the gneisses pass into a tectonic schist. This schist marks the contact between the eastern and western parts of the Glenelg Lewisian Inlier. This zone of high deformation encloses blocks of undeformed gneiss and lenses of Moinian metasediments along its course, and represents a slide zone analogous to other internal shear zones which have been described elsewhere from the Caledonides of Northern Scotland.

Further to the northwest, as far as the village of Dornie road-cuts display migmatitic gneisses and amphibolites, the latter with relict granulite facies mineralogy and textures, as well as more homogeneous granitic gneisses of the western part of the Glenelg Lewisian Inlier. To the west of the road near Eilean Donan Castle good views can be obtained down Loch Alsh to the Cuillin Hills on Skye, formed of Tertiary granitic intrusions. At Dornie the A87 crosses the bridge over the tidal Loch Long to Ardelve. A good view may be obtained down Loch Duich with the castle set against the background of the Five Sisters of Kintail at the head of the loch. The road crosses the raised beach by the Loch Duich Hotel. Road-cuts on the by-pass around Ardelve village are largely in massive speckled amphibolite.

North of Ardelve the road follows an east-west valley marking a
major fault zone, as far as the village of Balmacara. The floor of
this valley is covered by glacial deposits, often reworked into raised
beach terraces during high sea-level stands in the early Post-Glacial.
   As far as the village of the Auchtertyre, hills to the north and
south of the road are composed of Lewisian rocks in the western part of
the Glenelg Inlier. Basic pyroxene-granulites form the prominent crag
to the north of the road by Nostie. At Nostie Bridge a pipeline comes
down the hill from the hydroelectric dam and reservoir in Glean Udalain
to the turbine house which stands on the south side of the road.
   At Auchtertyre the road enters the Moine Thrust Zone marking the
northwestern margin of the Caledonian Mountain Belt in NW Scotland. In
Lochalsh the thrust zone consists of a series of thrust slices (nappes)
separated by thrust planes. The dip of the thrust planes and also of
the dominant planar structures in the rocks forming the thrust slices
is at about $20^{\circ} - 30^{\circ}$ towards the east.
   From east to west the sequence in the Moine Thrust Zone of
Lochalsh is:

Moine Nappe. Acid gneisses and amphibolites of the western part of
the Glenelg Lewisian Inlier rest on a thin zone of mylonitised Moinian
feldspathic sandstone in Maol Beag and Maol Mor to the north of
Auchtertyre. The mylonite is underlain by a major structural
discordance seen at Hangman's Bridge, locality 12.4.

----------------------------------Moine Thrust--------------------------

Balmacara Nappe. Complexly folded mylonitised Lewisian gneiss and
amphibolite. This unit is confined to the southern part of Lochalsh
and forms Auchtertyre Hill. The underlying thrust is seen in Ard
Hill, Balmacara, locality 12.5.

---------------------------------Balmacara Thrust-----------------------

Kishorn Nappe. Arkosic sandstones of the Applecross and Diabaig
Groups affected by low grade metamorphism and folded into the large
scale Lochalsh syncline. The axial trace of the syncline passes
through Kyle of Lochalsh so that only the western margin of the
Lochalsh peninsula shows rocks which are right-way-up, seen in the
Plock of Kyle (locality 12.1). To the east Torridonian sandstones of
the inverted limb of the Lochalsh syncline develop a mylonitic
foliation beneath the Balmacara thrust seen at Ard Hill, Balmacara
(locality 12.5). The underlying thrust plane extends from Loch Kishorn
to Skye and its outcrop lies beneath the Inner Sound to the west of the
Lochalsh peninsula.

---------------------------------Kishorn Thrust-------------------------

Caledonian Foreland. Undeformed Lewisian Gneiss of northern Raasay,
the Torridonian sandstones of Raasay and Applecross and the Cambrian
limestones of Broadford on Skye represent the Caledonian Foreland in

THE CALEDONIAN GEOLOGY OF THE SCOTTISH HIGHLANDS

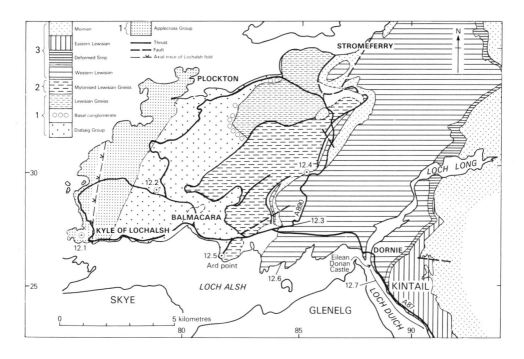

Fig. 4.8. Geological map of the Moine thrust zone in the Lochalsh area. The numbers in the key are as follows; 1= Kishorn nappe, 2= Balmacara and Sgurr Beag nappes, 3= Moine nappe. (after Barber and Soper, 1973).

this area and may be seen in the view from the Plock of Kyle (locality 12.1).

Continuing westwards along the A87 from Auchtertyre a drumlin occupies the floor of the valley to the north of the road at Kirkton. The prominent west-facing escarpment of Sgurr Mor marked by a conspicuous orange-weathered scree to the north of the road at Balmacara indicates the outcrop of the Balmacara thrust plane where mylonitised Lewisian Gneiss overlies mylonitised Torridonian sandstones. Eastward dipping deformed Torridonian sandstone is seen in the cliff on the north side of the road by the War Memorial about 1 km west of the Balmacara Hotel. These rocks form part of the inverted limb of the Lochalsh syncline. Inverted Torridonian sandstones are exposed in road cuts all the way from Donald Murchison's monument, erected by Sir Roderick Impey Murchison, Director of the British Geological Survey towards the end of the last century, to the village of Kyle of Lochalsh. At Scalpaidh Bay (NG 774 273) a lamprophyre dyke intruded into the Torridonian is seen to be cut by a thin basaltic sill. In Kyle itself a fine ripple-marked sandstone surface is exposed by the bridge leading to the railway station.

Locality 12.1. Plock of Kyle. Viewing point 500 m west of the village of Kyle of Lochalsh (NG 755 273).

From the ferry terminal follow the road towards Plockton. A few hundred metres north of the centre of the village take the first road on the left immediately south of the railway bridge. By bearing right at the top of the hill and driving up the steep narrow road to the right, past the houses it is possible to reach the viewing point on the Plock, where there is a parking place and turning room. Coaches should be left at the houses. It is also possible to reach this location directly from the centre of the village by the footpath immediately behind the Lochalsh Hotel.

From this point a panoramic view of the Isle of Skye, Raasay and the mainlaind from Applecross to Glenelg may be obtained. The view encompasses the Tertiary igneous complexes of the Cuillins on Skye; the Tertiary lavas in the northern part of Skye, Lewisian Gneiss and Torridonian Sandstone of the Caledonian foreland in Raasay and Applecross; and the Moine Thrust Zone, from the Kishorn thrust in Beinn Suardal behind Broadford on Skye, to the Balmacara thrust and the Lewisian and Moinian of the Mountain Belt on the mainland near Dornie and in Glenelg. Glacial features and extensive raised beaches may also be seen on the adjacent parts of Skye.

Torridonian sandstones forming part of the right-way-up (western) limb of the Lochalsh syncline are exposed in a group of rock outcrops a few metres to the north of the track about 50 m east of the viewing point. Here the sandstones have lost the characteristic red colour seen in the Torridonian of the foreland as a result of the reduction of iron in low grade metamorphism. Way-up evidence may be seen in a vertical rock face in the form of cross-bedding and cleavage/bedding relationships. Glacial striae may be seen on the same rock face. Neighbouring outcrops show spectacular sets of <u>en echelon</u> tension gashes and sedimentary slump structures.

Locality 12.2.   Loch Iain Oig.

Vehicles may be parked in the lay-by opposite the loch (NG 791 292).

Two small lochans, Loch Palascaig and Loch Ian Oig lie to the north of the old road from Kyle of Lochalsh to Balmacara, 11 km from Kyle of Lochalsh. Low exposures of Torridonian sandstone on the northern side of the depression between the two lochans show inverted cross-bedding, indicating that these outcrops are on the inverted limb of the Lochalsh syncline. The sandstones here are more highly deformed than these at the Plock to the west, with foliation, lineations, minor folds and quartzofeldspathic segregations in thin interbedded pelitic bands. All these features may be seen in the crag immediately to the south of Loch Iain Oig.

Locality 12.3.   (NG 845 275)

This locality comprises road-cuttings 400-500 m long above Auchtertyre on the road to Strome Ferry. Mylonitised striped and banded feldspathic and basic Lewisian gneisses of the Glenelg inlier within the Moine Nappe. Mylonitic surfaces dip $15°$ SE and an LS shape fabric plunges about $10°$ towards the east-southeast. Virtually isoclinal folds have a mylonitic fabric passing around them, but also have an intense axial planar mylonitic fabric. A zone of ultramafic rock occurs 100 m from the west end of the section. This has been intensely sheared and now consists of talc and actinolite. Zones of 1m thick of calc-silicate gneiss occur and spectacular occurrences of randomly orientated needles and blades of actinolite are recorded at the NE end of the section. Dykes of brick-red minette of Devonian age are approximately 2m thick and trend north-south.
   Parking places for coaches and cars are available up the new road above the section.

Locality 12.4.   Hangman's Bridge.

West of the A890 between Auchtertyre and Stromeferry, about 7 km north of Auchtertyre. A gate on the western side of the road opens on a track which leads to an abandoned section of the road and the old bridge. Vehicles can be parked on the new road or taken through the gate to the bridge (NH 856 298).
   The Moine thrust plane is exposed in the bed of a small stream about 100m upstream from the bridge. Leave the bridge heading northwards following the stream. After 100 m cut down into the bed of the stream. The thrust plane is exposed in the bed of the stream margined by small cliffs 2 - 3 m. high. The stream bed is composed of foliated and lineated mylonitised Lewisian Gneiss, the foliation dipping at $20°$ to the SE. The stream flows over a surface of structural discontinuity dipping due south at $14°$ and cutting across the foliation in the underlying Lewisian mylonites. At the base of

the cliffs forming the banks of the stream is a thin zone of clay gouge, only a few centimetres thick which separates the mylonitised Lewisian from flaggy mylonitised Moine schists with foliation surfaces dipping southwards parallel to the underlying thrust plane. The foliation is folded into a large number of small scale brittle monoclinal folds which do not occur in the Lewisian at this locality.

Locality 12.5.   Balmacara Foreshore.

Park opposite the Balmacara Hotel (NG 815 272). Southwards across the shingly and marshy foreshore on the east side of Balmacara Bay mylonitised Torridonian sandstones separated from mylonitised Lewisian Gneiss by the Balmacara thrust are exposed in a raised cliff.

The Torridonian seen here in the reefs and at the northern end of the cliff section is much more highly deformed than the equivalent rocks to the west, with an easterly dipping mylonitic foliation, an ESE mineral lineation, asymmetrical folds with axial plane cleavage which fold the earlier foliar and linear structures, sigmoidal tension gashes and a variety of monoclinal folds and kink bands representing several phases of deformation.

The Balmacara thrust plane is exposed on a grassy ledge in the cliff towards the point (NG 818 263). Mylonitised Torridonian sandstone dipping east at $20°$ is overlain by 3 - 5 m of gouge and breccia containing fragments of mylonite, which passes into mylonitised Lewisian gneiss above. The degree of structural discordance represented by the thrust can be appreciated from the neighbourhood of a large rock on the foreshore a short distance to the south. Uniformly dipping Torridonian mylonites are cut by the plan of discontinuing at a slight angle. On the other hand the foliation of the Lewisian mylonites forming the cliff above is highly folded and generally vertical, making a large angle with the thrust plane. This large angular discordance may account for the relatively thick zone of breccia along the thrust, in contrast to the thin zone of gouge along the Moine thrust plane at Hangman's Bridge, where the foliation in the overlying Moine is concordant with the thrust plane.

Highly folded Lewisian mylonites can be examined at the point of Ard Hill.

Locality 12.6.   Avernish shore section, 2.5 km SW of Nostie (NG 846 260 to 834 261).

This shore section shows quartzo-feldspathic gneisses and amphibolite lenses of the western part of the Glenelg Lewisian inlier cut by amphibolite dykes (? Scourie dykes). An interbedded garnetiferous siliceous schist at Avernish Point (NG 834 257), identified as part of the Moinian cover interbanded with the Lewisian basement, indicates that the Moinian sediments were metamorphosed to the amphibolite facies before the formation of the mylonites of the Moine Thrust Zone immediately to the west.

The coastal exposures on the northern shore of Loch Alsh may be reached by way of the minor road running southwards from the A87 at

Nostie. The road crosses the bridge and follows the shore of Loch Alsh to a fork in the road at (NG 846 262), where there is space for parking cars or a minibus.

The shore may be reached by heading due south across rough ground. Excellent exposures of amphibolite, quartzo-feldspathic gneiss and migmatite form eroded surfaces on the foreshore. The gneisses and amphibolites show by their macroscopic appearance and by their textures in thin section, that they have been derived from granulite facies rocks by retrogression in the amphibolite facies (Barber and May, 1976). The amphibolites form flat-lying lenticular pods, several metres in thickness, separated by screens of quartzo-feldspathic gneiss. The gneiss shows prominent rodding lineation of quartz and feldspar aggregates plunging at a low angle to the south-east. Close inspection of the surfaces of these rods show that individual mineral grains are elongated across the trend of the rods, in an east-south-easterly orientation (Barber and May, 1976, p. 40).

To the west at (NG 847 259) the gneisses are cut by vertical brick red lamprophre dykes some of which contain feldspathic xenoliths. These dykes are considered to be related to igneous activity of Lower Old Red Sandstone age in the Highlands. In Lochalsh they have a roughly radial relationship to the Ratagan granite of the same age, at the head of Loch Duich. Black basaltic dykes occur in the same shore section and near (NG 845 259), one of these basalts is seen to cut an earlier lamprophre dyke. These basalts are monchiquites related to east-west dykes of Permian age seen throughout the Northern Highlands Richey, 1939).

Migmatites are well exposed in the rock platform and raised cliff north of the point, which quartzo-feldspathic and hornblendic materials mixed together in a complex fashion. Pods of hornblendite are generally enclosed in a biotite-sheath where they are in contact with the surrounding quartzo-feldspathic gneiss. In the same area are amphibolites with relict igneous textures, and folded mineral banding which passes into a shear zone with increasing deformation.

Continuing westwards a series of amphibolite dykes are encountered which are believed to be equivalent to the Scourie dyke suite of the foreland. Those dykes can locally be seen to swing into parallelism with the regional foliation. To the west of this locality on Avernish Point (NG 835 257) a small strip of Moine metasediments tectonically interleaved with the Lewisian can be examined, details of this latter locality are given in Barber et al. (1978).

Locality 12.7. Dornie Road Section (NG 907 231 to NG 884 258).

This section is provided by road cuts on the main trunk road (A87). Care should be taken because of fast traffic. Hard hats should be worn because of danger from falling rock on parts of the section. Parking places occur at intervals along the section.

An almost continuous road section through the eastern and western parts of the Glenelg Lewisian inlier is exposed along the A87 between An Leth Allt (NG 907 231) and Eilean Donan Castle (NG 884 258) at Dornie.

At the southern end of the section for about 1 km north of the bridge over the An Leth Allt the road-cut shows pelitic garnet-biotite-gneiss, sometimes showing macroscopic kyanite which is commonly seen in thin sections. In places the pelites are interfolded with coarse migmatitic gneisses and cut by veins of white pegmatite. Spectacular examples of folds and associated structural features such as rodding, boudinage and mullions occur throughout the section. The relationships between these features allows a structural sequence to be established.

Beyond the first cliff immediately to the north of the landslip retaining wall is a small outcrop of forsterite-marble containing nodules of diopside. The next few outcrops along the road to the northwest consist of amphibolites and migmatitic gneisses with occasional small eclogite pods. Bands of pink marble and green diopside in the gneisses indicate the reaction of limestones with the surrounding siliceous gneiss. In the next major cliff 1.5 km northwest of An Leth Allt fine banded hornblende-and biotite-gneisses with some bands of massive amphibolite and pink feldspathic veins are highly folded. In the face of the cliff is a 4 m long fold mullion 0.5 in diameter plunging at $45°$ crossed by folded rods making a small angle with the axis of the mullion.

Further north at the end of the cliff above the retaining fence is a small adit opened in the early years of this century for copper. Chalcopyrite can be found in chloritic schist on the waste tip below the adit.

Two kilometres northwest of An Leth Allt the road-cut exposes a limonite-stained fault plane, part of a series of faults parallel to the trend of Loch Duich, which has been eroded along a major fault zone. Further northwest the road sections show fine-grained and finely banded amphibolite facies blastomylonites. The mylonitic banding, including highly deformed eclogites drawn into thin bands a few centimetres wide, are folded into large scale upright folds in the cliff section (NG 888 248), earlier small scale intrafolial folds can be seen in the limbs of the larger folds.

Three kilometres from An Leth Allt the blastomylonites pass into a tectonic schist containing lenses of hornblende-rock and small (2 - 3 cm) white feldspar augen (NG 887 252). Northwards along strike the schist includes large (20 - 30 m) lenses of Lewisian gneiss and Moine psammite.

This tectonic schist marks a major zone of deformation which separates the eastern from the western part of the Glenelg Lewisian inlier. This shear zone is analogous to the other major slides zones, such as the Sgurr Beag slide, described from other parts of the Caledonian mountain belt (Tanner, 1970; Rathbone and Harris, 1979).

From the tectonic schist to the village of Dornie the road cuttings are in hornblende and biotite migmatitic gneisses with homogeneous granodioritic gneiss, lenses of amphibolite and smaller actinolite pods. One of the latter is seen at the northern end of the car park opposite the turning to Eilean Donan Castle (NG 883 260). Several basalt dykes are also seen in this section.

Locality 12.8.   (NH 120 104)

Locality 12.8 is a road cutting about 100m long on the north side of the A87, 5.4 km east along the road from the Cluanie Inn. The psammite is coarse-grained and is striped and laminated containing cross-bedding more-or-less deformed. Many fold pairs are present which are probably of similar age to the Loch Quoich line and are therefore probably of similar age to the folds seen on the previous day at locality 11.8. The structures are cut by thin veins leucocratic pegmatite. This is an excellent locality at which to discuss the problems of interpreting sedimentary structures in deformed terranes and the possibility that the cross-bedding is mimicked by truncated folds limbs.

Locality 12.9.   (NH 184 104).

Hard hats should be worn, as the face is now locally unstable.

Large quarry in hornblendic granite cut by a sheet of mafic microdiorite dipping south at approximately $60°$. The granite which contains mafic patches of hornblende rich diorite is characteristically composed of quartz, oligoclase (with well defined oscillatory zoning), microcline, microperthite, hornblende and sphene. Abundant myrmekite is apparent in thin section. This granite is a typical Caledonian I-type granite which has been tentatively dated at 425 ± 4 Ma. The mafic microdiorite sheet is typical of the suite of the late orogenic microdiorites described by Smith (1979). Parking near the Cluanie Dam is easy.

ITINERARY 13.   THE MOINE THRUST ZONE.   PART 1.

This itinerary illustrates the geology of the NW foreland to the Caledonides and introduces some of the problems associated with the Moine thrust zone.

Introduction

The Moine thrust zone marks the NW boundary of the Caledonide Orogen and extends from Eriboll in the north to Sleat on Skye in the south, a distance of over 150 km (see Coward, this volume, Fig. 1). Internally deformed Proterozoic sediments, (the Moine schists) deformed during the Caledonian orogeny (500 - 400 Ma) were thrust over a foreland of early Proterozoic Lewisian gneiss, unmetamorphosed late Proterozoic – (Torridonian) and Cambro-Ordovician sediments (Peach et al., 1907; MacGregor and Phemister, 1972). There are several thrust sheets stacked on top of each other. The lowermost thrust sheet contains an imbricated sequence of Cambrian rocks and the base of this nappe is termed the Sole thrust. This thrust climbs down the stratigraphy from NW to SE; in the NW it lies in or above the Cambro-Ordovician sediments but to the east it must lie within the Lewisian complex as these gneisses are involved in the imbricate sequence. Above this imbricate zone are several large thrust sheets of Lewisian and Cambrian rocks. The thrusts which bound these large sheets are not parallel but converge and diverge and the sheets are lens shaped in plan. The most easterly thrust carries mylonites and the Moine schists. This is structurally the highest thrust though locally small faults thrust quartzites over the Moines from the underlying thrust nappes. The transport direction is to the NW, as seen from slickensides, sheared worm tubes and frontal/lateral ramps. The sequence of faults is from East to West, towards the foreland as the eastern faults are folded above steps in the underlying thrusts. However, the eastern thrusts were also the last to move, slicing through structures in the underlying sheets.

Stratigraphy

The Lewisian Gneiss. The Lewisian rocks were first described in detail by Peach et al. (1907). They noted that, around and to the south of Scourie, gneisses were cross cut by a suite of NW trending metamorphosed dolerite dykes. North of the Laxford River there are no such discordant dykes but gneisses and amphibolites are cut by considerable amounts of granite and pegmatite. Peach et al. (1907) noted a similar change in Lewisian intrusive rocks in the Glencoul nappe where the most northerly recognised dolerite dyke was mapped on the southern slopes of Beinn Aird da Loch, Glencoul, much farther south than the most northerly dykes on the Lewisian foreland. They (op.cit.) considered this offset to be due to the movement of the Glencoul sheet.

Sutton and Watson (1951) further defined the chronology of the Lewisian rocks and considered the gneisses north of Laxford Bridge to be the reworked equivalents of the gneisses and dykes to the south. This period of reworking was termed the Laxfordian and dated at between 1850 Ma and 1575 Ma (Giletti et al., 1961; Lambert and Holland, 1972). The easterly period of gneiss formation was termed the Scourian and subsequently dated at between 2900 Ma and 2600 Ma (Giletti et al., 1961; Moorbath et al., 1969). Park (1970) has subsequently divided the Scourian into an early Scourian (or Badcallian) granulite facies episode and a later Scourian or (Inverian) amphibolite facies episode (cf. Evans and Lambert, 1974). Bowes (1969) and Bowes and Khoury (1965) have disputed Sutton and Watson's (1951) claim that there was only one set of dykes, separating the Scourian and Laxfordian events but many authors (Park, 1970; Beach et al., 1974; Davies and Watson, 1977) consider that the field evidence supports the model of Sutton and Watson rather than that of Bowes. More recently Davies (1976) has observed an early set of dykes predating most of the Scourian deformation but post-dating an early gneiss forming event. He suggested that these early gneisses underlie a series of layered basic rocks and schists and gneisses with metasedimentary affinities (Davies, 1974; Beach et al., 1974). Davies (1977) proposed that these early gneisses might be of similar age to the Amitsok (3700 Ma) gneisses of West Greenland (MacGregor, 1973) but from these early gneisses Chapman and Moorbath (1977) obtained a Pb/Pb age of 2860 ± 60 Ma similar to those obtained from the surrounding gneisses.

Torridonian. The "Torridonian" is the name long given to the thick sequence of arkosic red-beds which rest unconformably on the Lewisian gneisses of the foreland and thrust zone.

Detailed field-mapping, palaeomagnetic studies and radiometric dating has revealed the "Torridonian" of Peach et al. (1907) comprises two quite distinct successions. The Stoer Group and Torridon Group are separated by a $25°$ angular unconformity, a $50°$ change in palaeomagnetic pole position and approximately 180 my. as determined by Rb/Sr whole rock isochrons. As Stewart (1969, 1975) has indicated, the term "Torridonian" has historical but little stratigraphical significance.

The older Stoer Group (995 ± 24 Ma) comprises 2.3 km of red arkosic sandstones and siltstones with locally derived breccias. These sediments were deposited on a hilly Lewisian land surface as bajada and playa deposits which are overlain by fluvial sands which are often pebbly. Local lake cycles and a volcanic mudflow occur. The Stoer Group crops out in a narrow belt between Stoer and Loch Maree.

The more extensive and younger Torridon Group (810 ± 17 Ma) oversteps the Stoer Group onto a rugged Lewisian topography (see Stewart, 1969, fig. 2; 1975, fig. 9). Red and purple arkosic sandstones, often pebbly, predominate but red and grey shales occur throughout the 7 km thick succession. Four formations are recognised (in ascending order, the Diabaig, Applecross, Aultbea and Cailleach Head Formations). Sedimentological analysis indicates that the bulk of the Torridon Group was deposited by river on large alluvial pans draining an upland region which lay to the north-west (Williams, 1969a and b). The thicker grey shale units in the Diabaig and Cailleach Head Formations probably were associated with a marine influence although some of the shales elsewhere in the sequence may represent local lakes.

The Diabaig Formation is considerably thicker in the south and absent in the north of the region. Such facies variations may have influenced the style of structures in the thrust zone (Barton, 1978). Following the deposition of the Torridon Group and prior to the deposition of overlying Lower Cambrian quartzites the "Torridonian" was gently folded about fold axes trending approximately north-south with a fold wavelength of approximately 50 km. There is a large syncline in the west, while in the east near the thrust zone, Torridon Group sediments and the underlying Gneisses form a large Precambrian anticline. Peneplanation of these structures preceded the deposition of the Cambro-Ordovician sequence which oversteps from the Torridon Group into the Lewisian.

The Cambro-Ordovician Sediments. These sediments were deposited on a plane erosion surface, very different to the older irregularly dissected surface beneath the Torridonian. The Cambrian base rests sometimes on Torridonian, sometimes on Lewisian. The lowermost Cambrian rocks are quartz-rich sandstones and grits (the Basal Quartzite, 75 - 125 m) sometimes with a conglomerate at the base.

These pass up into the well bedded quartz-sandstones with vertical bioturbation structures (the 'Pipe Rock', 75 - 100m). The burrows are of the genus skolithus (Hallam and Swett, 1966). In undeformed rock, the pipes are normal to bedding and circular on the bedding plane and make ideal strain markers for estimating layer parallel strain ratios and layer parallel shear. The quartz sandstones are followed by rusty weathering shales with hard dolomitic beds (the Fucoid Beds, 20m). Abundant flattened worm casts on the bedding planes were originally mistaken for the remains of fucoids and so give the rock its name. They contain the earliest shelly fossils in this sequence and are characterised by the trilobite Olenellus. The Fucoid Beds are followed by a 10 m thick grit (the 'Serpulite Grit', 10m) which takes its name from small conical fossil shells, originally named Serpulites, but now known as Salterella. There are vertical worm burrows in the Fucoid Beds and Serpulite Grit as well as Pipe Rock. The highest Cambro-Ordovician sediments are a thick sequence of limestones and dolomites (the Durness limestone). This 'Durness Group' has been subdivided into 4 sub-divisions, the two lowest, the Ghrudaidh and Eilean Dubh Formations are widely exposed in the Ullapool and Assynt district. The dolomites and limestones of the Ghrudaidh Formation are usually dark or grey coloured; those of the Eilean Dubh are white and fine grained with occasional layers of chert.

Post-Cambrian Intrusives. These include a suite of felsite, porphyrite and lamprophyre sills in the Assynt area and also large alkaline intrusives and syenites in SE Assynt. The large intrusives form the Loch Borralon and Loch Ailsh complexes, both generally of laccolithic shape. They intruded Cambrian sediments during the thrusting and so could date the main thrust episode (Sabine, 1953; Woolley, 1970; Parsons, 1979). A U-Pb age of 430 ± 4 Ma (van Breemen et al., 1979) dates the Loch Borralan intrusive.

Minor intrusives occur as sills sometimes exceeding 30 m thickness and include aegrine-felsites, quartz-microsyenite (eg. Canisp porphyry), hornblende-lamprophyres and nepheline-syenites. The aegrine-felsites are restricted to the higher thrust sheets within the thrust zone and hence help thrust correlation.

Structures of the Moine Thrust

Between the foreland and the Moine thrust there is what the Geological Survey called "the belt of complication". This belt varies from 0 to 11 kms wide and consists of numerous thrust faults, including several thick thrust sheets containing Lewisian as well as Cambrian or Torridonian rocks. Examples of the major sheets include the Arnaboll sheet of the north coast of Scotland, the Glencoul and Ben More sheets of Assynt and the Kinlochewe-Kishorn sheets of Lochalsh and Skye.

Cambrian rocks are carried by all the major thrust nappes except those of Kishorn and Tarskavaig. The sediments show variable amounts of deformation. Caledonian shear zones in the underlying Lewisian gneiss pass up into the steep limbs of asymmetric inclined folds in the Basalt Quartzite and Pipe Rock. The amplitude of these folds decreases upwards in the Cambrian sediments.

The quartzites and Pipe Rocks are strained and the pipes make elliptical sections on the bedding. The ellipse long axes are not always parallel to the fold axes but trend more N-S and thus change orientation slightly around the folds. This discordance between ellipse axes and fold axes suggest a rotational component to the strain in the bedding plane section. Coward and Kim (1981) have described how these strains may be separated into two components; (1) layer-parallel shortening and (2) shear strains on planes normal to the main thrust plane but with the same movement direction as the main bedding-parallel thrust shear. A combination of these two components results in strain ellipses with long axes oblique to the general thrust transport direction. It is suggested that the same sinistral shear strains shown by the pipes were also responsible for the development of the oblique trending folds; the shear strain would cause buckling of the beds and the trend of the hinges would depend on the combination of shear strain and layer parallel shortening. Oblique trending folds are common in Assynt as far north as Glencoul, suggesting a wide zone of differential sinistral movement (Coward, 1984, fig. 1).

The Moine thrust exemplifies several concepts of thin-skinned tectonics, applicable not only to high-level thrust zones but also to deeper-level ductile shear zones. In many thrust zones, the faults develop and move in piggy-back fashion, the lower faults being youngest (Dahlstrom, 1970; Elliott and Johnson, 1980). Thus tip strains and folds in lower faults affect the higher fault sheets. This may lead to fold interference and the production of several phases of cleavage development. If lateral folds related to ramps or tip zones interfere, areas of cross folding may be produced. Sometimes the folds and minor thrusts develop out of a tip region, with the opposite sense to that of the main fault, giving rise to back-folds and back-thrusts. These structures may have only local importance, related to tip zones in some underlying thrust or shear zone; the structures may have no regional significance, that is fold phase numbers may have only local significance.

From restoration of balanced cross-sections and a correlation of Lewisian structures in the thrust sheets with those of the foreland, displacement estimates of 30 - 50 Ma have been made for the major thrust sheets (Coward, 1982, fig. 1).

Not all this displacement has taken place on contractional faults. Northeast of Inchnadamph, extensional faults cut through the earlier folds and thrusts (Fig. 4.9; Coward, 1982, fig. 6). The extensional faults can be traced into strike-slip and then contractional faults and together bound a large scoop-shaped surge zone (cf. Coward, 1982) which has moved the western part of the Glencoul sheet some 1.75 km further to the northwest. The extensional faults probably flatten upwards to join reactivated, higher-level, low-angle faults (Fig. 4.10). Thus the Moine thrust in eastern and southern Assynt has been reactivated by extensional fault movement.

At Assynt the frontal part of the surge zone is defined by the arcuate trend to the imbricate faults. Similar arcuate trends occur along the length of the Moine thrust belt though the majority of these

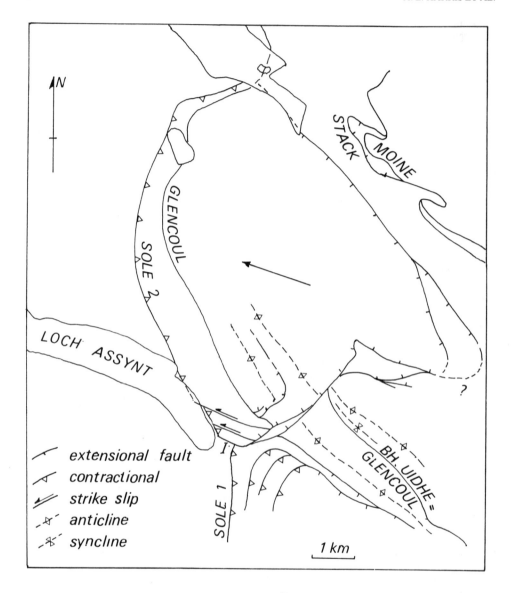

Fig. 4.9. Map of northern Assynt (box d of Fig. 4.16) showing the form of a surge zone defined by a closed system of contractional and extensional faults (after Coward, 1982, 1983).

are not part of closed systems of contractional and extensional faults as at Assynt. The real dimensions of the surge zones are not known. Those at Assynt may mark the keels of much larger eroded structures.

The surge zones obviously lead to some anomalous thrust tectonic situations. Though the normal thrust sequence is from east to west, in the transport direction, low-angle extensional faults may cut back to join and reactivate earlier eastern faults. This may explain some of the controversial fault sequences proposed for the Moine thrust zone; at Eriboll for example, the Upper Arnaboll thrust has moved later than the underlying and more westerly thrusts (Soper and Wilkinson, 1975; Coward, 1980). Between Kishorn and Kinlochewe the Moine thrust may have been reactivated by extensional flow. Similarly in southern Assynt, the Moine thrust has moved later than the lower Ben More and Glencoul faults. This reactivation may be one explanation for the range in textures along particular thrust planes; Christie (1960) records two distinct mylonite types in the Moine thrust zone, early ductile phyllonitic mylonites and later cataclastic brecciated mylonites.

The mylonites were first described by Lapworth (1885) from the Eriboll district and later by Peach et al. (1907), Peach and Horne (1930) and more recently by Christie (1960), Johnson (1961), McLeish (1971) and Rickels and Baker (1977). The mylonites lie immediately beneath the overthrust slab of Moine schists; some mylonites owe their origin to intensely deformed Moines, others to intensely deformed Cambrian rocks below.

Originally the term mylonite was proposed by Lapworth (1885) for microscopic breccia with fluxion structure in which the interstitial grains have only partially recrystallised. Later authors have applied the term to a variety of other rocks along the thrust zone. Here the term mylonite is applied to rocks in which the grains show evidence of plastic deformation leading to change in grain shape as well as grain size reduction by dynamic recrystallization as a result of intense strain within the thrust zone.

Christie (1960) examined the textures of mylonites and divided them into two groups: primary and secondary mylonites. According to Christie the primary mylonites are characterised by crystalloblastic textures and grade with increasing grain size into the Moine schists above. The secondary mylonites are characterised by cataclastic breakdown of all the minerals. Christie thus concluded that there were at least two periods of mylonitisation, an earlier one contemporaneous with deformation in the Moine schists and a later post-metamorphic deformation. Both groups of mylonites were formed in the movement zone between the Moine nappe and the foreland. Barber and Soper (1973) and Soper and Brown (1971) and Soper and Wilkinson (1975) have proposed a larger and more complex structural history of the mylonites with several fold phases affecting an early mylonitic fabric followed by late movement on the thrust faults.

Johnson (1960), Christie (1963) and Barber (1965) have shown that many of the folds on the mylonite zone have axes which plunge to the SE close to the transport direction. Indeed, Christie (1963) proposed a NE-SW transport direction parallel to the srike of the Moine thrust

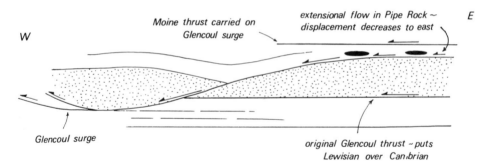

Fig. 4.10. Possible cross-section through the Glencoul/ N Assynt surge zone suggesting that the normal faults flatten to the east to cause extensional flow in the Moine rocks and/or high level thrusts.

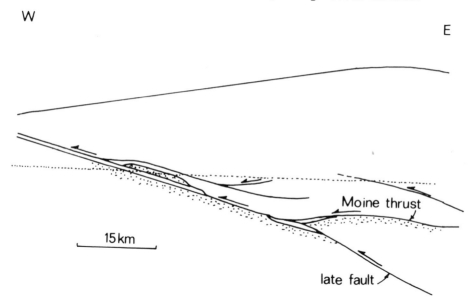

Fig. 4.11. Simplified section through the Moine Thrust Zone, showing the form of the Moine Thrust wedge and late extensional flow reactivating the Moine Thrust. This late extension may be due to gravitational collapse of the wedge, due to either a change in shear strength at the base or to uplift from a late fault at depth causing changes in the surface slope.

zone, on the basis of these folds. Many of the folds may have been initiated by layer parallel shortening or by shear where the layering has been deflected from the shear plane and originally many would have their axes normal to the transport direction. Their axes may have been rotated towards the X-direction during later shearing. However, there are many kink bands and chevron type folds which plunge to the SE. It is unlikely that these open folds have been rotated into this direction by intense deformation. In the Glencoul nappe, folds in Cambrian quartzites formed obliquely to the transport direction where the rocks carried a component of shear on the plane containing the normal to the layering. Thus if the mylonites formed at an early stage of thrust movement, differential movement of the Moine nappe would cause folds to form with hinges which plunge to the E or SSE.

Recent textural studies in deformed quartzites (Law et al., in press) show that the primary mylonites often show symmetric c-axes fabrics suggesting coaxial deformation and extensional flow parallel to the major thrust zone. The rocks with symmetric fabrics grade down to zones of asymmetric fabric suggesting dominantly simple shear deformation, along the thrust zone itself. Variable amounts of extensional flow may lead to differential movement and hence to the many phases of oblique folds. Note this represents an early phase of extensional flow structures before the later phases recorded by surge zones, late discordant low angle faults and secondary cataclastic mylonites on the Moine thrust zone.

A model for thrust development in the Assynt area has to include the following.
(1) Some of the Moine thrust zone has moved forward by gravitational spreading (Fig. 4.11; Coward, 1983) as it generally involves thrusts which cut up section from basement to cover in the transport direction, but also involves localised extensional flow and thinning of the thrust sheets.
(2) In northern Assynt, the thrust zone involves a wide zone of sinistral shear about a vertical plane where the northern zone has moved further than that in the south. A differential displacement of about 3 km has been estimated for this zone, though this is small compared to the total displacement of well over 70 km for the Moine thrust zone as a whole (Elliott and Johnson, 1980).
(3) More ductile deformation structures occur in the north compared with less ductile, often cataclastic faults structures in the south. These suggest that the northern zone has moved further but more slowly, and probably under a thicker cover, while the southern zone has moved more intermittently, probably under a thinner cover. Soper and Barber (1982) estimate a cover thickness of over 10 km for the northern area. This variation in cover thickness may be due to a change in thrust geometry east of the Moine thrust outcrop.

Locality 13.1.

From Kyle of Lochalsh follow the A87 and A890 through Stromeferry along the side of Lochcarron to the east end of the avalanche tunnel at (NG

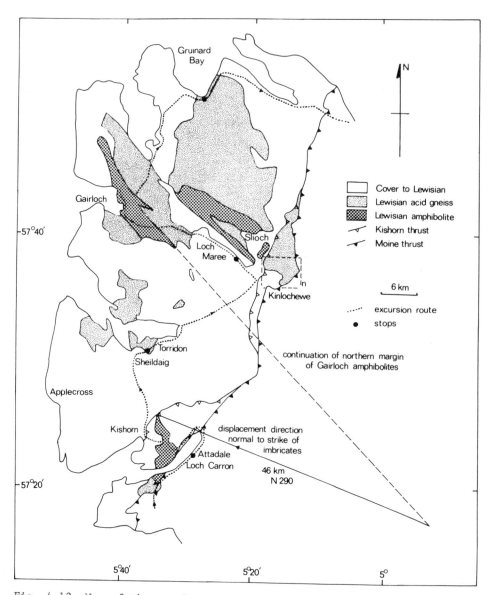

Fig. 4.12. Map of the southern part of the Moine Thrust Zone and the Lewisian rocks of the foreland, showing the route of itinerary 13. The offset of the Lewisian amphibolites suggests a thrust displacement of some 46 km.

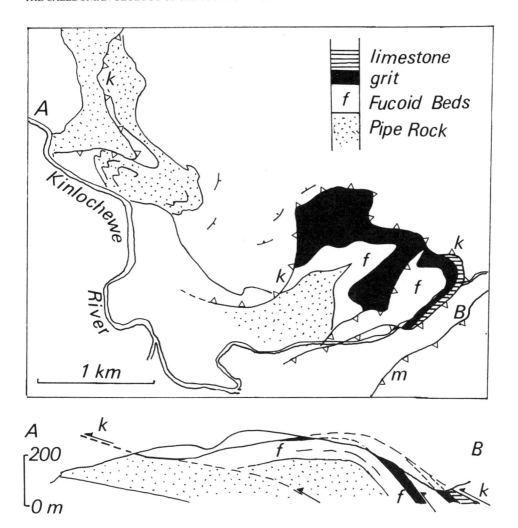

Fig. 4.13. Simplified map of the hills NE of Kinlochewe (box n in Fig. 4.12) and a cross-section showing the Kinlochewe thrust bulged by imbricates beneath but also cutting down stratigraphy in the thrust movement direction (N290) from limestone in the east to Pipe Rock in the west.

923 377) where there are exposed the basal conglomerates to the Moines. The conglomerates are overlain to the east by acid migmatites and amphibolites of the Lewisian, suggesting that the original sedimentary contact was inverted. The pebbles within the basal Moine include a variety of rock types but are mainly quartzo-feldspathic. They are flattened to define the foliation. The matrix is rich in ferromagnesian constituents with biotite and acicular hornblende.

Route from Attadale to Gruinard.

From Attadale follow the road (A890) through Strathcarron and turn right on the A896 to Lochcarron village. At the south end of the village, the peninsula Slumbay Island is made up of Moine mylonites. The Moine thrust is crossed near the main bend in the road across to Kishorn, but no topographic feature is produced. To the west are Lewisian rocks, often intensely sheared and mylonitic, consisting of large tracts of deformed basic rocks as well as acid gneiss. To the west of the narrow pass the Lewisian rocks overlie inverted Torridonian rocks. The contact is sheared; the original basal Torridonian being removed by thrusting. The Torridonian rocks lie in the inverted limb of the Kishorn syncline and show an intense fabric as in Loch Alsh.

Near the Kishorn lochside the road crosses the Kishorn thrust which also does not make much of a feature. By the lochside the road passes an extensive outcrop of Eilean Dubh limestone with distinctive vegetation, different to that of the Torridonian hills.

The route continues to the head of Loch Kishorn and then onto Shieldaig. To the west of the road there should be good views of the Applecross hills, made of foreland Torridonian. The road between Shieldaig and Torridon crosses the unconformity between Torridonian and Lewisian several times. North of Torridon are the mountains of Beinn Eighe and Liathach. On Beinn Eighe and south of the road on Sgurr Dubh the thick sequence of Torridonian sandstones are overlain by Cambrian quartzites showing several thrusts and associated folds.

From Kinlochewe the position of the Moine thrust can be seen making the escarpment to the SE of the village and to the NE along the Kinlochewe valley. North of the village the Cambrian quartzites make prominent escarpments; they are repeated several times by thrusting. The top of the ridge is made of Torridonian of the Kinlochewe thrust which is folded by a thrust culmination north of the road to the Heights of Kinlochewe (Fig. 4.13). A klippe of the Kinlochewe thrust occurs on Beinn Eighe, south-west of the village.

To the west, the route follows the shores of Loch Maree on the A832. To the north rises the hill Slioch; its upper part comprises horizontally folded Torridonian of the foreland, resting with a highly irregular unconformity on Lewisian gneiss. Here the pre-Torridonian Lewisian topography had a range of nearly 300 m.

From Loch Maree the route continues through Gairloch, crossing the Tollie antiform - a late large Laxfordian fold structure (Park 1970), and on to Inverewe and Gruinard.

THE CALEDONIAN GEOLOGY OF THE SCOTTISH HIGHLANDS

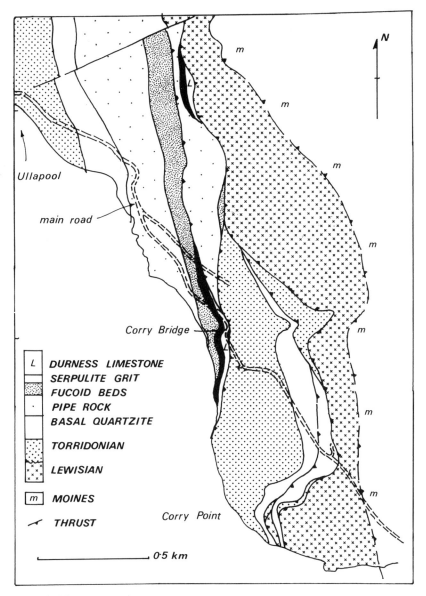

Fig. 4.14. Map of the thrust zone geology SE of Ullapool (box a on Fig. 4.16).

Locality 13.2 (NG 940 901).

The Lewisian can be examined in road cuts at the top of a steep hill, where there are well preserved Scourian migmatites. In road cuts to the north, these Scourian structures are shown to be cut by members of the Scourie dyke suite.

Route from Gruinard to Ullapool

Continue along Little Loch Broom following the Dundonnel River. At (NH 104 903), north of the road, there is the Dundonnel thrust structure, an antiformal stack of Lower Cambrian rocks (Elliott and Johnson, 1980; Boyer and Elliott, 1982). Unfortunately this area is very badly exposed and better examples of thrust structures can be seen more easily in Assynt. At Braemore, follow the road to Ullapool, passing road cuts in Moinian quartzofeldspathic psammites. At (NH 148 923) these road cuts show fine banded mylonitised Moines.
    The Moine thrust is crossed some 3.5 km E of Ullapool but is badly exposed here. Beneath the Moine thrust are coarse Torridonian sandstones showing varying degrees of deformation. The better exposures are south of the road (see Fig. 4.14) where Torridonian and Lewisian rocks are imbricated.

Locality 13.3

The best road cuts in this part of the thrust zone occur near the bridge over Corrie Burn (NH 135 929) where sheared Torridonian rocks are thrust over Durness Limestone, enclosing small thrust horses of quartzite between two major units.
    West of Corrie Burn the lower part of the Cambrian succession from Fucoid Beds to Basal Quartzites is well exposed in road cuts and at (NC 135 940), at the southern end of Ullapool the unconformity between the Torridonian sandstones and Cambrian quartzites occurs in a back garden.

Locality 13.4

Continue through along the road to the north to Ardmair Bay 3 km of Ullapool where there are good exposures of gritty Torridonian rocks with cross cut contorted bedding and good views of the Torridonian Hills to the NW.

Locality 13.5

Return to Ullapool and follow the track up the east side of the Ullapool River. After a few hundred metres the road is private, used by estates and for access to limestone quarries. The geology of the Ullapool valley is summarised in Figure 4.15.
    At locality 13.5A, Torridonian sandstones and quartzites occur in road cuts, repeated by normal faults downthrowing to the west. The quartzites are not well exposed by the roadside but at locality 13.5B a quarry south of the road shows a Lewisian quartzo-feldspathic gneiss

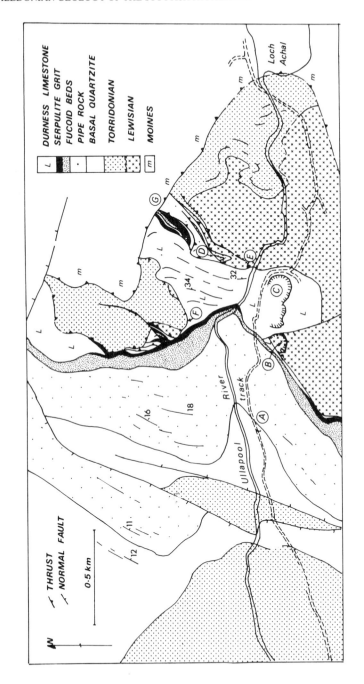

Fig. 4.15. Geology of the Ullapool Valley area, NE of Ullapool (box b on Fig.4.16). Letters A-G refer to locality 13.5.

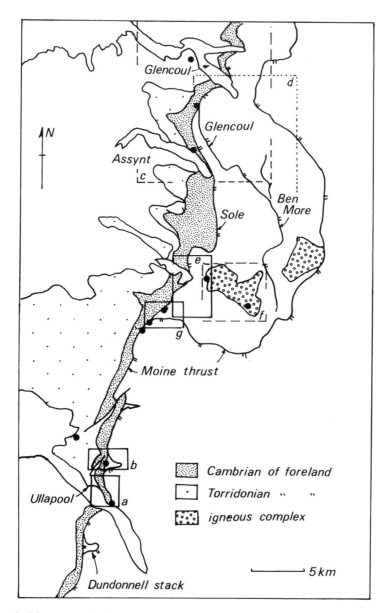

Fig. 4.16. Map of the Ullapool to Assynt area showing the figure locations. Locality stops shown by solid circles.

lying above the Sole thrust. At the east edge of the quarry there is a steep fault with fine black cataclastic rock, bringing the gneiss into contact with limestone. This fault does not cut the Sole thrust nor apparently the Moine thrust and thus predates both. East of the Lewisian gneiss quarry there is a large quarry in the limestone (locality 13.5C). The bedding is disrupted by several low angle normal faults, with downthrow to the west and the limestone generally shattered with thin extensional vein systems.

The easiest place to cross the Ullapool River is by bridge at the west end of Loch Achal where to the north and west the outcrops are of Lewisian and Torridonian which have been thrust over the limestone. At Locality 13.5D there is a fold of Torridonian rocks in the hanging wall of this thrust, while to the north the Lewisian and Torridonian are thrust over quartzites which are in turn thrust over Fucoid Beds/ Serpulite Grit imbricates. At locality 13.5E the thrust plane of Lewisian over limestone occurs in the East Dubh waterfall and adjacent cliffs.

To the west at locality 13.5F there is a north-westward dipping fault carrying Torridonian over limestone, but carrying them down to the NW. The fault plane can be easily examined in overhanging cliffs. This fault is considered to be a low angle normal fault dropping the Torridonian with some underlying Lewisian basement down to the NW from an originally over-riding thrust sheet.

The Moine thrust is a planar feature slicing off underlying thrust structures and also normal faults. It produces only a narrow zone, a few metres thick mylonitic rock with flaggy but folded Moines above. The thrust plane can be seen in the stream section near locality 13.5G.

The sequence of events in the Ullapool valley may be summarised as follows:

(i) Lewisian gneiss, with its cover of Torridonian, was folded and thrust over Durness Limestone, collecting numerous small thrust horses of Cambrian quartzite, Fucoid Beds, Serpulite Grit and Limestone. The thrust direction was to the WNW, the folds and thrust ramps are generally perpendicular to this transport direction but these are also oblique to lateral ramps. The Sole thrust may develop in this thrust sequence.

(ii) Normal faults cut across the thrust sequence, generally dropping it down to the west. These faults either flatten into or are cut by movement on the Sole thrust.

(iii) The Moine thrust moves last slicing through all earlier structures.

ITINERARY 14.   THE MOINE THRUST ZONE.   PART 2.

This itinerary examines the large thrust sheets of Assynt, the rocks of the foreland and the Sole thrust, the Moine thrust and mylonites and the syn-tectonic Borolan igneous complex (Fig. 4.16).
Locality 14.1.   The north side of Loch Assynt, by Skiag Bridge (Fig. 4.17).

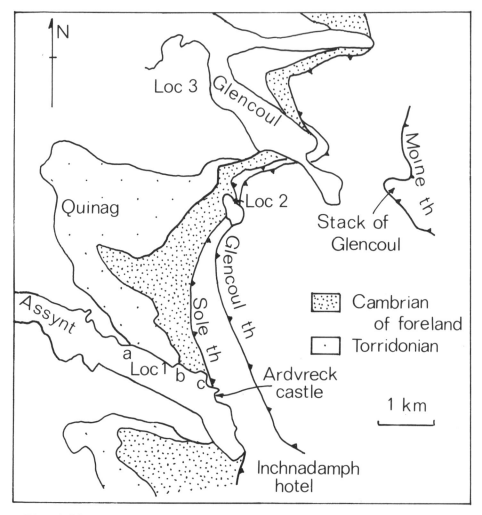

Fig. 4.17. Map of N Assynt, showing localities 14.1 and 14.2.

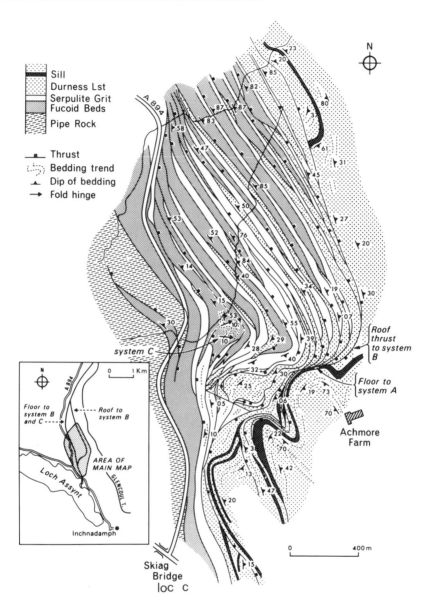

Fig.4.18. Detail of the imbricate structures, NE of the A894, Skiag Bridge area, N Assynt.

Locality 14.1A. Lewisian gneiss are exposed on the shore of Loch Assynt (NC 213 251), with a Scourian gneiss banding cut by strongly discordant ultrabasic (medium-grained olivine gabbro-peridotite) and basic (epidiorite) dykes. These rocks show almost no effects of Laxfordian deformation.

On the roadcuts to the north (NC 215 253) there is a long and good section of the unconformable contact between Torridonian sandstones and highly weathered Lewisian gneiss. The Torridonian passes up into course cross bedded sandstones and grits.

The views across the Loch shows the irregular unconformity between flat lying Torridonian and Lewisian and the planar but E-dipping unconformable contact at the base of the Cambrian.

Locality 14.1B. At (NC 228 247) roadcuts show the Cambrian basal cross bedded quartz-sandstones while at (NC 231 246) there is an exposure of a small-scale duplex zone with floor, roof and several small imbricate thrusts well preserved often with slickensides on the bedding related both to folding and thrust slip.

To the east the quartz-sandstone become finer grained and are cut by small worm tubes: the base of the Pipe Rocks. Some higher units of the Pipe Rock, are exposed in the moorland between the road and the Allt Skiathaigh river.

Locality 14.1C. Near Skiag Bridge (NC 235 243) the upper part of the Pipe Rock shows a characteristic red stained horizon where the white pipes stand out clearly (the 'Bird-Shit Pipe Rock').

This passes up into white Pipe Rock with blotchy purple pipes and then to dolomitic sandstones and shales of the Fucoid Beds overlain by careous weathering Serpulite (or Salterella) Grit. On the promontary at (NC 237 239) there is a good section through the grits with numerous thick worm burrows and concentrations of the conical shells of Salterella.

In the roadcuts the grits are overlain unconformably by basal Ghrudaidh limestone, with several lamprophyric sills. To the east these become involved in the imbricate faults above the Sole thrust, the Sole must lie parallel to bedding somewhere in the Durness limestone. Limestone and lamprophyre are more steeply tilted and highly fractured on the hanging walls to the imbricate ramps.

A more detailed analysis of the faulting can be made in roadcuts at (NC 244 234) where imbricate thrusts cut a lamprophyre sill, at (NC 238 237), at the north end of the peninsula of Ardvreck Castle where Fucoid Beds and Serpulite Grits are imbricated and in roadcuts at (NC 243 235), where limestones and sills are intensely fractured by approximately E-W trending fractures related to late N-S extension and in the hills NE of Skiag Bridge (Fig. 4.18).

Locality 14.2

At the north edges of Sandy Loch (Loch na Gainmhich, (NC 243 293)) there are good examples of Pipe Rock with large 'trumpet pipes'. The cliffs on the east side of the Loch are of Lewisian gneisses of the

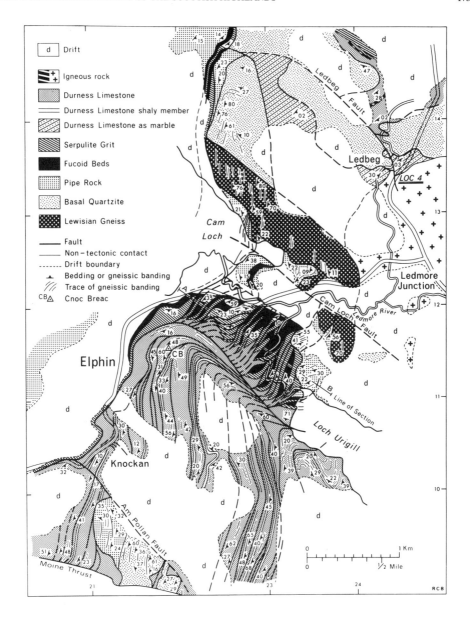

Fig. 4.19. Map of the geology between Ledmore junction and Knockan (box e on Fig. 4.16).

Glencoul thrust sheet.

In the road cuts west of this locality are good exposures of cross-bedded grits of the Basal Quartzite unit.

Locality 14.3

This stop gives a panoramic view of the Moine thrust zone from Kylesku to the Stack of Glencoul. North of the Loch, Cambrian Quartzites unconformably overlie Lewisian gneiss of the foreland. The Quartzites and overlying Pipe Rock are overthrust by the Glencoul sheet with its thick slab of Lewisian gneiss at the base and a thin cover of Cambrian sediments, which is itself overthrust by the Moines at the Stack of Glencoul. A simplified geological map of the area is given in Fig. 4.17. Note that at the east end of Loch Glencoul the Glencoul sheet lies on Durness Limestone while to the north-west it lies on Pipe Rock, that is the Glencoul thrust sheet appears to cut down stratigraphy in the transport direction. This implies that the Glencoul fault cut across more steeply dipping strata to give an apparent extensional geometry or it may have had a real component of extensional flow and cut down section into the foreland.

Roadcuts show Lewisian gneisses of the foreland largely unaffected by Laxfordian events, with a Scourian gneissic banding and early tight folds associated with high grade amphibolite facies metamorphism, folded by large open late Scourian structures. The hill Quinag shows a thick sequence of purple-brown weathering Torridonian sandstones and grits. The irregular boundary between these flat lying sandstones and the grey Lewisian gneisses can be traced along the hillside S of Glencoul.

The excursion continues south past Inchnadamph where there are views of the Glencoul thrust sheet. Basal Quartzites and Lewisian gneisses overlying Durness Limestone. The hills and cliffs near Inchnadamph are made out of limestone and dolomite, with prominant lamprophyric sills. The floor thrust to the imbricate system can be seen in the cliff at Stronchaibie with several imbricate faults in its hanging wall. The route continues past the Allt nan Uamh. About 1 km E of the road there are caves in the limestone which were excavated at the end of the last century and again in 1926 and 1927 producing the remains of bear, lynx, reindeer and lemmings. The hills to the N and S of the river: Beinn nan Cuaimhseag and Beinn an Fhurain, carry klippes of the Ben More thrust sheet, the highest of the major Assynt thrust slices. These thrusts carry Basal Quartzite, Torridonian and Lewisian in a large recumbent fold.

Locality 14.4

South of the Ledbeg River, (at NC 238 140) roadside outcrops show coarsely crystalline white brucite-marble, where the limestone has been intruded by the Loch Borolan complex. The metamorphosed Cambrian strata and the igneous complex occupy the Ledbeg thrust sheet lying beneath the over-thrust quartzites of the Ben More sheet. Later normal faulting has dropped the quartzites down to the NW, so that they

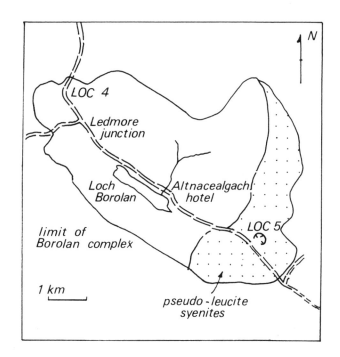

Fig. 4.20. Simple map of the Borrolan complex (box f on Fig. 4.16).

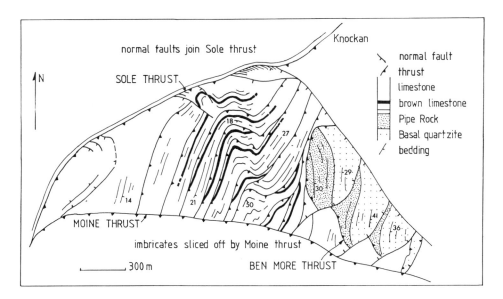

Fig. 4.21. Map showing the relationship between thrusts and later extensional structures at Knockan (box g on Fig. 4.16).

occupy the tops of the low hills W of the road. The badly exposed hill, Cnoc na Sroine, east of the road is composed mainly of the igneous material which is easiest seen in road cuts to the south, at (NC 244 133), south of the track to Ledbeg farm. Here a red augite-nepheline-syenite enclosed xenoliths of marble. Both are locally sheared. At both ends of the outcrop the syenite is faulted against quartzite.

Locality 14.5

Continue along A837 past Ledmore Junction and the Altnacealgach Hotel to a small quarry (NC 293 090), N of the road (Fig. 4.20). This is the type locality of 'borolanite' a melanite-pyroxene-nepheline-syenite which contains large white spots of feldspar, muscovite and nepheline, thought to be pseudomorphs after leucite. These 'pseudoleucites' show varying degrees of deformation, most are ellipsoidal and define an east dipping fabric approximately parallel to the Moine thrust. The fabric is considered to be related to early Moine shearing while the borolanite was still hot. Later pegmatite dykes cut the fabric and both dykes and borolonite are affected by later shears.

Locality 14.6

The excursion returns to Ledmore junction and then along the A835 towards Ullapool passing the Cam Loch klippe of the Ben More sheets to the west and imbricates of the Glacbain and Elphin duplexes east of the road (Fig. 4.19). At the roadside at (NC 212 106), at Knockan village, there are exposures of limestone with clearly listric normal faults. The fault cut-offs suggest a foreland propagating sequence while at the west end of the crags there is a steep tear or transfer fault, which together with the slickenside orientations, suggests a WNW extension direction, approximately parallel to that of the Moine thrust.
 At (NC 207 103), immediately west of the last house in Knockan there are exposures of dark Durness limestone with a basic sill, deformed into an extensional imbricate zone (Fig. 4.22). The floor fault to these normal faults lie just above the Serpulite Grit and is probably the same as for the original Sole thrust. This floor fault or an extensional splay from it, is exposed at the west end of the outcrops as a fine white brecciated limestone, locally as gauge. A map showing the relationship of these normal faults to the nearby thrust imbricates and Moine thrust is given in Figure 4.21.
 This location gives good views of the eastward dipping Cambrian rocks of the foreland, the Cam Loch klippe of the Ben More sheet to the north and the Assynt hills, with the large hill of Breabag forming a thrust culmination bulging up the Ben More thrust sheet.

Locality 14.7

The Knockan Crag section at (NC 189 094) forms part of the Inverpolly Nature Reserve. Follow the nature trail from the car park to the Moine

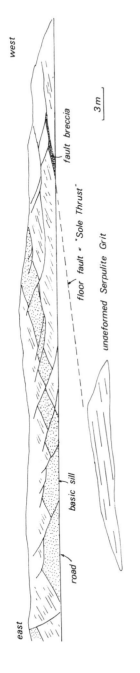

Fig. 4.22. Road section through faulted sill in limestone at Knochan, locality 14.6.

Fig. 4.23. Cross-section through the southern part of the Knockan area, along the southern boundary of Fig. 4.21, showing the relationship of the Moine 'thrust' to earlier imbricates (after Coward, 1983).

thrust plane which lies above a small thickness of Gradaidh dolomite and crushed and brecciated Eilean Dubh white dolomite. The Moine thrust is clearly exposed at several localities along the cliff and forms a clean cut surface. The Moines show an intense ductile mylonite fabric, defined by isoclinal folds with hinges plunging E or ESE parallel to the mineral lineation. The basal Moines have been re-brecciated and are cut by the clean cut break of the Moine thrust.

Continue along the track to the top of the cliff, crossing the Moines and then follow the outcrop of the mylonite schists to the NE and E for about 1 km, where the Moine thrust outcrops in a small stream at (NC 202 094). Here the easterly dipping dolomites and limestones are abruptly truncated by the Moine thrust plane indicating that the Moine thrust slices off the top of the under-lying imbricates and hence is a late feature. It also slices off the continuation to the normal faults seen at locality 14.6, that is it post-dates some extensional fault movement and may itself locally be a low-angle extensional fault.

Weather permitting there are good views from this cliff section of the Assynt hills and the imbricate sequences shown in Figures 4.19 and 4.21.

Locality 14.8

A roadside quarry and road cuts at (NC 186 090), south of Knockan show excellent exposures of the mylonites with isoclinal folds, later crenulations and small thrust ramps.

The return to Ullapool is made generally following the line of the Moine thrust which here lies directly on the middle Cambrian succession of the foreland with no underlying imbricates. At Strath Kanaird there is a large NE-SW trending normal fault which drops the Moine thrust down to the NW. Between Strath Kanaird and Ullapool the route lies in foreland Torridonian.

References

ANDERSON, J.G.C. 1935. 'The Arrochar Intrusive Complex'. Geol. Mag., 72, 263-83.

BAILEY, E.B. 1934. 'West Highland Tectonics: Loch Leven to Glen Roy'. Q. J. Geol. Soc. London, 90, 462-523.

_____ 1960. In: BAILEY, E.B. & MAUFE, H.B., The Geology of Ben Nevis and Glen Coe (Explanation of Sheet 53). Mem. Geol. Surv. UK.

BARBER, A.J. 1965. 'The history of the Moine thrust zone, Loch Carron and Lochalsh, Scotland'. Proc. Geol. Assoc., 76, 215-42.

_____ & SOPER, N.J. 1973. 'Summer field meeting in the north-west of Scotland'. Proc. Geol. Assoc., 84, 207-35.

_____ & MAY, F. 1976. 'The history of the western Lewisian in the Glenelg inlier, Lochalsh, northern Highlands'. Scott. J. Geol., 12, 35-50.

_____ et al. 1978. 'The Lewisian and Torridonian rocks of North-West Scotland'. Geol. Assoc. Guide, 21.

_____ et al. 1978. 'Metamorphic Caledonides and the Precambrian Foreland in the Highlands of Scotland and the Hebrides'. Int. Geol. Congr. Excursion 002. In: OWEN, T.R. (ed.) United Kingdom Guidebook. Inst. geol. Sci. London.

BARTON, C.M. 1978. 'An Appalachian view of the Moine thrust'. Scott. J. Geol., 14, 247-57.

BEACH, A., COWARD, M.P. & GRAHAM, R.H. 1974. 'An interpretation of the structural evolution of the Laxford front'. Scott. J. Geol., 9, 294-308.

BOWES, D. 1969. 'The Lewisian of Northwest Highlands of Scotland'. In: KAY, M. (ed.), North Atlantic - geology and continental drift. Mem. Am. Assoc. Petrol. Geol., 12, 575-94.

_____ & KHOURY, S.G. 1965. 'Successive periods of basic dyke emplacement in the Lewisian complex, south of Scourie, Sutherland'. Scott. J. Geol., 1, 295-9.

_____ & WRIGHT, A. 1973. 'Early phases of Caledonian deformation in the Dalradian of the Ballchulish district, Argyll'. Geol. J., 8, 333-44.

BOYER, S.E. & ELLIOT, D. 1982. 'Thrust systems'. Bull. Am. Assoc. Petrol. Geol., 66, 1196-230.

CHAPMAN, H.J. & MOORBATH, S. 1977. 'Lead isotope measurements from the oldest recognised Lewisian gneisses of north-west Sutherland'. Nature, 268, 41-2.

CHRISTIE, J.M. 1960. 'Mylonitic rocks of the Moine Thrust Zone in the Assynt region, north-west Scotland'. Trans. Geol. Soc. Edinburgh, 18, 79-93.

_____ 1963. 'The Moine thrust zone in the Assynt region, north-west Scotland'. Univ. Calif. (Berkeley) Publ. Geol. Sci., 40, 345-440.

CLIFFORD, T.N. 1957. 'The stratigraphy and structure of part of the Kintail district of Southern Ross-shire; its relation to the Northern Highlands'. Q. J. Geol. Soc. London, 113, 57-8.

COWARD, M.P. 1980. 'The Caledonian thrusts and shear zones of NW Scotland'. J. Struct. Geol., 2, 11-7.

_____ 1982. 'Surge zones in the Moine thrust zone of NW Scotland'. J. Struct. Geol., 4, 247-56.

_____ 1983. 'The thrust and shear zones of the Moine thrust zone and the Scottish Caledonides'. J. Geol. Soc. London, 140, 795-812.

_____ 1984. 'The strain and textural history of thin-skinned tectonic zones: examples from the Assynt region of the Moine thrust zone, NW Scotland'. J.Struct. Geol., 6, 89-99.

_____ & WHALLEY, J.S. 1979. 'Texture and fabric studies across the Kishorn Nappe, near Kyle of Lochalsh, western Scotland'. J. Struct. Geol., 1, 259-73.

_____ & KIM, J.H. 1981. 'Strain within thrust sheets'. In: McCLAY, K. R. & PRICE, N. J. (eds.), Thrust and Nappe Tectonics. Spec. Publ. Geol. Soc. London, 9, 275-92.

CURRY, G.B., BLUCK, B.J., BURTON, C.J., INGHAM, J.K., SIVETER, D.J. & WILLIAMS, A. 1984. 'Age, evolution and tectonic history of the Highland Border Complex, Scotland'. Trans. R. Soc. Edinburgh, 75, 113-33.

DAHLSTROM, C.D.A. 1970. 'Structural evidence of the eastern margin of the Canadian Rocky Mountains'. Bull. Can. Petrol. Geol., 18, 332-406.

DAVIES, F.B. 1974. 'A layered basic complex in the Lewisian south of Loch Laxford, Sutherland'. J. Geol. Soc. London., 130, 279-84.

_____ 1976. 'Early Scourian structures in the Scourie-Laxford region and their bearing on the evolution of the Laxford front'. J. Geol. Soc. London, 132, 543-54.

_____ & WATSON, J.V. 1977. 'The early history of the type Laxfordian complex, NW Sutherland'. J. Geol. Soc. London, 133, 123-31.

ELLIOT, D. & JOHNSON, M.R.W. 1980. 'The structural evolution of the northern part of the Moine thrust zone'. Trans. R. Soc. Edinburgh, 71, 69-76.

EVANS, C.R. & LAMBERT, R.St.J. 1974. 'The Lewisian of Loch Inver, Sutherland; the type area for the Inverian metamorphism'. J. Geol. Soc. London, 130, 125-50.

FETTES, D.J. & McDONALD, R. 1978. 'Glen Garry vein complex'. Scott. J. Geol., 14, 335-38.

GILETTI, B., MOORBATH, S. & LAMBERT, R.St.J. 1961. 'A geochronolgical study of the metamorphic complexes of the Scottish Highlands'. Q. J. Geol. Soc. London, 117, 223-64.

HALLAM, A. & SWETT, K. 1966. 'Trace fossils from the Lower Cambrian Pipe Rocks of the north-west Highlands'. Scott. J. Geol., 2, 101-6.

HARTE, B., BOOTH, J.E., DEMPSTER, T.J., FETTES, D.J., MENDUM, J.R. & WATTS, D. 1984. 'Aspects of the post-depositional evolution of Dalradian and Highland Border Complex rocks in the Southern Highlands of Scotland'. Trans. R. Soc. Edinburgh, 75, 151-63.

HENDERSON, W.G. & FORTEY, N.J. 1982. 'Highland Border rocks at Loch Lomond and Aberfoyle'. Scott. J. Geol., 18, 227-45.

_____ & ROBERTSON, A.H.F. 1982. 'The Highland Border rocks and their relation to marginal basin development in the Scottish Caledonides'. J. Geol. Soc. London, 139, 433-50.

HICKMAN, A.H. 1978. 'Recumbent folds between Glen Roy and Lismore'. Scott. J. Geol., 14, 191-212.

IKIN, N.P. 1983. 'Petrochemistry and tectonic significance of the Highland Border Suite mafic rocks'. J. Geol. Soc. London, 140, 267-78.

JOHNSON, M.R.W. 1960. 'The structural history of the Moine thrust zone at Loch Carron, Wester Ross'. Trans. R. Soc. Edinburgh, 64, 139-68.

_____ 1961. 'Polymetamorphism in movement zones in the Caledonian Thrust Belt of north-west Scotland'. J. Geol., 69, 417-32.

JOHNSTONE, G.S., SMITH, D.I. & HARRIS, A.L. 1969. 'The Moinian Assemblage of Scotland'. In: KAY, M. (ed.), North Atlantic - geology and continental drift. Mem. Am. Assoc. Petrol. Geol., 12, 159-80.

KELLEY, S.P. & POWELL, D. 1985. 'Relationships between marginal thrusting and movement on major, internal shear zones in the Northern Highland Caledonides, Scotland'. J. Struct. Geol., 7, 161-74.

LAMBERT, R.St.J. & HOLLAND, J.G. 1972. 'A geochronological study of the Lewisian from Loch Laxford to Durness, Sutherland, NW

Scotland'. Q. J. Geol. Soc. London, 128, 2-19.

LAPWORTH, C. 1885. 'The Highland controversy in British Geology'. Nature, 32, 558-9.

LAW, R. KNIPE, R. & DAYAN, H. in press. 'Strain path partitioning within thrust sheets: microstructural and petrofabric evidence from the Moine thrust zone at Loch Eriboll, NW Scotland'. J. Struct. Geol.

LEEDAL, G.P. 1952. 'The Cluanie igneous intrusion, Inverness-shire and Ross-shire'. Q. J. Geol. Soc. London, 108, 35-63.

McGREGOR, M. & PHEMISTER, J. 1972. Geological Excursion Guide to the Assynt district of Sutherland. Edinburgh Geological Society.

McGREGOR, V.R. 1973. 'The early Precambrian gneisses of the Godthaab district, West Greenland'. Phil. Trans. R. Soc. London, 273, 343-58.

McLEISH, A.J. 1971. 'Strain analysis of deformed pipe rock in the Moine thrust zone, north-west Scotland'. Tectonophysics, 12, 469-503.

MOORBATH, S. 1969. 'Evidence of the age of deposition of the Torridonian sediments of NW Scotland'. Scott. J. Geol., 5, 154-70.

_____ WELKE, H. & GALE, N.H. 1969. 'The significance of lead isotope studies in ancient, high grade metamorphic basement complexes, as exemplified by the Lewisian rocks of north-west Scotland'. Earth Plan. Sci. Lett., 6, 245-56.

NICHOLLS, G.D. 1950. 'The Glenelg-Ratagan Igneous Complex'. Q. J. Geol. Soc. London, 106, 309-44.

PARK, R.G. 1970. 'Observations on Lewisian chronology'. Scott. J. Geol., 6, 379-99.

PARSONS, I. 1979. 'The Assynt alkaline suite'. In: HARRIS, A.L., HOLLAND, C.H. & LEAKE, B.E. (eds.), The Caledonides of the British Isles - reviewed. Spec. Publ. Geol. Soc. London, 8, 677-80.

PEACH, B.N., HORNE, J., GUNN, W., CLOUGH, C.T., HINXMAN, L.W. & TEALL, J.J.H. 1907. 'The geological structure of the NW Highlands of Scotland'. Mem. Geol. Surv. UK.

_____, _____ and others 1910. 'The geology of Glenelg, Lochalsh and southeast part of Skye'. Mem. Geol. Surv. UK.

_____ & _____ 1930. Chapters on the Geology of Scotland. Oxford University Press, London.

POWELL, D., BROOK, M. & BAIRD, A.W. 1983. 'Structural dating of a Precambrian pegmatite in Moine rocks of northern Scotland and its bearing on the status of the 'Morarian orogeny'. J. Geol. Soc. London, 140, 813-24.

RAMSAY, J.G. 1957. 'Moine-Lewisian relation at Glenelg, Inverness-shire'. Q. J. Geol. Soc. London, 113, 487-523.

_____ 1963. 'Structure and metamorphism of the Moine and Lewisian rocks of the northwest Caledonides'. In: JOHNSON, M.R.W. & STEWART, F.H. (eds.), The British Caledonides. Oliver & Boyd, Edinburgh.

RATHBONE, P.A. & HARRIS, A.L. 1979. 'Basement cover relationships at Lewisian inliers in the Moine rocks'. In: HARRIS, A.L., HOLLAND, C.H. & LEAKE, B.E. (eds.), The Caledonides of the British Isles - reviewed. Spec. Publ. Geol. Soc. London, 8, 101-8.

_____, COWARD, M.P. & HARRIS, A.L. 1983. 'Cover and basement: a contrast in style and fabrics'. Mem. geol. Soc. Am., 158, 312-23.

RICHEY, J.E. 1939. 'The dykes of Scotland'. Trans. Geol. Soc. Edinburgh, 13, 402-35.

RIEKELS, L.M. & BAKER, D.W. 1977. 'The origin of the double maximum pattern of optic axes in quartzitic mylonite'. J. Geol., 85, 1-14.

ROBERTS, A.M. & HARRIS, A.L. 1983. 'The Loch Quoich line - a limit of early Palaeozoic reworking in the Moine of the Northern Highlands of Scotland'. J. Geol. Soc. London, 140, 883-92.

ROBERTS, J.L. 1976. 'The structure of the Dalradian rocks in the North Ballachulish district of Scotland'. J. Geol. Soc. London, 133, 139-54.

SABINE, P.A. 1953. 'The petrography and geological significance of the post-Cambrian minor intrusions of Assynt and the adjoining districts of NW Scotland'. Q. J. Geol. Soc. London, 109, 137-71.

SANDERS, I.S. 1979. 'Observations on eclogite- and granulite-facies rocks in the basement of the Caledonides'. In: HARRIS, A.L., HOLLAND, C.H. & LEAKE, B.E. (eds.), The Caledonides of the British Isles - reviewed. Spec. Publ. Geol. Soc. London, 8, 97-100.

SMITH, D.I. 1979. 'Caledonian minor intrusions of the N Highlands of Scotland'. In: HARRIS, A.L., HOLLAND, C.H. & LEAKE, B.E. (eds.), The Caledonides of the British Isles - reviewed. Spec. Publ. Geol. Soc. London, 8, 683-97.

SOPER, N.J. & BARBER, A.J. 1979. 'Proterozoic folds on the northwest Caledonian foreland'. Scott. J. Geol., 15, 1-11.

_____ & BROWN, P.E. 1971. 'Relationship between metamorphism and migmatization in the northern part of the Moine Nappe'. Scott.J.Geol., 7, 305-25.

_____ & WILKINSON, P. 1975. 'The Moine thrust and Moine nappe at Loch Eriboll, Sutherland'. Scott.J.Geol., 11, 339-59.

STEWART, A.D. 1969. 'Torridonian rocks of Scotland reviewed'. Mem. Am. Assoc. Petrol. Geol., 12, 595-608.

_____ 1975. 'Torridonian rocks of western Scotland'. In: HARRIS, et al. (eds.), A correlation of the Precambrian rocks in the British Isles. Spec. Rep. Geol. Soc. London, 6, 43-52.

SUTTON, J. & WATSON, J.V. 1951. 'The pre-Torridonian metamorphic history of Loch Torridon and Scourie areas on the north-west Highlands, and its bearing on the chronological classificationn of the Lewisian'. Q. J. Geol. Soc. London, 106, 241-307.

_____ & _____ 1958. 'Structures in the Caledonides between Loch Duich and Glenelg'. Q. J. Geol. Soc. London, 114, 231-58.

TALBOT, C.J. 1983. Microdiorite sheet intrusions as incompetent time- and strain- markers in the Moine assemblage NW of the Great Glen fault, Scotland. Trans. R. Soc. Edinburgh, 74, 137-152.

TANNER, P.W.G. 1970. 'The Sgurr Beag Slide - a major break within the Moinian of the western Highlands of Scotland'. Q. J. Geol. Soc. London, 126, 435-63.

_____, JOHNSTONE, G.S., SMITH, D.I. & HARRIS, A.L. 1970. 'Moinian stratigraphy and the problem of the Central Ross-shire inliers'. Bull. Geol. Soc. Am., 81, 299-306.

TILLEY, C.E. 1936. 'Eulysites and related rock types from Loch Duich, Ross-shire'. Mineralog. Mag., 24, 331-42.

TREAGUS, J.E. 1964. 'Notes on the structure of the Ben Lawers Synform'. Geol.Mag., 101, 260-70.

_____ 1974. 'A structural cross-section of the Moine and Dalradian rocks of the Kinlochleven area, Scotland'. J. Geol. Soc. London, 130, 525-44.

VAN BREEMEN, O., AFTALION, M., PANKHURST, R.J. & RICHARDSON, S.W. 1979. 'Age of the Glen Dessary Syenite, Inverness-shire; diachronous Palaeozoic metamorphism across the Great Glen'. Scott. J. Geol., 15, 49-62.

_____, _____ & JOHNSON, M.R.W. 1979. 'Age of the Loch Borrolan complex, Assynt and late movements in the Moine thrust'. J. Geol. Soc. London, 136, 489-96.

WILLIAMS, G.E. 1969a. 'Characteristics of a Precambrian pediment'. J. Geol., 77, 183-207.

_____ 1969b. 'Petrography and origin of pebbles from Torridonian strata (late Precambrian) north-west Scotland'. Mem. Am. Assoc. Petrol. Geol., 12, 609-29.

WOOLLEY, A.R. 1970. 'The structural relationships of the Loch Borrolan complex, Scotland'. Geol.J., 7, 171-82.

A COMPARISON OF THE LOWER PALAEOZOIC VOLCANIC ROCKS ON EITHER SIDE OF
THE CALEDONIAN SUTURE IN THE BRITISH ISLES.

C J Stillman
Department of Geology
Trinity College
Dublin 2
Ireland

ABSTRACT. South of the suture Lower Palaeozoic volcanics are almost all subduction-related and are found on a plate margin zoned from a continental foreland in the southeast across the Welsh back-arc ensialic marginal basin, to the Irish and Lake District compound volcanic arc in the north-west, the inner part erupted onto continental crust, the outer less obviously so.
  North of the suture there is evidence of obduction of oceanic crust in the Ballantrae ophiolite and part of the Highland Border Complex and of oceanic components in the Southern Uplands accretionary prism. The only pre-Devonian subduction related magmatism is seen in S Mayo, though granitic clasts at Girvan and elsewhere suggest an early arc in the Midland Valley. It is not until the Siluro-Devonian post-orogenic period that widespread calc-alkaline volcanism is seen.

INTRODUCTION

Lower Palaeozoic volcanism in the British Isles represents an important aspect of Caledonian orogenic development and the volcanic rocks appear to have been erupted almost exclusively along continent/ocean boundaries. In particular many were erupted in island arcs, marginal basins and on continental margins, as products of subduction-related magmatism. Although volcanism occurred from time to time throughout the whole period, activity built up to major widespread climaxes in Cambro-Ordovician boundary times and again in the mid- to late-Ordovician. The volcanic climaxes presumably mark some change in the plate dynamics which appear to have affected both northern and southern margins of Iapetus. Nevertheless the responses of the two margins were sufficiently different to produce and preserve quite dissimilar suites of volcanic rocks.
  Crucial factors governing the differences between volcanics erupted on northern and southern flanks of the Iapetus, which are now juxtaposed across the Caledonian suture, include initially the type of eruption generated in response to the geometry and kinetics of the

respective subduction systems and subsequently the nature of changes brought about by the type and intensity of deformation, dislocation and metamorphic grade which are quite different on either side of the ocean. As a result of these differences, the volcanics south of the suture are today preserved on a crustal substrate which may well be the original one through which they were erupted and in something resembling the same relative positions as when they were formed, whilst those to the north are sometimes allochthonous, seldom found above their own basement and quite commonly not even in their own primary depositional setting. Many indeed exist only as components of clastics sediments. Nevertheless the northern margin has preserved slices of crust containing components of juvenile Iapetus lithosphere, whilst such ocean crustal rocks are yet unknown south of the suture. The Shetland rocks, including the Unst Ophiolite, have not been included in this account as it seems possible that the extension of the 'Tornquist Line' may separate the Shetlands from Scotland.

This account lays emphasis on differences which are based not just on the petrography of the volcanic rocks but also on their presumed tectonomagmatic origins. The vast majority of Lower Palaeozoic volcanics are low-grade metamorphic rocks, often referred to simply as greenstones and though careful petrographic examination in field and laboratory has led to widespread recognition of environments of eruption and emplacement, the plate-tectonic settings are deduced largely from their geochemical characteristics. It is obvious that these rocks have all undergone chemical alteration which, for many, began at the moment of their eruption. For submarine volcanics, which include the majority of the rocks under consideration, chemical exchange involved substantial alteration in both major element and LIL trace elements, an effect enhanced by sub-sea floor metamorphism and metasomatism, locally at fairly high temperatures. However, these effects can be calibrated by reference to more recent submarine occurrences (eg. Saunders et al., 1979, 1982) and subsequent orogenic metamorphism appears to have had little effect on overall chemistry, though localised effects in regions of high tectonic strain may be considerable.

It has thus been possible for many modern workers to develop techniques for presenting geochemical evidence which seems to give reliable plate-tectonic indications; these include the avoidance of LIL elements and the use of 'immobile' trace elements as discriminators and comparisons using rock/MORB or rock/mantle normalised diagrams. Nevertheless such techniques provide clues, not absolute proofs, and it is only a combination of palaeoenvironmental evidence with consistent petrographic and geochemical indicators that should be regarded as reliable.

VOLCANICS SOUTH OF THE SUTURE

The Caledonian Suture is considered to be the junction between terranes marginal to the northern flank of the Iapetus and those along its southern edge. Its presumed course (Phillips et al., 1976) is shown in Figure 1. South of this line it appears that late-Proterozoic

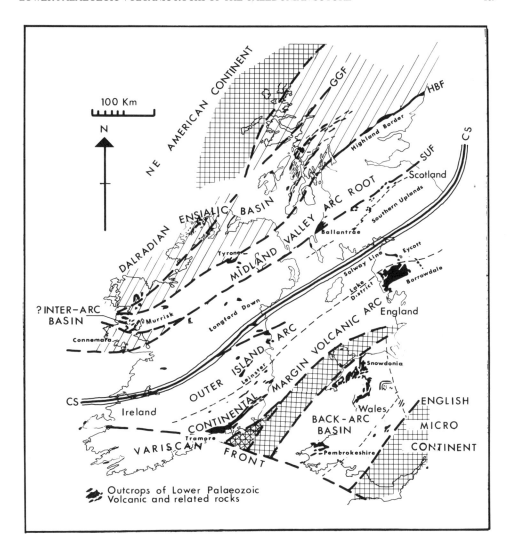

Figure 1. Location map for Lower Palaeozoic Volcanic Rocks in the British Isles and Ireland, in a Caledonide volcano-tectonic zonal framework. (Outcrops from Moseley 1982).

Cross-hatched areas — Pre-Caledonian continental crust.
Diagonal line pattern — Dalradian supracrustal metasediments on continental basement.
Unshaded areas — Caledonian volcano-sedimentary accretions onto the continental margins.

continental crust - possibly an Avalonian Terrane similar to that of south-eastern Newfoundland (Kennedy, 1979) - has been added to by a sequence of Lower Palaeozoic marine sediments and igneous rocks, by procedures apparently similar to those operating in active continental margins at the present day.

Starting within the continental plate, at Tortworth (Gloucestershire) and in the Mendips (Somerset) outcrops expose Llandovery submarine lavas and pyroclastics, apparently situated close to their source vents. They range from basaltic andesite to rhyodacite and have sub-alkaline chemistry transitional from tholeiitic to calc-alkaline (Van de Kamp, 1969). They may be the products of island arc volcanoes in a transitional stage of evolution, and are almost unique in the British Silurian.

Closer to the Welsh basin are the Shelve-Breiden Hills and Builth Wells Ordovician inliers of the Welsh Borderlands. Here rather earlier (Llanvirn-Caradoc) volcanism is seen. At Builth, Upper Llanvirn volcanoes were situated near the shores of a sea which had, by Caradoc times, deepened so that shallow intrusions and tuffs of basic composition were emplaced in deep-water sediments. At Shelve, Llanvirn volcaniclastics indicate proximity to similar eruptive centres which, by the Caradoc, were erupting magmas of basic, intermediate and acid compositions. (Kokelaar et al., 1984). Kokelaar and his co-authors believe the Builth rocks to show a calc-alkaline fractionation continuum from basic through intermediate to acid, consistent with ensialic island arc evolution, a model initially proposed by Furnes (1978). Silurian volcanic products are found here in a very reduced form with thin bentonitic tuff beds in the Llandovery and Wenlock, but activity terminates with a bang - a massive plinian explosive eruption in Lower Old Red Sandstone times, which produced the Townsend Tuff (Allen and Williams, 1981) - an ash fall deposit which is recorded around the whole of the southern edge of the Welsh Basin, from Milford Haven to the Clee Hills north of Ludlow. The source volcano for this tuff has not yet been located.

The Welsh Basin itself apparently came into being when extensional tectonics led to the foundering of blocks of the Proterozoic crust in Cambrian times and volcanism appears to be associated at an early stage with faulting (Reedman et al., 1984). Widespread discontinuities indicate sporadic subsidence of the basin - the most extensive producing the Tremadoc/Arenig marine transgression seen over much of the basin. This transgression was preceded by major volcanic activity which Kokelaar et al. (op.cit.) consider to mark the early stages of south-easterly subduction of Iapetus lithosphere beneath the English continental plate, following the reversal in plate polarity which initiated the closure of the Iapetus. In mid-Wales, well-developed arc volcanism is seen at Rhobell Fawr where Tremadoc arc-tholeiites occur, deformed by tectonism along the Rhobell fracture and eroded before the Arenig transgression. Arc volcanism is also seen in Pembrokeshire, where the Trefgarn Volcanic Group, of probable Tremadoc to lowermost Arenig age, comprises a calc-alkaline fractionation continuum from basalt through andesite to dacite and rhyolite, resembling that of modern volcanic arcs (Bevins et al., 1984). This phase of arc volcanism may well have been much more widespread than the present outcrops would indicate, as clasts of these lithologies have been found in the Arenig transgressive strata through much of Wales.

Volcanic activity subsequent to the Arenig transgression was also widespread, but different in character – no longer a calc-alkaline continuum but a bimodal basalt-rhyolite association in which the basic rocks resemble those of modern ensialic marginal basins and the acid magmas were most probably generated, at least in part, by fusion of underlying continental crust. This changed, believe by Bevins et al. (op.cit.) to mark the transition from arc to marginal basin environment, may well reflect a significant change in plate kinematics associated with an effective northward shift of the subduction zone (Fitton et al., 1982), since from this time onwards subduction-related volcanism only appears further to the north-west in the Lake District and in eastern Ireland. In the Welsh basin, Arenig to Llandeilo volcanism is seen in southern Snowdonia and in Pembrokeshire. In the former, the Aran Volcanic Group crops out round the Harlech Dome, and comprises rhyolitic ash-flow tuffs and domes, with an intercalated unit of submarine basaltic composition consisting of pillowed and sheet lavas and hyaloclastites. In Pembrokeshire two volcanic centres produced the Fishguard Volcanic Complex and the Ramsay Island rhyolites respectively. Here again the sequence is bimodal and largely submarine. The bimodality, together with the trace-element geochemistry of the basic rocks appear to preclude an arc origin and to indicate a mantle source-magma modified possibly by metasomatism by fluids from dehydrating subducted oceanic lithosphere beneath.

In the Welsh Basin in general, Kokelaar et al. 1984, have shown that the basalts have ocean floor basalt affinities, with transitional to arc-like compositions, characteristic of basalts erupted in back-arc marginal basins. Minor quantities of intermediates do show a fractionation continuum to rhyodacite which is typically tholeiitic, but in both Snowdonia and Pembrokeshire the large volume of acid rocks relative to basic rocks, the scarcity of intermediate compositions and the absence of extensive hidden basic bodies makes it unlikely that these rocks are linked solely by fractionation; more probably the acid magmas were generated largely by crustal fusion (Dunkley, 1978; Kokelaar et al., 1984. This appears to have been localised and to have produced large 'blisters' of acid magma. The basic rocks are more widespread, though volumetrically less abundant, and together with sheet intrusions seem to have been localised by deep crustal fractures, presumably tapping a basic-wide zone of potential partial melting of the mantle.

The significance of crustal fracturing in controlling the eruptive magmatism is apparent especially in the subsequent activity in Snowdonia. This is essentiallty Caradocian in age and again largely acid and basic in composition. At an earlier phase, the Llewelyn Volcanic Group, produced ash-flow tuffs which transgressed from land to sea to produce the submarine ash-flow tuffs of the Capel Curig Volcanic Formation (Francis and Howells, 1973) – tuffs which have gained notoriety as demonstrably illustrating submarine welding. Minor intermediate and basic eruptives are also seen, with distribution apparently controlled by active faulting. The later sequence, the Snowdon Volcanic Group, erupted from a number of centres with alternating acid and basic cycles; in Central Snowdonia the basic compositions erupted in a shallow marine environment, whilst in the east, acid eruptives were emplaced in a deeper marine setting. Howells (1977) proposed that these were separated by an active ridge controlled by a deep-seated fracture.

The production of magma beneath the Welsh Basin seems to have ceased by the end of the Ordovician as, with the exception of the Skomer volcanics, only the wind-blown representatives of plinian activity, eg. the numerous bentonite horizons in the Silurian of the Welsh Borderland, and the Old Red Sandstone, the Townsend Tuff are seen in the subsequent Lower Palaeozoic strata. The Llandovery Skomer volcanics are alkaline hawaiites, mugearites and rhyolites which appear to represent an oceanic volcanic island but - according to Sanzen-Baker (1972) - are found in an allochthonous block of crust which may perhaps have been transported from the south, to its present situation on the edge of the Welsh Basin.

By analogy with modern continental margins, oceanward of the Welsh Basin should lie a volcanic island arc. Such an arc may indeed be found in the Leinster-Lake District zone, but between it and the Welsh basin exists the Irish Sea Horst. By the time of the initiation of the Welsh Basin, the block was acting as a positive crustal feature standing between depositional basins on its northern and southern sides. The Ordovician hiatuses and transgressions of the two areas do not appear to match across the horst. Whilst little is seen of Cambro-Ordovician sequences in Ireland, so that the presence or absence of the Tremadoc/Arenig break cannot be verified, the Irish succession is dominated by a mid-Ordovician (Llanvirn/Llandeilo) hiatus which is not seen in the Welsh basin.

The Leinster-Lake District zone is dominated by subduction-related magmatism and clearly demonstrates a northward transition from continental margin magmatism, isotopically very similar to the present-day Java Arc (Davies, 1983) to volcanics produced at the outer limits of subduction-related magmatism (Davies, op.cit.). This transition is seen especially well in south-east and east Ireland (Stillman and Williams, 1979) but is also represented along strike to the north-east in the Lake District, with the Borrowdale Volcanics and the Binsey-Eycott and Cross Fell rocks - a distinction whose significance was early recognised by Fitton and Hughes (1970).

There is a notable absence of Tremadocian rocks throughout the zone and late Cambrian ocean-floor basalts seen in County Wicklow, Ireland, are followed by Arenig turbidites that contain but a few andesitic ash bands and occasional lavas, apparently of calc-alkaline chemistry. It is not until the end of the Arenig that subduction-related volcanism is recorded on a large scale. As might be expected, if the magma genesis is related to the reactivation of the subduction system after it had shifted substantially to the north during and after the Arenig transgression seen in Wales, volcanism starts first in the more northerly parts. In the Binsey-Eycott region of the Lake District, late Arenig island arc tholeiites evolve, by mid-Llanvirn times, to calc-alkaline basalts and basaltic andesites. A widespread hiatus then occurred, as referred to above, which may again have been caused by a change in plate motion for it is followed by the most extensive and voluminous phase of volcanism in the Caledonides of the British Isles. Repeated eruptions occurred in an arc which extended from the Lake District to Tramore on the south-east coast of Ireland, and continuing for almost 20 million years built up a plateau on the continental margin, so that towards the end of the Ordovician, stratiform composite volcanoes were to a large extent subaerial. This phase of activity evolved diachronously along the arc. The earliest representatives are found immediately south of the Eycott area, where the Borrowdale Volcanic Group built up evolved andesite-rhyolite stratovolcanoes similar to those of the present day Cascade Range of Northwestern America (Moseley, 1983). Subaerial ash-fall tuffs, some with accretionary lapilli, and ash flows with welded zones, are commonly andesitic to dacitic and often highly brecciated. Widespread plinian ash falls are commonly reworked in an aqueous environment, and volcanic mudflows testify also to the local presence of water; probably in intermontane or crater lakes. An Sm-Nd garnet age of 457+/-4 Ma has been obtained from garnets crystallised from andesites and dacites of these calc-alkaline volcanics (Thirlwall and Fitton, 1983), the lower part of which is geologically constrained as Llandeilo in age.

Across the Irish Sea, south-west along the arc, volcanoes of this age were submarine, erupting basalts and basaltic andesites of tholeiitic and transitional calc-alkaline chemistry (Stillman and Williams, 1979). The earliest are known from south Leinster, and are well exposed in the sea cliffs west of Tramore, where pillow lavas and shallow intrusions abound in sea floor sediments of mudstone and volcaniclastic debris. A local but extensive pause in volcanism, accompanied by carbonate sedimentation, was followed by abundant Caradoc volcanism which was dominantly rhyolitic, with subordinate andesites and basalts. Rhyolite domes and brecciated lava flows formed cores to volcanic piles (O. Swennen pers.comm., 1984) which built up by sub-aqueous debris flows, mass flows, and ash flows, sometimes welded, to form locally subaerial volcanoes which do not seem to have been as high or as extensive above sea level as the Borrowdale volcanoes. At least one rhyolite dome was associated with sulphide mineralisation in a Kuroko-style deposit (Sheppard, 1980). During this Caradocian volcanic climax, in the region north of Dublin on the outer, seaward edge of the arc, less mature magmas were erupting as island arc tholeiites and transitional calc-alkaline basalts and basaltic andesites, with compositions similar to the earlier Eycott volcanics. These eruptions formed a series of island volcanoes built up from an ocean floor which was receiving only pelagic sediment, and were often fringed by local limestone reefs, commonly brecciated and mixed with volcanic debris by the abundant volcano-seismic activity. The volcanic pedestals were constructed of submarine lavas and hyaloclastites, which only occasionally built up to sea level and sealed their vents to the water, to enable the production of subaerial lava flows. The compositional range was restricted and no acid rocks are known

(Stillman and Williams, 1979).

Petrochemical and isotopic studies have given supportive evidence of the zonation from outer to inner arc (Davies, 1983). In the northern, outer arc, basalts and basaltic andesites persist to the end with transitional tholeiitic to calc-alkaline chemistry, though with anomalously high LIL element concentrations. In the southeastern, inner arc, early basalt dominance gave way to abundant rhyolite, and the few basic rocks have calc-alkaline chemistry with significantly different rare-earth element patterns to the northern rocks. In the northern suite the trace elements are suggestive of derivation near the extreme outer limit of subduction-related volcanism, (Davies, op.cit.) In the southeast, on the other hand, the basalt and andesites have the rare-earth element patterns and trace element characteristics of subduction-related magmatism on continental margins well back from the edge of the plate. The abundant acid volcanics show evidence of both fractionation from primary subduction-related magma and contamination by the addition of some 30% of partial melt from the lower crust - specifically from crust of the Anglesey or Rosslare Complex type, not upper crustal sediments.

The volcanism of the zone ends in the Ashgill with a late phase of alkali gabbros and rhyolites, some per-alkaline, and a suite of tholeiitic dolerite intrusions. Subduction-related magmatism seems to have come to an end and the Silurian is largely deficient in volcanic rocks, except in the far west, on the Dingle Peninsula in Kerry, where some Wenlock rhyolites and andesites are known.

In previous discussions of the eastern Irish Ordovician volcanic rocks the regularity of the geochemical pattern was complicated by the presence of Llanvirn to Llandeilo alkali basalts, mugearites and rhyolites at Grangegeeth, Collon and Bellewstown, north of the normal outer arc volcanoes. Like Skomer, these are more appropriately designated as products of oceanic island volcanism, and now evidence has been produced that they may be situated on a block of crust which seems to be exotic, and to have been tectonically emplaced at the end of the Ordovician (Murphy, F.C., pers.comm., 1984).

VOLCANICS NORTH OF THE SUTURE

The northern continental margin presents a radically different picture; no well-preserved island arc and marginal basin but a series of tectonic zones shuffled by strike-slip movements, in which there is overthrusting, obduction, and the build up of an accretionary wedge in which both sediment and slivers of the ocean-floor basalts are stacked on the continent edge. There is very fragmentary evidence of an early marginal arc with a plutonic core, stripped and exposed by the Arenig, and of marginal basins which were closed by overthrusting and their contents sliced up, strongly deformed and metamorphosed. A substantial amount of the volcanic material is preserved only as clastic components of sediments after the original rocks have been deformed, uplifted and eroded. It is not until the Lower Devonian, after the main Caledonide collision, that regionally extensive volcanicity is seen in-situ, and this, situated as it is across an orogenic system which appears to have

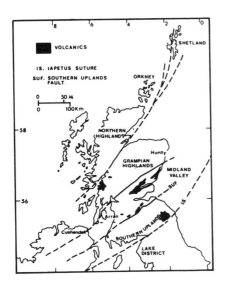

Figure 2. Distribution of Lower and Middle Old Red Sandstone volcanic rocks in Northern Britain (after Thirwall (1982) Fig.1).

IS= Iapetus Suture
SUF= Southern Uplands Fault
HBF= Highland Boundary Fault
GGF= Great Glen Fault

terminated movement, curiously presents a spatial geochemical variation which might be expected across an active subduction-related volcanic arc.

Starting again within the continental plate, this time the Northern American plate with its Archean cratonic core; isotopic studies on Caledonian granites (Halliday, 1984) suggest that the sources material for the more southerly granites contains a younger crustal component derived from younger continental crust thrust under the southern side of the Grampian highlands. This younger crust, which is believed to exist under the Midland Valley and Southern Uplands is inferred by Halliday (op.cit.) to be in part of Lower Palaeozoic age and include arc and oceanic crustal material, whilst the remainder is of late Proterozoic, post-Grenvillian age.

North of the Highland Boundary Fault, Late Proterozoic to Lower Palaeozoic Dalradian sedimentation was accompanied by early Cambrian volcanism, seen principally in the Tayvallich volcanics of the Southern Highland Group. Basaltic pillow lavas, hyaloclastites and tuffs were probably associated with a suite of tholeiitic dykes and sills (Graham, 1976). These have been interpreted as indicating an extensional regime, perhaps accompanying rift faulting and the incipient development of an early marginal basin on the Iapetus northern continental margin.

The Dalradian and older rocks were then involved in the Grampian Orogeny, a Lower Ordovician event which imposed the highest grade metamorphism and intensity of ductile deformation seen in the Caledonide orogeny. There is some evidence that this orogenic event was diachronous and that various depositional and volcanic events were taking place at the same time, the products of which were not involved in the Grampian orogeny. Most of these are now found immediately to the south of the Grampian 'orthotectonic' region, though one orthotectonic segment, Connemara, appears to have become detached and migrated to a more southerly position where it 'docked' some time after the Arenig. The Grampian orogenic event took place only on the northern continent and did not affect either the southern margin of this continent, the Iapetus itself, or the continental plate south of the Iapetus. The tectonic junction between the Dalradian orthotectonic region and the paratectonic zone to the south of it is the Highland Boundary Fault, along which is found the Highland Border Complex. The group of rocks includes a suite of spilites and amphibolites, with cherts, black shales, arenites and limestones from which Curry et al. (1983) have reported an Arenig fauna. Longman et al. (1979) have postulated that these rocks were deposited in a marginal basin to the south of the Highland Boundary Fault, between it and a magmatic arc thought to be located beneath what is now the southern part of the Midland Valley. Closure of this basin by subduction-related underthrusting in Llanvirn to Caradoc times preserved the ophiolite sliver along the fault. Henderson and Fortey (1982) and Graham and Bradbury (1981) would have this marginal basin on the northern side of the fault zone.

Slightly older volcanics, of apparently pre-Grampian age, are seen on Arran and Bute as greenschist and epidote-amphibolite facies basic metavolcanics, apparently originally submarine eruptives which were spilitised. Two magmatic groups are represented; tholeiites of marginal basin ocean floor type, and more alkaline within plate basalts (Ikin, 1983). These have been interpreted by Ikin as either the remnants of an ocean basin crust on which alkaline ocean island lavas were erupted, or the products of a pre-Grampian marginal basin which, because of its location, avoided the orogenic event.

In Ireland, the Tyrone Igneous Complex may provide an analogy with the Highland Border Complex, and perhaps with Ballantrae. Hutton et al. (1985) now recognise it as an ophiolite formed and obducted onto the northern margins of the Iapetus close to 470 Ma (Arenig). Here is a jumble of tectonically juxtaposed segments including a calc-alkaline basic to acid series of eruptives and minor intrusions and a suite of gabbros, ultrabasic rocks and doleritic sheet intrusions, together with a metamorphic complex of unknown provenance, all faulted against the Dalradian metasediments to the north. The complex may contain components of island arc and ocean crust, plus a basement similar to that postulated for the Midland Valley of Scotland. Still further to the west, on the south shores of Clew Bay, Co. Mayo, the Deer Park Complex has been regarded as an Arenig ophiolite (Dewey 1982). Ryan et al. (1983) have suggested that it is now a northerly dipping melange unit, which would again suggest analogy with the Highland Border Complex.

The Caledonian rocks of the Midland Valley in Scotland are covered by Upper Palaeozoic strata and their nature is largely inferred, but in the westward extension of the structure into Ireland, Lower Palaeozoic rocks are seen in the Curlew Mountains, and in the South Mayo basin. On the south flank of this basin, subduction-related magmatism is known from the Tremadoc, when the Lough Nafooey island arc volcanism took place. Lower tholeiitic pillow lavas give way upwards to calc-alkaline

pillowed basalts and andesites (Ryan et al., 1980). The overlying
sediments contain granite clasts, and by mid Arenig times, volcanism
was largely acidic. Succeeding Arenig-Llanvirn rhyolitic ignimbrites
are interbedded with sediments which contain abundant granitic clasts
(Graham, J.G., pers. comm.,1984) which appear to be possibly arc-
derived. Clast varieties in the sediments suggest that their sources
was not the now-adjacent Dalradian terrane of Connemara but may be an
unroofed arc core - a situation entirely analogous to that postulated
by Bluck (1982) for the Midland Valley in Scotland.

The southern margin of the Midland Valley in Scotland also
contains the Girvan basin and the Ballantrae Complex which is,
according to Bluck et al. (1980), an ophiolite slice comprising an
island arc/marginal basin assemblage that was thrust onto the
continental margin by obduction as a hot slab in Arenig times. The
components of the ophiolite have been apparently derived from an ocean
basin to the south - whether an open ocean or a marginal basin is still
debateable; indeed recent Sr-Nd isotope evidence (Thirlwall and Bluck,
in press) implies that at least four tectonically juxtaposed units are
involved. In the Girvan basin are seen granitic clasts, derived
apparently from some arc plutonism, plus proximal fore-arc effusives.

The whole of the Midland Valley zone, including the South Mayo
basin, was affected by widespread earth movement in mid-Ordovician
times, which between the Llanvirn and Caradoc brought an end to the
volcanic activity and uplifted the older sequences, which were then
eroded and incorporated in the subsequent sediments (Longman et al.,
1979). This movement appears to have been controlled by some change
in the dynamics of the subduction zone for it coincides with the
closure of the Highland Border Complex source basin and the deposition
of the Margie Series molasse (Bluck, 1983). It is noteworthy that
though the evolution of the northern margin of Iapetus seems
independant of that of the southern margin, this mid-Ordovician event
appears to have affected all areas north of the Welsh Basin, on both
sides of Iapetus.

Southward across the Southern Uplands Fault lies the Southern
Uplands and its western extension, the Longford-Down Inlier. This
tectonostratigraphic unit has been widely interpreted as an
accretionary prism built by a northward stacking of slices of sediment
and ocean floor volcanics from the south (Leggett et al., 1979).
Basalts are found at the bases of sedimentary wedges comprising cherts,
shales and greywackes of Arenig age. The basalts are submarine -
lavas and hyaloclastites, and in Scotland, two facies can be
determined. These are the Abingdon 'ocean-floor tholeiites' and the
Noblehouse alkaline, oceanic fracture type volcanics of Galapagos
affinities (Lambert et al.,1981). These are not subduction-related but
are of constructional margin origin, though their present position is
apparently the result of subduction-related processes. Later members
include the Llandeilo-Caradoc Bail Hill volcanics of calc-alkaline to
alkaline affinities (Lambert et al., op.cit.) though in addition there
are the Wrae Hill alkaline and peralkaline trachytes and rhyolites
(Thirlwall, 1981).

Similar volcanic facies to those of the Southern Uplands are seen to the west, in Ireland at Strokestown at the western end of the Longford-Down Inlier (Morris 1981), and in South Connemara, south of the Galway Granite (Ryan et al., 1983).

Throughout the zone, there is an apparent diminution of volcanic activity in late Ordovician and Silurian times — again possibly associated with some change in plate dynamics. Small occurrences of basic and intermediate submarine lavas and tuffs have been recorded in the Ashgill and Lower Llandovery, (Stillman and Francis, 1979) but the only widespread representatives are the bands of metabentonite in the Birkhill shales. Undoubtedly volcanic in origin (Cameron and Anderson, 1980), they indicate repeated plinian activity of volcanoes some considerable distance away and bear comparison with similar Silurian bentonites in the Welsh Borderlands. The tuffs are most probably distal, of ash-fall or submarine ash-flow origin, and their frequency implies very considerable repeated activity somewhere. As they appear to become thicker in Scotland than Ireland, perhaps the source was to the north-east. In the whole British Isles region the only significant Lower Silurian proximal volcanism is that of Skomer Island, the Mendips and west Kerry and these are all of Llandovery age. The Birkhill metabentonites range from Caradoc to Llandovery. They were probably derived volcanic glass of monzonitic composition (Cameron and Anderson, op.cit.) and it seems unlikely that their source can be found in these islands. It may not be outside the bounds of possibility that they represent plinian ash falls from the source region of volcanics of this age found in south-western Norway, for example the Langevåg Group volcanics which are late-Ordovician in age and have been likened to the products of Cascade Range volcanism in Northwestern America (Brekke et al., in press). On the basis of palinspastic reconstructions such as that of Cocks and Fortey (1982), a source region to the west of southern Norway would be near enough and given a suitable prevailing wind direction would be well within the dispersal range of plinian volcanoes.

The obducted volcanics and fragments of marginal basin ocean floor seen in the Ballantrae Complex and the Southern Uplands prism are the most southerly representatives of Iapetus margin volcanism seen north of the Suture, and it is not clear whether there was at one time an arc and fore-arc system on their oceanward side. Indeed the only true widespread calc-alkaline activity does not appear until late in the orogeny at a time when magmatism was dominated by granitic plutonism on both sides of the Suture. Then, in Northern Britain, there was a remarkable phase of eruptive volcanism which did not extend across the Suture but seems to be in some way related to a norther subduction zone which should have become dynamically inactive by this time. In the Cheviots, suites of calc-alkaline andesitic to dacitic lavas are of Middle Old Red Sandstone age (389-393 Ma; Thirlwall, 1982). However, recent studies (Thirlwall, 1981, 1982, 1983), indicate that the similar andesites and dacites of the Midland Valley, Scottish Highlands and Shetlands, may in fact be slightly older — late Silurian, or at youngest, Siluro-Devonian. The chemistry of all the supposedly Old Red Sandstone lavas north of the Southern Uplands Fault is consistent with a relationship to a subduction zone, and Thirlwall (1982) has now shown that the isotopic ratios cannot be explained simply, but require a mixed origin with two mantle sources of differing histories. He postulates the co-existance of enrich and depleted mantle sources as a vertical layering, which he suggests could provide the link between subduction and the spatial chemical variation.

Watson (1984) suggests a direct link with crustal structure. She suggests that the Lorne andesitic lavas (410 Ma) and the Lintrathen Ignimbrite (411 +/- 6 Ma; Thirlwall, 1983) and the Glen Coe and Lochnager complexes, which have subduction zone chemical characteristics, may owe their origin to a two-stage process. First, fluids from the subducted slab metasomatise the mantle wedge above. Subsequently this is tapped by deep-seated fault structures (Great Glen Fault, Highland Boundary Fault, etc.) inducing a partial melting, triggered from above by block movements or dislocations propagated downwards to the mantle. This would provide magmas with subduction zone chemistry which were actually erupted long after the end of active subduction. Whilst this model would overcome difficulties regarding the lack of synchroneity with subduction, it is not clear whether the mechanism would provide the degree of partial melting sufficient to derive the trace element variation noted by Thirlwall. However, it would provide a suitable means for tapping the heterogeneous mantle envisaged by Thirlwall and erupting the volcanics in their widely distributed localities scattered across a number of the Caledonide tectonic zones.

SUMMARY

Crucial factors governing the differences between volcanics on the northern and southern flanks of Iapetus, which are now juxtaposed across the Caledonide Suture, include initially the type of eruption generated in response to the geometry and kinetics of the respective subduction systems and subsequently to the nature of changes brought about by the type and intensity of deformation and dislocation.

The southern volcanics are found on a plate margin zoned from the Welsh Borderland continental foreland across the Welsh back-arc ensialic basin to a southeast Ireland - Lake District compound volcanic arc; the inner arc erupted into continental crust, the outer arc less obviously so. Two suites are widely recognised: subduction-related volcanics ranging from arc-tholeiites to calc-alkaline andesites, dacites and rhyolites; and marginal basin bimodal assemblages of transitional tholeiitic basalts and calc-alkaline to alkaline rhyolites. The eruptives constructed volcanoes ranging from oceanic islands with associated volcaniclastic infill of back-arc basins, to subaerial cordilleran stratovolcanoes.

North of the Suture, such a unified plate margin model cannot be substantiated, as the volcanics occur in tectonostratigraphic units whose present juxtaposition is largely the result of late-Caledonide strike-slip movement. There is evidence of a compressive regime during the early development of the orogen which resulted in the obduction of the Ballantrae Ophiolite and perhaps some of the

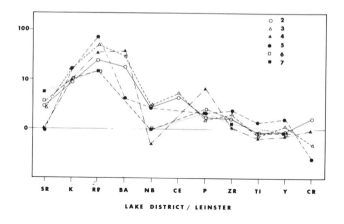
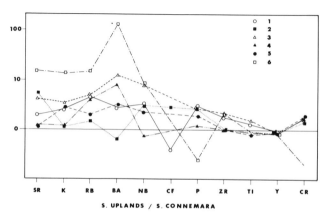
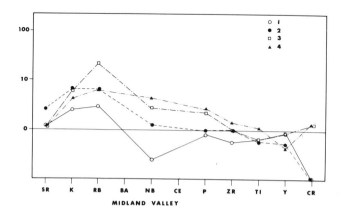

components of the Highland Border Complex and in the accretion of the
Southern Uplands prism which also contains volcanic components of
oceanic origin. There is little evidence of island arc or other
subduction-related volcanism in Cambrian or Ordovician strata except
for the South Mayo basin where island arc tholeiites and calc-alkaline
basalts are seen. However, evidence of early arc plutonism is provided
by the granitoid clasts in the Girvan basin and in the Lower Ordovician
sediments of the South Mayo basin. Some authors suggest that these
clasts are derived by the mid-Ordovician erosion which uncovered a
Cambro-Ordovician arc root situated in the Midland Valley. It is not
until the Siluro-Devonian that widespread regional calc-alkaline
volcanism is seen.

ACKNOWLEDGEMENTS

The author owes a considerable debt of gratitude to his colleagues who
have made many helpful suggestions and have allowed the use of much new
and hitherto unpublished data. In particular, acknowledgement is due
to Messrs Bevins, Davies, Jackson, Kokelaar, Thirlwall and Thorpe. It
is hoped that much of the work which is currently 'in press' will be
published before this paper.

Figure 3.   Geochemistry of representative sample of averages, plotted
on multi-element diagrams; Normalised to MORB.

MORB normalisation values (ppm for trace elements, % for oxides)

Sr 120, K_2O 0.15,   RB 2,   Ba 20,   Nb 4,   Ce 10, P_2O_5 0.12, Zr 90, TiO_2 1.5, Y 30,   cr 250.

3(a).   Lake District/Leinster :

2.   Borrowdale Volcanics (average basalt).   Fitton, (1972)
3.   Borrowdale Volcanics (average basaltic andesite).   Fitton, (1972)
4.   SE Leinster   (Main phase, basalt S1)   Stillman & Williams (1979)
5.   SE Leinster   (Late alkaline basalt S3)   Stillman & Williams (1979)
6.   N Leinster   (Tholeiitic basalt L100)   Stillman & Williams (1979)
7.   N Leinster   (Calc-alkaline basalt L33) Stillman & Williams (1979)

3(b).   Southern Uplands/S Connemara:

1.   S. Connemara Series; Golam Fm.(basalt 43)   Ryan et al. (1983)
2.   S. Connemara Series; Gorumna Fm.(basalt 50)   Ryan et al. (1983)
3.   S. Uplands (Tectonic Slice 1 (35))   Lambert et al. (1981)
4.   S. Uplands (Tectonic Slice 2 (39))   Lambert et al. (1981)
5.   S. Uplands (Tectonic Slice 3 (5))   Lambert et al. (1981)
6.   S. Uplands (Bail Hill (10))   Lambert et al. (1981)

3(c).   Midland Valley:

1.   Lough Nafooey, Murrisk (Av. Ben Corragh Fm.)   Ryan et al. (1980)
2.   Lough Nafooey, Murrisk (Av. Knock Kilbride Fm.) Ryan et al. (1980)
3.   Highland Border Series, Bute (Gp.I; 463)   Ikin (1983)
4.   Highland Border Series, Arran (Gp.IIA; 435)   Ikin (1983)

REFERENCES

ALLEN, J.R.L. & WILLIAMS, P.J. 1981. 'Sedimentology and stratigraphy of the Townsend Tuff Bed. (Lower ORS) in S. Wales and the Welsh Borders'. J.geol.soc.London, 138, 15-29.

BEVINS, R.E., KOKELAAR, B.P. & DUNKLEY, P.N. 1984. 'Petrology and geochemistry of early to mid Ordovician igneous rocks in Wales; a volcanic arc to marginal basin transition'. Proc. Geol.Assoc., 95, 337-47

BLUCK, B.J. 1983. 'Role of the Midland Valley of Scotland in the Caledonian Orogeny'. Trans.roy.Soc.Edinburgh, Earth Sci., 74, 119-136.

─────────, HALLIDAY, A.N., AFTALION, M. & MACINTYRE, R.M. 1980. 'Age and origin of the Ballantrae ophiolite and its significance to the Caledonian Orogeny and the Ordovician time scale'. Geology, 8, 492-495.

BREKKE, H., FURNES, H., NORDAS, J. & HERTOGEN, J. (in press). 'Lower Palaeozoic convergent plate margin volcanism on Bømlo, SW Norwegian Caledonides'. J.geol.Soc.London.

CAMERON, T.D.J. & ANDERSON, T.B. 1980. 'Silurian metabentonites in Co. Down, N. Ireland'. Geol. J., 15, 59-75.

COCKS, L.R.M. & FORTEY, R.A. 1982. 'Faunal evidence for oceanic separation in the Palaeozoic of Britain'. J.geol.Soc.London, 139, 467-480.

CURRY, G.B., INGHAM, J.K, BLUCK, B.J. & WILLIAMS, A. 1982. 'The significance of a reliable Ordovician age for some Highland Border rocks in central Scotland'. J.geol.Soc.London., 139, 451-454.

DAVIES, G. 1983. 'The isotopic evolution of the British lithosphere'. Ph.D. Thesis, Open University (unpubl.)

DEWEY, J.F. 1982. 'Plate tectonics and the evolution of the British Isles'. J.geol.Soc.London, 139, 371-412.

DUNKLEY, P.N. 1978. 'The geology of the SW Arans with particular reference to the Igneous history'. Ph.D. Thesis, Univ. of Wales (unpubl.)

FITTON, J.G. 1972. 'The Genetic significance of Almandine-Pyrope phenocrysts in the calc-alkaline Borrowdale Group, Northern England'. Contrib.Mineral.Petrol., 36, 231-248.

───────── & HUGHES, D. 1970. 'Volcanism and plate tectonics in the

British Ordovician'. Earth Planet.Sci.Lett., 8, 223-228.

―――――――, THIRLWALL, M.F. & HUGHES, D.J. 1982. 'Volcanism in the Caledonian orogenic belt of Britain'. In THORPE, R.S. (ed.) Andesites: orogenic andesites and related rocks. Wiley, 611-636.

FRANCIS, E.H. & HOWELLS, M.F. 1973. 'Transgressive welded ash-flow tuffs amongst the Ordovician sediments of NE Snowdonia, N. Wales'. J.geol.Soc.London, 129, 621-641.

FURNES, H. 1978. 'Comparison of Caledonian volcanicity in Wales and West Norway'. Ph.D. Thesis, Univ. of Oxford (unpubl.).

GRAHAM, C.M. 1976. 'Petrochemistry and tectonic significance of Dalradian metabasaltic rocks of the SW Scottish Highlands'. J.geol.Soc.London, 132, 61-84.

―――――――― & BRADBURY, H.J. 1981. 'Cambrian and Late Precambrian basaltic activity in the Scottish Dalradian; a review'. Geol.Mag., 118, 27-39.

HALLIDAY, A.N. 1984. 'Coupled Sm-Nd and U-Pb systematics of late Caledonian granites and basement under N. Britain'. Nature, 307, 229-233.

HENDERSON, W.G. & ROBERTSON, A.H.F. 1982. 'The Highland Border rocks and their relation to marginal basin development in the Scottish Caledonides'. J.geol.Soc.London, 139, 443-450.

HOWELLS, M.F. 1977. 'The varying pattern of volcanicity and sedimentation in the Bedded Pyroclastic/Middle Crafnant volcanic formations in the Ordovician of central and eastern Snowdonia'. J.geol.Soc.London, 133, 401-411 (Conference Report p. 404).

HUTTON, D.H.W., AFTALION, M. & HALLIDAY, A.N. 1985. 'An Ordovician ophiolite in County Tyrone, Ireland'. Nature, 315, 210-212.

IKIN, N.P. 1983. 'Petrochemistry and tectonic significance of Highland Border Suite mafic rocks'. J.geol.Soc.London, 140, 267-278.

KENNEDY, M.J. 1979. 'The continuation of the Candadian Appalachians into the Caledonides of Britain and Ireland'. In HARRIS, A.L., HOLLAND, C.H. & LEAKE, B.E. (eds.) The Caledonides of the British Isles - reviewed. Spec. Publ. Geol. Soc, London, 8, 33-64.

KOKELAAR, B.P., HOWELLS, M.F., BEVINS, R.E., ROACH, R.A. & DUNKLEY, P. N. 1984. 'The Ordovician marginal basin of Wales'. In: (KOKELAAR, B.P. & HOWELLS, M.F.(eds.). Volcanic and associated sedimentary and tectonic processes in modern and ancient marginal basins. Spec.Publ.Geol.Soc.London.

LAMBERT, R. St J., HOLLAND, J.G. & LEGGETT, J.K. 1981. 'Petrology and tectonic setting of some Upper Ordovician volcanics from the' Southern Uplands of Scotland'. J.geol.Soc.London, 128, 421-436.

LEGGETT, J.K., McKERROW, W.S. & EALES, M.H. 1979. 'The Southern Uplands of Scotland, a Lower Palaeozoic accretionary prism'. J.geol.Soc.London, 136, 755-770.

LONGMAN, C.D., BLUCK, B.J. & Van BREEMAN, O. 1979. 'Ordovician conglomerates and the evolution of the Midland Valley'. Nature, 280, 571-581.

MORRIS, J.H. 1981. 'The geology of the western end of the Lower Palaeozoic Longford-Down inlier'. Ph.D. Thesis. Univ. of Dublin (unpubl.)

MOSELEY, F. 1982. 'Lower Palaeozoic Volcanic Rocks'. In SUTHERLAND, D.(ed.). Igneous Rocks of the British Isles. J. Wiley, London.

────────── 1983. 'The volcanic rocks of the Lake District'. McMillan, London.

PHILLIPS, W.E.A., STILLMAN, C.J. & MURPHY, T. 1976. A Caledonian Plate tectonic model'. J.geol.Soc.London, 132, 579-609.

RYAN, P.D., FLOYD, P.A. & ARCHER, J.B. 1980. 'The stratigraphy and petrochemistry of the Lough Nafooey Group, Tremadocian, W. Ireland'. J.geol.Soc.London, 137, 433-458.

────────── , MAX, M.D. & KELLY, T. 1983. 'Petrochemistry of basic volcanic rocks of the South Connemara Group (Ordovician) W. Ireland'. Geol.Mag., 120, 141-152.

────────── , SAWAL, V.K. & ROWLANDS, A.S. 1983. 'Ophiolite melange separates ortho- and paratectonic Caledonides in W. Ireland'. Nature, 302, 50-52.

SANZEN BAKER, I. 1972. 'Structural relations and sedimentary environment of the Silurian and early Old Red Sandstone of Pembrokeshire'. Proc. Geol. Assoc., 83, 139-169.

SAUNDERS, A.D., TARNEY, J., STERN, C.R. & DALZIEL, I.W.D. 1979. 'Geochemistry of Mesozoic marginal basin floor igneous rocks from Southern Chile'. Geol. Soc. Amer. Bull., 90, 237-258.

────────── , FORNARI, D.J. & MORRISON, M.A. 1982. 'The composition and emplacement of basaltic magams produced during the development of continent-margin basins: the Gulf of California, Mexico'. J.geol.Soc.London, 139, 335-346.

THIRLWALL, M.F. 1981. 'Implications for Caledonian plate tectonic modelling of chemical data for volcanic rocks of the British Old Red Sandstone'. J.geol.Soc.London, 138, 123-138.

——————, 1981. 'Peralkaline rhyolites from Ordovician Tweedale lavas Peebleshire, Scotland'. Geol.J., 16, 41-44.

——————, 1982. 'Systematic variation in chemistry and Nd-Sr isotopes across a Caledonian calc-alkaline volcanic arc; implications for source materials'. Earth & Planet.Sci. Letters, 58, 27-50.

——————, 1983. 'Discussion on Implications for Caledonian plate tectonic models of chemical data for volcanic rocks of the British Old Red Sandstone'. J.geol.Soc.London, 140, 315-318.

—————— & FITTON, J.G. 1983. 'Sm-Nd garnet age for the Ordovician Borrowdale Volcanic Group, English Lake District'. J.geol.Soc.London, 140, 511-518.

—————— & BLUCK, B.J. 1984. 'Sr-Nd isotope and chemical evidence that the Ballantrae "ophiolite", SW Scotland is polygenetic'. In SHELDON, A.W. (ed.), Ophiolites and Oceanic Lithosphere. Spec.Publ.geol.Soc., London, (in press).

VAN de KAMP, P.C. 1969. 'Silurian volcanic rocks of the Mendip Hills, Somerset, and Tortworth, Gloucestershire, England'. Geol. Mag., 106, 542-555.

WATSON, J. 1984. 'The ending of the Caledonian Orogeny in Scotland'. J.geol.Soc.London, 141, 193-214.

THE TECTONIC SETTING OF THE SOUTHERN UPLANDS

W. S. McKerrow
Department of Geology and Mineralogy
Parks Road
Oxford, OX1 3PR

ABSTRACT. The pattern of fault-bounded sedimentary sequences established in southern Scotland can be traced for almost 600 km to western Ireland. The sediments forming this accretionary prism were deposited, in part on the floor of the Iapetus Ocean, and in part in a trench, either by longitudinal fans (mostly from the north-east) or by fans and mass-flow deposits flowing from the north-west down the inner trench wall. In addition, a trench-slope basin has been recognised in eastern Scotland.

Recent work has shown that some of the original ten tracts can be subdivided into smaller fault bounded units. The detritus in the Ordovician greywackes of the Northern Belt (Tracts 1 to 3) is varied. Tracts 1 and 3 have sediments with a low quartz content and abundant igneous rocks, perhaps derived from an Early Palaeozoic arc; while in Tract 2, the greywackes are richer in quartz and contain metamorphic detritus. Rudites occur in all three tracts; they have been interpreted as fans fed by submarine canyons feeding into the trench from the north-west. As well as metamorphic and other clasts the rudites include some granites which give ages of around 800 to 1200 Ma. Granites of this age are not known in Scotland, but they do occur in western Newfoundland.

The Central Belt (Tracts 4 to 9) and the Southern Belt (Tract 10) include greywackes with both igneous and metamorphic detritus. In Tracts 9 and 10, the greywackes are more quartz-rich. The petrography suggests that these younger Silurian sediments are derived from pre-existing sediments, perhaps the elevated parts of the trench-slope break (known as Cockburnland). Geochemical studies suggest that some fine-grained sediments came neither from Scotland nor North America; perhaps Scandinavia may have been a source terrane for the axially-derived trench sediments. The presence of 460 Ma granite clasts in Early Llandovery greywackes of Tract 4, is consistent with the accretionary prism still lying south of Newfoundland or Rockall through the Ordovician period. Strike-slip movement is inferred to have taken place during the Silurian (perhaps mainly during the Late Silurian).

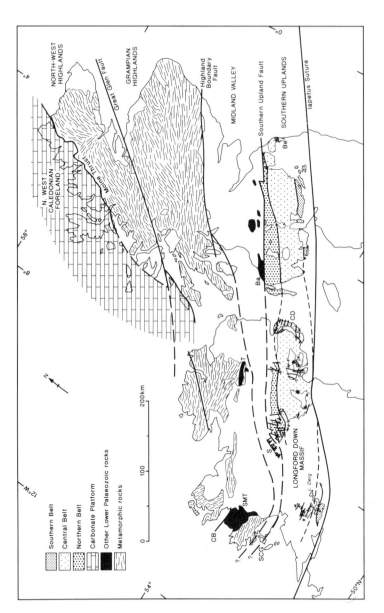

Fig. 1. The structural divisions of Scotland. The line of the Highland Boundary Fault is drawn to the south of all Dalradian rocks and to the south of central Arran, Kintyre, Cushendall and Tyrone, where Lower Old Red Sandstone rests unconformably on Dalradian rocks. The line of the Southern Upland Fault is drawn along the Stinchar Fault, and in Ireland to the south of Strokestown and the South Connemara Group which have sequences comparable with Girvan.
CB= Clew Bay Fault; SMT= South Mayo Trough; SCG= South Connemara Group; S= Strokestown; T= Tyrone; CD= County Down; Ba.= Ballantrae; Be= Berwickshire.

TECTONIC SETTING

The hypothesis that the Southern Uplands represents an Ordovician and Silurian accretionary prism on the northern margin of the Iapetus Ocean (Mitchell and McKerrow, 1975) has stood up to the tests of further work. Recent research on the Southern Uplands and its tectonic setting has developed in two ways: studies in the Midland Valley and the Scottish Highlands (Fig. 1) suggest the presence of large strike movements along both the Highland Boundary Fault and the Southern Upland Fault during the Ordovician and Silurian (e.g. Bluck, 1983), and secondly, detailed studies of the structure, stratigraphy, sedimentology, sedimentary petrography and geochemistry have provided more information on the sedimentary environments, structural development and the possible source regions of the Southern Uplands sediments and igneous rocks.

The Southern Uplands of Scotland was divided into three belts by Peach and Horne (1899): the Northern Belt, with entirely Ordovician rocks; the Central Belt, with Silurian greywackes above Ordovician and Early Silurian pelagic sediments; and the Southern Belt, with dominantly Wenlock sediments (Fig. 1). Leggett et al. (1979a, b, 1982) described ten tracts, each with its own distinctive stratigraphic sequence, and each bounded by strike faults. Tracts 1 to 3 correspond to the Northern Belt; Tracts 4 to 9 to the Central Belt; and Tract 10 forms the Southern Belt. Many of these tracts can now be split into smaller fault-bounded units with different (though not always quite so distinct) stratigraphic sequences (Table 1). So, while some tracts can still be traced along strike for over 200 km, many others only appear to extend for less than 20 km. (Fig. 2). For clarity, the original numbers of Leggett et al. (1979a) are retained.

The oldest sequence (exposed in Tract 1 just to the south of the Southern Upland Fault) contains fossils as young as the _gracilis_ Zone (which straddles the Llandeilo/Caradoc boundary), and it is inferred that accretion commenced in Caradoc time, perhaps around 455 Ma ago (McKerrow et al., 1985). The youngest sediments occur in the Southern Belt of both Scotland and western Ireland, and include Wenlock fossils up to the _lundgreni_ Zone (for zonal succession, see Fig. 3), which has an age of around 420 Ma. So accretion lasted for about 35 Ma.

Most of the subduction-related plutonism and volcanics of the Grampian Highlands and the Midland Valley are younger than the Wenlock (Thirlwall, 1981a), though there was some contemporary plutonism to the north of the Great Glen Fault (van Breemen and Piasecki, 1983). These anomalies can be resolved, in part, by invoking strike slip movements, but it is also necessary to conclude that subduction in Scotland continued through much of the Early Devonian, long after accretion ceased. The Lower Old Red Sandstone rests unconformably on the Dalradian rocks just north of the Highland Boundary Fault in several places, and similar Old Red Sandstone facies appear to occur on both sides of the Great Glen Fault, so these faults could not have had large horizontal movements after the end of the Silurian.

Sedimentological studies (Leggett, 1980a; Casey 1983; Floyd, 1982; Hepworth et al., 1982; Morris, 1983; Craig, 1984; Kemp and White, 1985) show that deposition of coarse clastic sediments took place by

three distinct processes.
  (a) deposition by axial currents along the trench, mostly from the north-east;
  (b) deposition by currents down the inner trench wall from the north-west, perhaps sometimes in submarine canyons, with rudites deposited by mass-flow, as well as greywackes by turbidity currents.
  (c) deposition in fans situated either out in the open ocean or near the plate margin, the latter being present perhaps after the trench-fill spreads oceanwards over the outer trench high (after the turriculatus Zone) and permitting more variable current directions.

In addition, a trench-slope basin has been recognised in Berwickshire (Casey, 1983), where the Late Wenlock or Early Ludlow Coldingham and Linkim Beds appear to lie unconformably on Early Silurian turbidites of the accretionary prism (Fig. 2).

The petrography of sediments reveals some indication of the provenance areas which provided sediment to the trench and ocean floor during the Late Ordovician and Silurian. Although at present our understanding is far from complete, it would appear that very different terranes were being eroded simultaneously. It is thus necessary to look at the petrography in some detail.

Many of the clastic sediments in Tracts 1 and 3 contain abundant basic igneous detritus derived from oceanic rocks and/or from an arc; the sediments generally have a very low content of quartz and metamorphic detritus, though some acid igneous material is present (Floyd, 1982; Craig, 1984). Tract 1 greywackes all occur above or with graptolitic beds of the gracilis Zone. In Tract 3 the age of the underlying graptolites is more variable (Table 1, Fig. 3) ranging from Early Caradoc to Early Ashgill; the Portayew Rocks in Tract 3d are distinct from the greywackes in Tract 1 and Tracts 3a, 3b and 3c in containing more quartz and being post-linearis (Floyd, 1982).

Tract 2 contains greywackes with more quartz and more metamorphic detritus than Tracts 1 and 3. It consists of two distinct divisions: Tract 2a occurs over the whole of the western half of the Southern Uplands (Fig. 2) and is post-gracilis Zone in age; Tract 2b is restricted to Nithsdale and Clydesdale and is of post-clingani age. In both 2a and 2b, the stratigraphic sequences include more than one coarse clastic formation (Table 1, Fig. 3). On the west coast of Galloway, the Kirkholm Group is present below the Galdenoch Group (Fig. 3); both have igneous and metamorphic detritus (Kelling 1961, 1962), but the latter is richer in volcanic material.

In Tract 2a, the Bail Hill Volcanic Group of Nithsdale has been interpreted as a mildly alkaline seamount (Hepworth et al., 1982) which provided detritus to the Kiln Formation, which also contains tuffs. Both the volcanics and the Kiln Formation are then overlain by rudites and sandstone - with dominantly sedimentary detritus - of the Spothfore Formation. Some shale clasts in the rudites have yielded gracilis Zone graptolites, and it has been suggested that they originated either from earlier accreted sediments or from lower slope deposits (Hepworth et al., 1982).

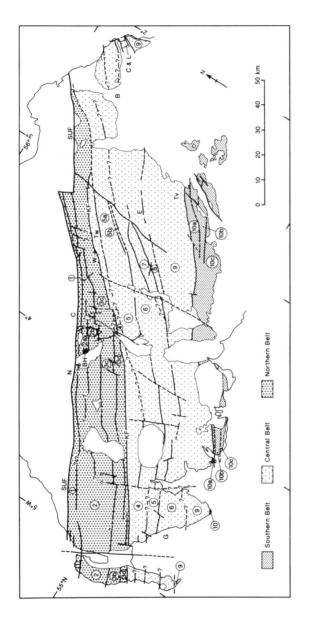

Fig. 2. The Tracts of the Southern Uplands. The Southern Uplands Fault (SUF) is drawn along the Stinchar Fault to the north of the Glen App and Downans Point rocks, which are correlated with the Tract 1 sequence further east. Sources as in Table 1.
G= Galloway; KF= Kingledores Fault; N= Nithsdale; BH= Bail Hill; C= Clydesdale; W= Wrae Hill; Tw= Tweedale; E= Ettrickbridgend; Tv= Teviotdale; B= Berwickshire; C & L= Coldingham and Linkim Beds.

A second seamount was accreted in Tract 3c, where the Tweeddale Lavas are overlain by a small olistostrome with blocks of limestone (the Wrae Limestone) and lava (Leggett, 1980b; Thirlwall, 1981b). These rocks occur in the Lowther Beds, a unit of quartz-poor greywackes. So far it is not known how much of this detritus may have originated from the Tweeddale Lava seamount, and how much came from elsewhere. Olistostromes are very rare in the Southern Uplands. Leggett (1980a) uses this as an indication that the Southern Uplands accretionary prism may have developed over a trench with a slow convergent rate. Clearly the Tweeddale olistostrome could have developed on the slopes of the seamount, and would thus have had no direct relation to the convergence rate.

The arc-derived material of Tract 1 and Tract 3a was deposited in the Early Caradoc, and that of Tracts 3b and 3c in the Late Caradoc (Fig. 3).

The greywackes with more quartz and more metamorphic detritus of Tracts 2a, 2b and 3d appear to have been deposited at about the same times as those with abundant igenous material (Table 1), so perhaps different terranes were providing sediments simultaneously to different parts of the trench system.

Whole-rock Rb-Sr dating of granite clasts from mass-flow conglomerates has yielded some interesting results. Mr C.F. Elders of Oxford University has obtained Rb-Sr whole-rock isochron ages of 1,200 Ma from clasts in the Corsewall Conglomerate in Tract 1 of western Galloway, and ages between ca. 800 and ca. 1,100 Ma from granites in similar conglomerates in the Blackcraig Formation of Tract 2b (Table 1, Fig. 3); and Bluck (1983) reports ages of 595 ± 28 Ma from granites in the Corsewall Conglomerate. Both these formations include conglomerate lenses in unbedded sandstones, which have been interpreted as mass-flow deposits on the inner trench wall or as the proximal parts of fans feeding south-eastwards into the trench (Casey, 1983; Floyd, 1982; Kelling, 1962). These results suggest that, whatever the provenance of the axial turbidites in the trench may prove to be, the terrane on the continental side of the trench was similar to Grenville Province of western Newfoundland and other parts of eastern Canada.

The detritus in Tracts 2, 3d, 4, 5, 6, 7 and 8 shows mixtures of igneous and metamorphic material in various proportions. Thus during the Ashgill and Llandovery the provenance of most of the coarse clastic sediments included both metamorphic and igneous (acid and basic) terranes. Granite clasts in rudites of Tract 4 have ages of 450 Ma - these could originate in Newfoundland or Scotland, but they are older than the majority of Scottish granites. After the end of the Ordovician, from Tract 4 southwards, there is a progressive increase in the quartz content of the greywackes. It is in the Early Silurian that the trench slope break (known in Scotland as Cockburnland) rose above sea level (Leggett et al., 1982), and it is reasonable to deduce that it provided progressively more detritus to the trench (Fig. 3). It is also noticeable (Casey, 1983) that during the Silurian the sedimentary petrography becomes more uniform along the length of the accretionary prism. With further development of the prism, Casey recognised a second Silurian stage (from the _gregarius_ to the _turriculatus_ Zones) when the trench was still separated by an outer trench high from the ocean floor, where, apart from sandstones in the Ashgill and gregarius Zone of Tract 8, only pelagic deposits were accumulating. After the turriculatus Zone, Casey considers that the outer trench high was swamped, and that a major fan system with variable current directions spread outwards from the trench over the ocean. From this time onwards, the only pelagic sediments occur interbedded with turbidites (and not below them as before). During this time (latest Llandovery and Wenlock) erosion of the trench-slope break introduced reworked sediments onto the ocean to give the quartz rich greywackes of the Hawick Rocks and the Riccarton Group (tracts 9 and 10).

Erosion of a sedimentary terrane is also indicated during the Late Wenlock/Early Ludlow deposition of the Coldingham and Linkim Beds in Berwickshire (Casey, 1983). Uplift of the Southern Uplands may thus have continued after accretion ceased in the late Wenlock lundgreni Zone. This uplift may have been related to post-Wenlock subduction of English continental crust below the Southern Uplands as the calc-alkaline igneous rocks continue in Scotland well after the start of the Devonian (Leggett et al., 1983).

A recent development in the study of the provenance of fine-grained sediments in the Southern Uplands has been the application of Sm-Nd isotope geochemistry (O'Nions et al., 1983; Miller and O'Nions, 1984). Crustal residence ages (Sm-Nd model ages calculated on the basis of De Paolo's (1981) depleted mantle model give an indication when, on average, the material examined was incorporated into the Earth's crust. Sixteen samples from the Southern Uplands (Tracts 1, 3b, 3c, 6 and 10) show (with one exception: 1.08 Ga from Tract 1) a restricted range from 1.43 Ga to 1.74 Ga (O'Nions et al., 1983). This range differs markedly from metasediments of the Lewisian, Moine, Torridonian and Dalradian of the Scottish Highlands, all of which include detrital components with a substantially greater crustal residence age: Lewisian up to 2.49 Ga; Torridonian up to 2.33 Ga; Dalradian up to 2.69 Ga (O'Nions et al., 1983). Hence it has been deduced that the Ordovician and Silurian sediments of the Southern Uplands had a different provenance than the Precambrian and Early Cambrian rocks of the Scottish Highlands. In addition, several of the Southern Uplands samples, especially those from Tract 1, have significantly higher $^{147}Sm/^{144}Nd$ ratios (0.120) than any metasediments sampled from the Highlands ($^{147}Sm/^{144}Nd$ 0.120) implying a large detrital component derived from basaltic source rocks. This is particularly clear for samples from Tract 1, but is less clear in the samples from Tracts 3, 6, and 10.

Miller and O'Nions (1984) argue that the basaltic components in the Southern Uplands detritus are not derived from penecontemporaneous igneous activity. Their data indicates that the source terrane may have had considerable petrographic diversity (granite to basalt ?) but limited geochronological variety, with crust formation at ca. 1.6 - 0.15 Ga. Miller and O'Nions (1984) suggest Baltica as a plausible candidate, and exclude the Scottish Highlands and the Canadian Shield as viable source areas. However, it is important to realise two points before making any definite conclusions. First, during the Ordovician there were several point sources of sediments transported down the inner trench wall, as well as the sediments deposited by axial currents along the trench. Secondly the Grenville ages on granite clasts from mass-flow deposits in

TABLE 1

Selected greywackes and rudite formations, with sedimentological, stratigraphical and petrographical data.

Tract 1. Coulter-Noblehouse
Galloway: Corsewall Group

Age of greywackes: post-*gracilis* Zone
Igneous detritus; some mass-flow rudites with 1,200 Ma and 600Ma granite clasts.

(Kelling 1961, 1962; Bluck, 1983; Elders, pers. comm. 1984).
Nithsdale: Marchburn Formation

Igneous detritus; quartz-poor; rudite fans building out to south-east.

(Floyd, 1982).
The rudite fans appear to be local in their distribution.
Tract 1 has not been recognised in Ireland.

Tract 2. Abington
Galloway: Galdenoch Group

(a) post-*gracilis* Zone
Turbidites from south-east; more volcanic detritus than Kirkholm Group.

Kirkholm Group

Turbidites from north-east; more metamorphic detritus than Corsewall Group (Tract 1).

(Kelling, 1961, 1962).
Nithsdale: Spothfore Formation

Laterally transported greywackes and rudites with sedimentary clasts.

Kiln Formation and Bail Hill Volcanic Group

A mildly alkaline seamount covered by sediments partly derived from its erosion.

(Hepworth et al., 1982).
Clydesdale: Guffock Formation

Mid-fan turbidites, flow to south-west.

(Hepworth et al., 1982).

(b) post-*clingani* Zone

Nithsdale: Blackcraig Formation

Rich in ferromagnesium minerals; fans fed by canyons from the north-west; rudites include 800 Ma and 1,100 Ma granite clasts.

Afton Formation

Quartz-rich; no rudites.

(Floyd, 1982; Elders, pers. comm. 1984).
Clydesdale: Abington and Glenflosh Formations

Quartzose greywackes; currents to south-west.

(Hepworth et al., 1982).

Tract 3. Tweeddale
County Longford: Carrickateane Formation

(a) post-*gracilis* Zone
Greywackes with tholeiitic volcanics.

(Leggett et al., 1979b; Morris, 1983)
County Down; Ballygrot Group

Rich in igneous clasts.

(Craig, 1984)

(b) post-*wilsoni* Zone

Galloway: Portpartick Formation (basic clast)

Quartz-poor; andesitic and metamorphic detritus; fans building south-east into trench.

(Kelling, 1962; Floyd, 1982)
Clydesdale: Glencaple and Elvan

(c) post-*clingani* Zone
Sediment gravity flows to south-west; Glencaple Formation more distal; quartz-poor.

(Hepworth et al., 1982)
Tweeddale: Lowther Beds, Wrae Limestone and Tweeddale Lavas

Quartz-poor greywackes, with lavas and olistostrome interbedded; olistostrome with blocks of limestone and lava.

(Leggett, 1980; Leggett et al., 1979a; Thirlwall, 1981b).

(d) post-*linearis* Zone

County Down: Ballymacormick Group

Rich in igneous and metamorphic detritus.

(Craig, 1984)
Galloway: Portayew Rocks

Quartz-rich greywackes from north-west, from an igneous and metamorphic terrane.

(Floyd, 1982).
Nithsdale: Shinnel Formation

Quartz-rich greywackes with rudites on mid-fan lobes fed by canyons from north-west.

(Floyd, 1982).

Tract 4. Talla
Galloway: Kilfillan Formation

(Gordon, 1962)
Tweeddale: Pyroxenous Group

(Walton, 1955; Casey, 1983; Elders pers. comm. 1984)

Tract 5. Hartfell
Galloway: Garheugh Formation
(Gordon, 1962; Casey, 1983).
Tweeddale: Intermediate Group

(Walton, 1955; Casey, 1983)

Tweeddale: Craigierig Sandstone
(Casey, 1983)

Tract 6. Dobbs Linn
Galloway: Craignell Formation (D)
(Cook and Weir, 1980)
Tweeddale: Gala Group and Grieston Shales

(Walton, 1955; Casey, 1983)

Tract 7. Craigmichen
Tweeddale: Gala Group
(Casey, 1983).

Tract 8. Ettrickbridgend
Ettrickbridgend: Abbotsford Flags

                 Upper Birkhill Shales
                 Hartfell Shales
(Casey, 1983)
Berwickshire: Westloch Formation

(Casey, 1983).
Tract 9. Hawick
Galloway: Hawick Rocks

(Rust, 1965a, b).
Teviotdale: Hawick Rocks

(Casey, 1983)

Tract 10. Riccarton
Galloway: Riccarton Group
(Rust, 1965a, b; Casey, 1983).
Teviotdale: Stobs Castle Beds
(Warren, 1963, 1964; Casey, 1983)

Galloway: Ross Formation
         Gipsy Point Formation
(Casey, 1983; Kemp, pers. comm. 1984).
Teviotdale: Penchrise Burn Beds
            Shankend Beds
(Warren 1963; Lumsden et al., 1967).

Galloway: Raeberry Castle Formation
          Mullock Bay Formation
(Kemp, pers. comm. 1984; Casey, 1983).
Teviotdale: Caddround Burn Beds
(Warren, 1963, 1964).

post-*vesiculosus* or *gregarius* Zones
    Greywackes with detritus from an island arc and metamorphic
    minerals up to garnet grade.

    High content of ferromagnesian minerals; metamorphic, basalt and
    granite detritus; rudites with 450Ma granite clasts.

(a) post-*gregarius* Zone
    Quartz-rich with granite and metamorphic detritus.

    Quartz variable; igneous and metamorphic detritus; some black
    shale and greywacke clasts.

(b) post-*convolutus* Zone (includes *gregarius* and
    sedgwickii Zones)
    Richer in quartz than Intermediate Group.

    Quartz-rich; acid igneous clasts. Post-*sedgwickii* Zone.

    Quartz-rich; metamorphic and igneous detritus. Turbidites spread
    out over ocean. Post-*maximus* Zone; includes *turri-
    culatus*, *crispus* and *griestoniensis* Zones.

post-*maximus* Zone.
    Quartz-rich; metamorphic and igneous detritus.

    Quartz-rich; igneous detritus; includes *turriculatus* and
    crispus Zones.
    Thin quartz-rich greywacke; includes *maximus* Zone.
    Quartz-rich; currents to SE; on ocean floor; Ashgill age.

    Quartz-dominant; mid-fan flow from north-west; post-*gregarius*
    Zone; includes graptolites up to *crispus* Zone.

?Late Llandovery/Early Wenlock.
    Quartz-rich, with some volcanic and metamorphic detritus;
    turbidites from north-east.

    Quartz-rich, with granite and metamorphic detritus; carbonate in
    matrix; recycled sediments; turbidites from north-east.

(a) Lower Wenlock
    Quartz-rich; derived from Northern Belt.

    Quartz-dominant

(b) Lower Wenlock
    Quartz-rich.
    Proximal Fan rudites.

    Quartz-dominant
    Quartz-dominant

(c) Middle and Upper Wenlock
    Quartz-rich
    Quartz-rich

    Quartz-rich

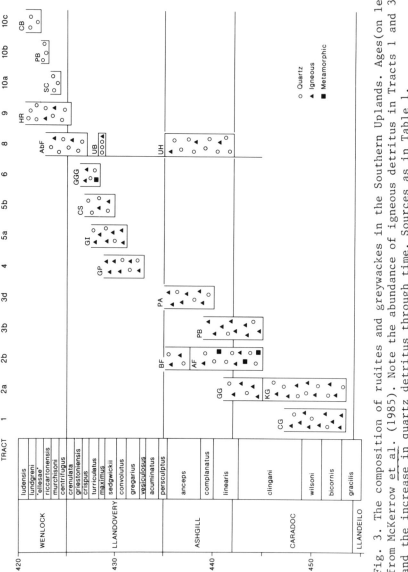

Fig. 3. The composition of rudites and greywackes in the Southern Uplands. Ages (on left) from McKerrow et al. (1985). Note the abundance of igneous detritus in Tracts 1 and 3b, and the increase in quartz detritus through time. Sources as in Table 1.
BF= Blackcraig Formation; PB= Portpatrick Formation (basic clast division); PA= Portpatrick Formation (acid clast division); KG= Kirkholm Group; GP= Gala Group (Pyroxenous); GI= Gala Group (Intermediate); GS= Craigirig Sandstone; GGG= Gala Group (Garnet); UH= Upper Hartfell sandstones; UB= Upper Birkhill sandstones; AbF= Abbotsford Flags; HR= Hawick Rocks; SC= Stobs Castle Beds; PB= Penchrise Burn Beds; CB= Caddroun Burn Beds.

Tracts 1 and 2b, show that these south-eastwardly transported deposits could be derived from nearby parts of North America, while many of the fine-grained sediments analysed by O'Nions et al. (1983) could well have been deposited by far-travelled axial currents flowing south-west along the trench. It seems that the Ordovician beds of the Northern Belt may have had two quite distinct sources.

If Grenville rocks were situated to the north-west of the Southern Uplands during the Late Ordovician, then it is probable that the accretionary prism was developing off a Grenville terrane like western Newfoundland or the Rockall Bank (Miller et al., 1973). We also have to explain tracts in the Northern Belt which are rich in basic igneous rocks. In the Northern Appalachians there is evidence that island arc rocks resting on oceanic crust (Bronson Hill, Tetagouche, Lushs Bight) collided with the passive margin of eastern North America in the Llandeilo (Humberian Orogeny of Newfoundland) and the Caradoc (Taconic Orogeny of New Brunswick and New England) (Hall and Robinson, 1982; Lyons et al., 1982; Hatch, 1982; Doolan et al., 1982; McKerrow, 1982). It is possible that a north-eastern extension of this arc (Dewey, 1982; Mitchell, 1984) existed to the north of the Southern Uplands.

The north-west source could thus be expected to contain Early Ordovician arc and oceanic tholeiitic basalts as well as Grenville and other continental rocks. Contrary to the conclusions of Miller and O'Nions (1984), it can be calculated (P. N. Taylor, personal communication, 1984) that it is possible to include up to around 70% basaltic detritus of Lower Palaeozoic age in the Southern Uplands sediments without greatly modifying the crustal residence model ages of Miller and O'Nions (1984), since oceanic basalts typically have Sm-Nd isotope compositions today similar to the model mantle of Miller and O'Nions. It is not likely that the abundant basaltic detritus in Tracts 1 and 3 travelled very long distances; derivation down the inner trench wall would seem more probable than long-distance transport by turbidity currents along the trench.

If an arc did exist to the north of the accretionary prism, subsequent sinistral strike-slip movements shifted both the island arc and the Southern Uplands to the north-east.

The presence of 450 Ma granite clasts in Early Llandovery greywackes of Tract 4, suggests that the accretionary prism was still south of Newfoundland or Rockall throughout the Ordovician period. Strike-slip movement is inferred to have taken place during the Silurian (perhaps mainly during the Late Silurian), as similar Lower Old Red Sandstone facies cross both the Highland Boundary Fault and the Southern Upland Fault, and Early Devonian 'greywacke' conglomerates of the Midland Valley have been derived from the Southern Uplands (Bluck, 1983). Strike-slip may also have taken place between the Northern and Central Belts along the line of the Kingledores Fault, and possibly on other strike faults in the Souther Uplands.

ACKNOWLEDGEMENTS.

Thanks are due to David Casey for producing an excellent D.Phil. thesis

and allowing me to use its most useful products; to Chris Elders for
allowing me to quote his unpublished work on Rb-Sr ages from granite
clasts; to Paul Taylor for advice on age dates and geochemistry and
for improving the typescript; Jeremy Leggett has discussed many aspects
of the Southern Uplands with me over the past decade; I will always be
in his debt. Clare Pope drew the diagrams, and Andria Fowler prepared
the manuscript.

REFERENCES

BLUCK, B.J. 1983. 'Role of the Midland Valley of Scotland in the Caledonian Orogeny'. Trans. R. Soc. Edinburgh : Earth Sciences, 74, 119-136.

CASEY, D.M. 1983. 'Geological studies in the Central Belt of the eastern Southern Uplands of Scotland'. D.Phil. thesis (unpubl.) Univ. of Oxford.

COOK, D.R. & WEIR, J.A. 1979a. 'Stratigraphy of the aureole of the Cairnsmore of Fleet pluton, southwest Scotland'. In: HARRIS, A.L., HOLLAND, C.H. & LEAKE, B.E. (eds.), The Caledonides of Britain - reviewed. Geol. Soc. London, Spec. Publ., 8, 489-493.

CRAIG. L.E. 1984. 'Stratigraphy in an accretionary prism: the Ordovician rocks of North Down, Ireland'. Trans. R. Soc. Edinburgh: Earth Sciences, 74, 183-191.

DE PAOLO, D.J. 1981. 'A neodymium and strontium isotopic study of the Mesozoic calc-alkaline granite batholiths of the Sierra Nevada and Peninsular Ranges, California'. J. Geophys. Research, 86, 10470-10488.

DEWEY, J.F. 1982. 'Plate tectonics and the evolution of the British Isles'. J. Geol. Soc. London, 139, 371-412.

DOOLAN, B.L., GALE, M.H., GALE, P.N. and HOAR, R.S. 1982. 'Geology of the Quebec Re-entrant: possible constraints from early rifts and the Vermont-Quebec Serpentine Belt'. Geol. Assoc. Canada, Special Paper, 24, 87-115.

FLOYD, J.D. 1982. 'Stratigraphy of a flysch succession: the Ordovician of W Nithsdale, SW Scotland'. Trans. R. Soc. Edinburgh: Earth Sciences, 73, 1-9.

GORDON, A.J. 1962. 'The Lower Palaeozoic rocks around Glenluce, Wigtownshire'. Ph.D. thesis (unpubl.), Univ. of Edinburgh.

HALL, L.M. & ROBINSON, P. 1982. 'Stratigraphic-tectonic subdivisions Southern New England'. Geol. Assoc. Canada, Special Paper, 24, 15-41.

HATCH, N.L. 1982. 'The Taconian Line in Western New England and its implications to Palaeozoic tectonic history'. Geol Assoc. Canada, Special Paper, 24, 67-85.

HEPWORTH, B.C., OLIVER, G.J.H. & McMURTRY, M.J. 1982. 'Sedimentology, volcanism, structure and metamorphism of a Lower Palaeozoic accretionary complex; Bail Hill - Abington area of the Southern Uplands of Scotland'. In: LEGGETT, J.K. (ed.), Trench - Forearc Geology. Geol. Soc. London, Spec. Publ.,

10, 521-34.
KELLING, G. 1961. 'The Stratigraphy and structure of the Ordovician rocks of the Rhinns of Galloway'. Q. J. Geol. Soc. London, 117, 37-75.
——— 1962. 'The petrology and sedimentation of the Upper Ordovician rocks of the Rhinns of Galloway, south-west Scotland. Trans. R. Soc. Edinburgh, 65, 107-37.
KEMP, A.E.S. & WHITE, D.E. 1985. 'Silurian trench sedimentation in the Southern Uplands, Scotland: implications of new age data'. Geol. Mag., 275-277.
LEGGETT, J.K. 1980a. 'The Sedimentological evolution of a Lower Palaeozoic accretionary forearc in the Southern Uplands of Scotland'. Sedimentology, 27, 401-17.
——— 1980b. 'Palaeogeographic setting of the Wrae Limestone: an Ordovician submarine slide deposit in Tweeddale'. Scot. J. Geol., 16, 91-104.
———, McKERROW, W.S. & CASEY, D.M. 1982. 'The anatomy of a Lower Palaeozoic accretionary forearc: the Southern Uplands of Scotland'. In: LEGGETT, J.K. (ed.), Trench-Forearc Geology. Geol. Soc. London, Spec. Publ., 10, 495-520.
———, ——— & EALES, M.H. 1979a. 'The Southern Uplands of Scotland. A Lower Palaeozoic accretionary prism'. J. Geol. Soc. London, 136, 755-70.
———, ———, MORRIS, J.H., OLIVER, G.J.H. & PHILLIPS, W.E.A., 1979b. 'The north-western margin of the Iapetus Ocean'. In: HARRIS, A.L., HOLLAND, C.H. & LEAKE, B.E. (eds.), The Caledonides of the British Isles - reviewed. Geol. Soc. London, Spec. Publ., 8, 499-511
———, ——— & SOPER, N.J. 1983. 'A model for the crustal evolution of Southern Scotland'. Tectonics, 2, 187-210.
LUMSDEN, G.I., TULLOCH, W., HOWELLS, M.E. & DAVIES, A. 1967. 'The geology of the neighbourhood of Langholm'. Mem. Geol. Surv. Scotland, Sheet 11, 225pp.
LYONS, J.B., BOUDETTE, E.L. & ALEINIKOFF, J.N. 1982. 'The Avalonian and Gander Zones in Central Eastern New England'. Geol. Assoc. Canada, Special Paper, 24, 43-66.
McKERROW, W.S. 1982. 'The Northwest margin of the Iapetus Ocean during the Early Palaeozoic'. In: DRAKE, C.L. & WATKINS, J.S. (eds.), Studies in continental margin geology. Am. Assoc. Petrol. Geol. Memoir, 34, 521-533.
———, LAMBERT, R.St.J. & COCKS, L.R.M. 1985. 'The Ordovician, Silurian and Devonian Periods'. In: SNELLING, N.J. (ed.), Geochronology and the Geological Record. Geol. Soc. London, Spec. Publ., 20, 73-80.
MILLER, J.A., MATTHEWS, D.H. & ROBERTS, D.G. 1973. 'Rock of Grenville age from Rockall Bank'. Nature, Phys. Sci., 246, 61.
MILLER, R.G. & O'NIONS, R.K. 1984. 'The provenance and crustal residence ages of British Sediments in relation to palaeogeographic reconstructions'. Earth and Planet. Sci. Letters, 68, 459-470.

MITCHELL, A.H.G. 1984. 'The British Caledonides: interpretation from Cenozoic analogues'. Geol. Mag., 121, 35-46.
───────── & McKERROW, W.S. 1975. 'Analogous evolution of the Burma orogen and the Scottish Caledonides'. Bull. Geol. Soc. Am., 86, 305-315.
MORRIS, J.H. 1983. 'The stratigraphy of the Lower Palaeozoic rocks in the western end of the Longford-Down Inlier, Ireland'. J. Earth Sci. R. Dublin Soc., 5, 201-218.
O'NIONS, R.K., HAMILTON, P.J. & HOOKER, P.J. 1983. 'A Nd isotope investigation of sediments related to crustal development in the British Isles'. Earth and Planet Sci. Letters, 68, 459-470.
RUST, B.R. 1965a. 'The stratigraphy and structure of the Whithorn area of Wigtownshire, Scotland'. Scot. J. Geol., 1, 101-131.
───────── 1965b. 'The sedimentology and diagenesis of Silurian turbidites in south-east Wigtownshire, Scotland'. Scot. J. Geol., 1, 231-246.
THIRLWALL, M.F. 1981a. 'Implications for Caledonian plate tectonic models of chemical data from volcanic rocks of the British Old Red Sandstone'. J. Geol. Soc. London, 138, 123-38.
───────── 1981b. 'Peralkaline rhyolites from the Ordovician Tweeddale lavas, Peebleshire'. Geol. J. Chicago, 16, 41-4.
van BREEMEN, O. & PIASECKI, M.A.J. 1983. 'The Glen Kyllachy Granite and its bearing on the nature of the Caledonian Orogeny in Scotland'. J.Geol. Soc. London, 140, 47-62.
WALTON, E.K. 1955. 'Silurian Greywackes in Peebleshire'. Proc. R. Soc. Edinburgh, B65, 327-57.
WARREN, P.T. 1963. 'The petrography, sedimentation and provenance of the Wenlock Rocks near Hawick, Roxburghshire'. Trans. Geol. Soc. Edinburgh, 19, 225-55.
───────── 1964. 'The stratigraphy and structure of the Silurian rocks south-east of Hawick, Roxburghshire'. Q. J. Geol. Soc. London, 120, 192-222.

THE CONTRIBUTION OF THE FINNMARKIAN OROGENY TO THE FRAMEWORK OF THE SCANDINAVIAN CALEDONIDES

Donald M.Ramsay[1] and Brian A.Sturt[2]
1) Department of Geology, University of Dundee, Dundee,Scotland
2) Geological Institute, University of Bergen, Bergen,Norway

ABSTRACT. Researches in the last decade have demonstrated that the Scandinavian Caledonides are a complex montage built-up during three successive and distinctive orogenic cycles. The earliest of these – the Finnmarkian Orogeny had a unique significance in the evolutionary pattern. In Late Precambrian times the Laurento-Baltic super-craton was rifted and the opening and spreading of the Iapetus Ocean initiated. Clearly marked continental miogeoclines were developed on either sides of the ocean during the Cambrian, and major subduction was initiated, seaboard of Scandinavia, in late Cambrian times. Ensimatic island arc complexes were developed at this stage and the subduction polarity was arguably westward. This major stage of oceanic contraction resulted in the destruction of vast quantities of ocean floor by subduction and the obduction of major ophiolites onto the deforming rise prism. This latter was subjected to polyphasal deformation and metamorphism culminating in continentwards thrusting over the western part of the Baltic Shield. The type development for the Finnmarkian Orogeny is in the northern part of Norway, though evidence can be found throughout the belt, where either Finnmarkian metamorphics or ophiolites are truncated by a major first-order unconformity capped by Ordovician sediments of varied age. The morphogenic stage of the Finnmarkian Orogeny was essentially a feature of the Lower and Middle Ordovician.

THE CONTRIBUTION OF THE FINNMARKIAN OROGENY TO THE FRAMEWORK OF THE SCANDINAVIAN CALEDONIDES

Introduction

Finnmarkian (U.Cambrian - L.Ordovician) orogenic deformation was initially envisaged and defined as a phase, albeit a major polycomponent phase, of the Caledonian Orogeny (Ramsay & Sturt,1977, Sturt et al., 1978). However, after two decades of tectonostratigraphic investigation in Scandinavia it emerges that three separate orogenic cycles have contributed to the build up of the major orogenic belt along the western edge of the Baltic Craton. The Finnmarkian Orogeny brought to a close the first cycle, a cycle initiated in late Riphean times and characterised by rifting of the Laurentian-Baltic super plate and the accumulation of thick developments of fluviatile and shallow marine

sediments in the resulting basins and troughs. Final separation of the
Laurentian and Baltic plates was effected towards the end of the Vendian and
sea-floor spreading came into operation with a progressive opening of the
Iapetus Ocean. Along both margins of this ocean well-differentiated sedimentary
prisms built up throughout Cambrian times, which, on the Baltic margin, became
involved in the major orogenic convolution during late Cambrian to early Arenig
times. This Finnmarkian Orogeny was distinguished by polyphasal deformation
and metamorphism and the formation of a thick sequence of nappes.

Physiographically the orogen was expressed in an emergent upland area
sited to the west of the present cratonic margin, separated from the main
craton by broad epicontinental seas.

Distribution and Characteristics of Finnmarkian Orogenesis.
A. Finnmark.

The largest intact remnant of the Finnmarkian Orogen is preserved in
the counties of Finnmark and Troms, where in an area of some 40.000 km^2
the structure is one of extensive, flat-lying, allochthonous nappes (Kalak
Nappe Complex) resting on a substructure of smaller, less-allochthonous and
parautochtonous nappes (Fig.1). The allochthonous nappes each comprise a
basal segment of Precaledonian ortho- or paragneiss with an unconformable
cover of Riphean-Cambrian metasediments (Table 1) (Sørøy Succession, Ramsay,
1971). The penetrative tectonothermal fabric of the Kalak Nappe Complex
displays an upward increase in the metamorphic grade through the nappe pile
and is in essence a transported metamorphism. Beneath the allochthon, as
seen along the orogenic front and in the prominent tectonic windows (e.g.
Komagfjord and Alta), the lower complex of thin and less extensive nappes
contain Karelian, Karelian and Archean gneisses or little metamorphosed
Riphean-Vendian sediments (Fig.2).

Geological confirmation of the age of the deformation and the basic morphology of the nappes emerged with the recognition of unconformities at the base
of the 'Caledonian' cover sequence (Ramsay & Sturt,1978; Ramsay et al.,1979)
and within the Lower Palaeozoic metasediments of the upper nappes in N.Troms
(Minsaas & Sturt,1984; Ramsay et al.,1984). Recognition of these first-order
unconformities (Ramsay & Sturt in prep.) within the metamorphic allochthon
provided the important key to unravelling the complex polyorogenic history of the
Scandinavian Caledonides. The lower unconformity established the tectonostratigraphic significance of the Precaledonian gneiss horizons in the allochthon
and in so doing delineated the constituent nappes of the Kalak Nappe Complex.
The consistent occurrence of gneisses of cratonic origin as the substrate to the
nappes and the repetition of the same cover stratigraphy reveals an encratonic
setting for late Precambrian sedimentation and emphasized the great width of
the sedimentary prism which flanked the cratonic margin.

At the upper unconformity in several of the post-Finnmarkian nappes
e.g. Vaddas and Lyngen Nappes, Upper Ordovician rocks truncate large-scale
Finnmarkian folds in the subjacent sediments (Fig.3)(Ramsay et al.,1984).
The unconformity separates rock sequences with different tectonothermal
histories, and therefore provides an upper limit for the timing of the
Finnmarkian Orogeny. The repetition of this unconformity in the aerially
extensive nappes of the Upper and Uppermost Allochthons, transgressing a
Finnmarkian orogenic substrate in each case, reveals the 4-500 km wide area
of Ordovician-Silurian sedimentation, following the deep erosion of the

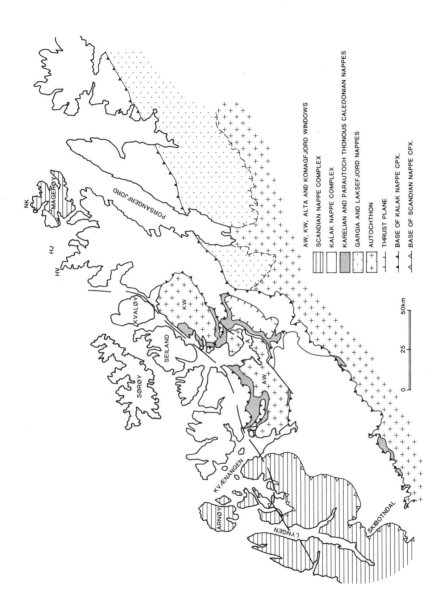

Fig. 1   Simplified map of the principal nappe complexes in Northern Norway.

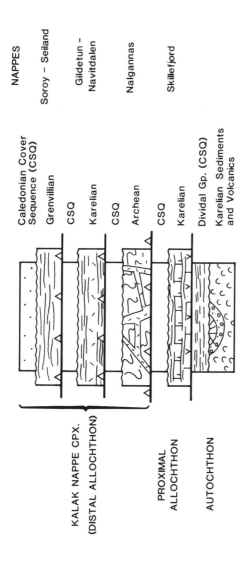

Fig. 2 Simplified tectonostratigraphy of the Finnmarkian nappes of northern Norway emphasizing the age of the pre-Caledonian substrate components.

Finnmarkian belt. More important perhaps, it provides an insight into the minimum dimensions of the original Finnmarkian Orogen along the western margin of the Craton.

B. South-west Norway

A comparable history of pre-Ashgillian deformation and metamorphism also characterises extensive tracts of S.W.Norway (Sturt et al.,1983). Within the Middle and Upper Allochthons mid-Ordovician to mid-Silurian rocks form unconformable cover sequences to more complexely deformed and metamorphosed substrates. Unlike Finnmark however, no obvious Finnmarkian thrusts or nappes have yet been identified. The most prevalent substrate in this region comprised various members of a major ophiolitic fragment(s) (Sturt & Thon, 1978; Sturt et al.,1979; Furnes et al.,1980) which are overprinted by a polyphase, pre-unconformity tectono-metamorphic fabric. Discussion of the regional significance of this will be deferred until later, but it is relevant to draw attention here to the pre-Ashgill West Karmøy Igneous Complex (WKIC), a sequence of post-orogenic acid plutons which intrude the ophiolite and its internal fabric on the island of Karmøy. Radiometric ages of ca. 460 m.a. (Mid-Ordovician) from these plutons indicates that the pre-intrusion orogenic fabric in the ophiolite is thus even older and probably Finnmarkian in age.

C. Central Norway

In the major segment of the belt between Bergen and Troms involvement of Ordovician-Silurian rocks in the major nappes of the chain, together with an absence of diagnostic older radiometric data, was thought to preclude the Finnmarkian as a significant contributor to the build up of the orogen. Consistent radiometric ages in the range 400-420 m.a. appeared to confirm a single late-Silurian deformation phase. As the Finnmarkian Orogeny was not anticipated evidence was not sought or discovered. However, in view of the nature and scale of observed Finnmarkian deformation in both northern and south-western Norway it is inconceivable that it should be confined to such restricted areas. It should be remembered also that events in these regions included regional metamorphism, polyphase folding and major nappe formation.

It is emerging however, that at least two and possibly three major orogenic cycles are involved in the tectonic framework of the whole belt; full cycles which included denudational, depositional and orogenic stages (Miyashiro,1982). In view of the role of nappe formation in the large-scale structural pattern of each orogeny it is not surprising that areas with well-preserved Finnmarkian fabrics are somewhat fragmentary and have been overlooked. The most fruitful approach to the identification of Finnmarkian activity depends heavily on finding direct or indirect evidence of unconformities. If, as in Finnmark and S.W. Norway, a major break does occur within the Lower Palaeozoic sequences, separating the effects of two discrete orogenic cycles, it is reasonable to expect it to be present throughout the belt. This first-order unconformity has indeed been identified in most of the nappes of the Helgeland Nappe Complex, (Oftedahl,1980) and the nappes of the Upper Allochthon in western Norway (Sturt & Thon,1976,1978) (Fig.4). As will be developed further this geological approach has proved to be most reliable approach in several of the nappe complexes and when used in conjunction with some key radiometric results has demonstrated

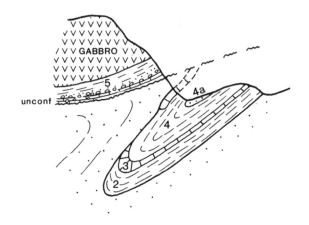

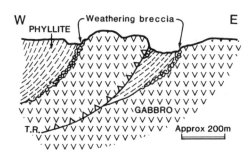

Fig. 3　The intra-Ordovician unconformity. (a) Structural relationships in the Vaddas area of N. Troms. Numbers 1-4a indicate members of the West Finnmark stratigraphy namely, Klubben, Stören, Falkenes and Aafjord Groups; 5 is the Upper Ordovician sequence. (b) Structural relationships at the upper surface of the Lyngen Gabbro.

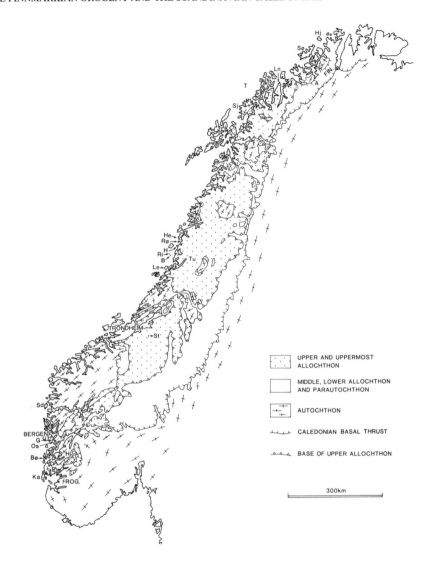

Fig. 4 Simplified tectonostratigraphic map of the
Scandinavian Caledonides. List of abbreviations:
A - Alta, B - Bolvaer, Bö - Bömlo, FIN - Finnmark,
G - Gullfjell, H - Hamnoen, He - Helgeland,
Hj - Hjelmsöy, Ka - Karmöy, K - Komagfjord, Le - Leka,
Ln - Lyngen, Ri - Risoen, ROG - Rogaland, Rö - Rödöy,
Sa - Samnanger, Se - Seiland, Sj - Senja, Sd - Solund,
Sö - Söröy, St - Stavfjord, St - Stören, T - Tromsö,
Tu - Tusenfjord.

the belt-length influence of Finnmarkian events.

Evidence of Finnmarkian Orogenesis

A. Geochronological

Finnmarkian tectonothermal activity was first identified on the basis of radiometric age dating. A suite of Rb/Sr isochron ages in the range 530-490 m.a. was obtained from members and products of the synorogenic Seiland Igneous Province and the metasedimentary cover sequence in the Kalak Nappe Complex, in addition to slates from the autochthon of East Finnmark (Sturt et al.,1978). The grouping of ages corresponds to the first (530 m.a.) and second (500-490 m.a.) phases of deformation in the metamorphic allochthon and D1 in the autochthon (505-485 m.a.), apparently similar to D2 in the allochthon and highlighting the diachronous nature of deformation between the interior and exterior segments of the belt. The later ages were confirmed by K-Ar ages on nepheline in the range 495-480 m.a. (Sturt et al.,1967) and recently $^{39}Ar/^{40}Ar$ mineral plateau ages for hornblende at ca. 510 m.a. have been obtained from shear zones in the western basement arch on Senja (Dallmeyer et al.,1984). Dallmeyer (pers. comm.1984) has obtained $^{39}Ar/^{40}Ar$ ages at c.475 Ma for slates from the autochthon.

Precaledonian ages obtained from gneisses in several of the Finnmarkian nappes, e.g. 2800 m.a. in the Skillefjord Nappe (Sturt & Austrheim, in press) and 1800 m.a. in the Carrovarre Nappe (Zwann & Quenardel,1980), confirmed the geological evidence of the basal Caledonian unconformity.

In the Sevé-Köli Nappe Complex of north-central Norway hornblendes from metabasic rocks have yielded $^{39}Ar/^{40}Ar$ plateau ages of 510-450 m.a. (Dallmeyer et al.,1984). These important results extend the influence of Finnmarkian activity to the largest nappe complex of the whole belt, and confirms the belt-long influence into the terrain which was subsequently to be overthrust by the Upper and Uppermost Allochthons, although the exact nature of the Finnmarkian framework in this sector is obscure. The full significance of these ages awaits some reappraisal of the tectonostratigraphy of the Sévé-Köli Complex. The Finnmarkian ages within the Köli are somewhat surprising and suggest that this nappe group is more complex than previously envisaged .

Finnmarkian ages (485 m.a.) have also been reported from mylonites in the Tennås Augen Gneiss, underlying the Särv Nappe of the Middle Allochthon (Claesson,1980). This is the lowest horizon in central Scandinavia from which evidence of the early orogeny has been obtained.

B. Geological Evidence

On the island of Seiland in West Finnmark (Fig.4) a major syn-D1 mylonite zone at the base of the gneissic substrate to the Söröy Nappe, e.g. the Eidvågeid Schist (Worthing,1971; Akselsen,1981; Ramsay & Sturt,1978) is intruded by a syn-D2 gabbroic member of the Seiland Igneous Province. Finnmarkian deformation therefore, must have involved large-scale thrusting, but the full extent of displacement is obscured by real or possible Scandian events. In central Norway the large-scale Finnmarkian framework is even more obscure, despite the growing bank of data on Finnmarkian events. For example, as mentioned earlier, mylonites (485 m.a.) beneath the Särv Nappe, at the base of the metamorphic allochthon underline Finnmarkian thrusting extending as far east as the provenance site of the Särv Nappe, i.e. a situation somewhat west of the present coastline.

The most widespread and convincing proof of Early Ordovician orogenesis in Norway however, is provided by the first-order unconformity within the Lower Palaeozoic metasedimentary sequences of the Upper and Uppermost Allochthons. The areal extent of this event is indicated by its recurrence in most of the constituent nappes of the afore-mentioned complexes. In several districts the surface of unconformity truncates large-scale orogenic folds in the Riphean-Vendian metasediments, or the ophiolitic suite of the substrate, confirming the presence of an earlier dynamothermal fabric in the older assemblages. This transgressive relationship has been observed, not only at widely separated points in the belt, but also in different nappes, e.g. Vaddas and Lyngen Nappes (Upper Allochthon) in Troms;Rödöy, Leka and Bolvaer in the Helgeland Nappe Complex (Uppermost Allochthon); the major Bergen Arc, and Bömlo and Karmöy in the Upper Allochthon of S.W.Norway (Fig.4).

The unconformity may be continental or marine in character. In the former case, where outcrop is sufficiently extensive, it may still be possible to distinguish some palaeorelief on the surface, as on the island of Leka in Helgeland and western Lyngen in Troms (Sturt et al.,1984;Minsaas & Sturt,1984). Marine unconfirmities on the other hand, are typically planar in form, succeeded by sediments which range in character from shallow marine to basinal.

The age of the surface is sometimes difficult to assess and a minimum age for the unconformity is provided only by age of the lowest members of the cover sequence. Conglomerates at or close to the base of the cover, whose pebbles may be closely linked to the substrate, suggest a close temporal connection between the unconformity and its immediate cover. Palaeontological evidence is not abundant but what does exist is fortunately widely distributed, confirming the belt-long development of this unconformity. In the central part of the belt closer stratigraphic control reveals a short time interval between orogeny, denudation and the lowest of the datable cover sequence, the Upper Middle Arenig Shales of the Hovin Gp. in the Trondheim Nappe Complex (Ryan et al., 1980). On Bömlo the folded ophiolitic substrate is truncated by Middle Ordovician bimodal continental volcanics which have been dated at 464 ± 7 and 467 ± 13 m.a. (Furnes et al.-1983). Elsewhere in the belt e.g. Finnmark-Troms, Helgeland, Rogaland (Fig.4), there may be a considerable time interval across the unconformity as the earliest faunally dated cover rocks are considerably younger i.e. Ashgillian. This age is provided by coral and brachiopod faunas in limestones, though these are at varying distances above the unconformity. Where similar relationships have been identified in the nappes of the Uppermost Allochthon however, but with an absence of fossils, the similarity of lithologies and relationships lead tentatively to similar conclusions and the suggestion of Upper Ordovician rocks characterising the cover. The different ages at different localities suggests a diachronous surface of unconformity.

In highly deformed orogenic terrains unconformities are frequently difficult to recognise and if not suspected may be overlooked entirely. This difficulty is especially acute when the tectonic fabrics on both sides of the structure are concordant, whether primary or by tectonic convergence. In less-deformed situations the original surface may be preserved, little modified from its original condition, and any original angular discordance between the two series is also preserved. While in many of the Scandinavian occurrences the ambiguous concordant relationship obtains there are a number of localities where the original relationships can be clearly seen. This is especially true in the situations of continental unconformity, e.g. Lyngen, Helgeland and Karmöy (Fig.4), where strain has not modified relationships significantly. At Lyngen and Leka it is possible to appreciate the irregularity of the original land surface with palaeorelief of several hundred metres at Lyngen (Minsaas & Sturt,1984; Sturt et al.1984). In addition to the fossil land surface in the Helgeland region there is a distinctive facies at the top of the substrate suggestive of a metamorphosed soil,(metasol). On the islands of Leka and Risoen and to a less well-defined extent on Rödöy and Hamnoen (Fig.4), this distinctive lithofacies is developed in the top 20 m of the substrate. In each case the protolith was ophiolithic, although drawn from different members of the pseudostratigraphy, e.g. greenstone, metagabbro and serpentinite.

The lithologies developed range from massive chlorite-chloritoid fels streaked with dolomite-quartz or dolomite-chlorite schists, to a very distinctive fabric characterised by a spheroidal fracture pattern. The inner cores of the latter have the character of core-stones in spheroidal weathering, varying from fresh to diapthorised protolith, while the enveloping skins are composed of soft chlorite schist or porphyroblastic chlorite-chloritoid schist. The tectonic effects of the Scandian Orogeny can be very pronounced in this zone of weak rocks and spheroidal cores may be distorted to flattened ellipsoids aligned parallel to the foliation.

Irregular patches, streaks or lenticular bands of dolomite are locally developed, some of which may contain ghosts of the original protolith. These carbonate bodies are reminiscent of caliche developments, implying an arid, subtropical climate. This deduction agrees with palaeolatitude determinations for Baltoscandia in L. Ordovician times deduced from the palaeontological record.

In these areas of continental unconformity the irregular land surface is progressively drowned beneath varying thicknesses of fluvial sand and gravel, which in turn pass upwards into shallow marine limestones and shales or proximal turbidites. In most of its preserved occurrences, including the higher nappes in the Helgeland Nappe Complex, the post-Finnmarkian unconformity in Norway is marine in character, succeeded by varied sequences of sandstones, calc shales and limestones, turbidites or polymict conglomerates.

Where the surface of unconformity is masked through the development of concordant fabrics it becomes necessary to contrast the lithologies or tectonothermal histories of the upper and lower sequences to distinguish any break in stratigraphic or tectonic continuity. Polymict conglomerates occur immediately above the surface of unconformity or a short distance up-sequence, at many locations between Vaddas in Troms and S.W.Norway. The fluvial conglomerates are normal clast-supported rocks with sandy matrix while, in the marine sequences these conglomerates may be clast or matrix supported often with a calcareous matrix.

These conglomerates provide important stratigraphic and tectonic data which not only confirm the primary structural relationships but may also, through their clast populations, provide the only evidence of the original stratigraphy in the substrate, now largely missing through erosion, and prove unequivocally the existence of preunconformity deformation in the substrate. In several nappes of the Helgeland Nappe Complex, but particularly well seen on the island of Leka and in Tosenfjord, the clast population includes representatives of the whole ophiolite pseudostratigraphy. This can also be seen in the Vaddas Nappe, the Stören Nappe of the Trondheim Nappe Complex and the Bergen Arcs of S.W.Norway, all members of the Upper Allochthon. In the Lyngen Nappe horizons of conglomerate, up sequence from the mafic basal conglomerates, contain pebbles of predominantly medium to high grade metasedimentary rocks, representative of the West Finnmark succession (Minsaas & Sturt,1984). Where tectonic convergence has produced concordant fabrics in both cover and substrate the actual surface of unconformity may be obscured and the polymict conglomerates may provide the only indication of the original stratigraphic relationships.

A high proportion of the clasts in these conglomerates display tectonic foliation (S_i) and sometimes folds, which may either be inherited Finnmarkian fabrics from the substrate or have been acquired in later Scandian deformation (Fig.5). The pebbles range in shape from irregular to ellipsoidal and in some situations these have been modified during later deformation to flattened ellipsoids and discs but in others they may be preserved unmodified. As foliation influenced many of the original pebble shapes it is frequently aligned parallel to the principal shape axes. With distortion of the pebbles in the Scandian foliation (S_e) the inherited S_i fabrics are concordant with S_e, although they may have been modified in the process. It is also possible of course that the pebble fabric is entirely a product of, if not merely modified by, Scandian strain. Concordant S_i fabrics therefore, are ambiguous in regard of pre-sedimentation tectonics. Where, however, the S_i fabric was originally oblique to the shape of the pebbles it may remain oblique during later deformation and still be recogniseable as an older fabric. In 'up sequence' conglomerates within the Lyngen Nappe for example, dimensionally orientated and flattened psammite clasts with transversely orientated banding (S_i) display folding of this banding with axial surfaces parallel to the plane of flattening and external foliation (S_e) (Fig.5b). These folds are geometrically compatible with the later, pebble flattening strain and are probably Scandian, although the S_i is pre-sedimentation in age. On the other hand, large irregular greenschist clasts within coarse breccias of the Havna Fm. on Leka preserve tight asymmetrical folds of an older foliation and these structures are partially truncated by the clast margins (Fig.5c). Both the schistosity and the folds are unrelated to the external Scandian fabric and indicate polyphasal, pre-unconformity tectonism. This is supported by the primary shape irregularities and lack of distortion of the clasts. In this deposit numerous pebbles with oblique foliation confirm the extent of the older fabric in the substrate.

In mafic conglomerates derived from the ophiolitic substrate care must be exercised in pebble analysis, however, especially with foliated plutonic rocks, as the pre-unconformity fabric could be the product of ocean-floor deformation rather than orogenic tectonism.

A strong line of evidence supporting the existence of a major post-orogenic unconformity and at the same time validating Finnmarkian orogenesis stems from comparative tectonothermal studies in the cover and substrate. In its most obvious form this is manifested by the truncation of large structures, primarily folds, in the substrate, and if the folds are associated with foliation and metamorphic reconstruction, then a significant tectonothermal event had occurred. This relationship has been reported from the Vaddas and Lyngen Nappes of Troms (Ramsay et al.,1984; Minsaas & Sturt,1984), Rödöy in the

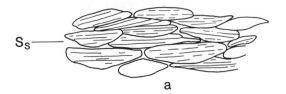

a

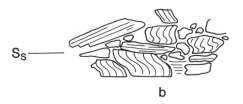

b

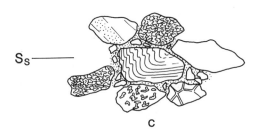

c

Fig. 5 Relationship of internal (possibly Finnmarkian) to external fabric in Upper Ordovician conglomerates at or close to the unconformity. (a) S_i is parallel to the principal shape fabric of the pebbles and the Scandian foliation (S_S). (b) S_i is oblique to S_S, although some of the intra-pebble folding is Scandian in age. (c) Folded foliation of pre-sedimentation age within a pebble in polymict mafic conglomerate.

Helgeland Nappe Complex and in the Upper Allochthon at Bömlo and Karmøy in Rogaland.

As the substrate has an independent as well as shared thermal history with the cover a distinctive hiatus may be identifiable where the two series are not isogradic. This thermal and tectonic distinction has now been recorded from widely separated localities along the length of the belt, e.g. Lyngen and Vaddas Nappes in Troms, Helgeland Nappe Complex, Solund in W.Norway and S.W.Norway (Karmøy and Samnanger).

Significance of the Unconformity

The tectonostratigraphic importance of the intra-Ordovician unconformity is its geological confirmation of the Finnmarkian Orogeny. Prior to this, as mentioned earlier, the only basis for the early orogeny was the radiometric ages from the Seiland Igneous Province and the autochthon of East Finnmark. The major stratigraphic break represented by this structure and the level of dedunation attained in the substrate affirm that the Finnmarkian Orogeny was the climax of a discrete orogenic cycle (Miyashiro,1982) and quite distinct from the more widely known Scandian (late Silurian) Orogeny.

The belt-long extent of the unconformity together with its repetition in a number of the major nappes make it a wide-ranging palaeogeographic and stratigraphic datum horizon. Although data is still sporadic and imperfect what exists can be assembled to distinguish broad palaeogeographic outlines, highlighted by local environmental and physiographic detail. The unconformity represents an interval of time during which a rising orogenic welt underwent deep dissection and the exposing of medium to high grade rocks at the ground surface. The relatively short time interval in which this was accomplished in central Scandinavia, i.e. before the upper Middle Arenig, gives a measure of the rapidity of uplift of the Finnmarkian Orogen. Elsewhere the oldest rocks of the cover are upper Ordovician (Ashgillian), suggesting an erosive surface of considerable longevity.

The role of block faulting in this uplift is suggested by the differences in the levels of related rock sequences which were exposed at the ancient land surface. On Karmøy, for example, erosion exposed granitoids of the WKIC, while on Bømlo 30 km to the north (Fig.4), a thick pile of Middle Ordovician acid-intermediate volcanics, the probable surface equivalents of the granitoids, occur immediately beneath the sub-Ashgillian unconformity, though themselves unconformable on the Lykling Ophiolite.

Repetition of the unconformity in so many nappes of the Upper and Uppermost Allochthons, e.g. Karmøy-Os, Stavfjord, Trondheim, Helgeland and Troms (Fig.4), make it an important marker horizon with great significance for inter-nappe correlation and lithostratigraphy. Within individual nappes the surface marks the base of the cover sequence and provides a sense of younging in the overlying assemblage. This has been especially valuable in the Helgeland Nappe Complex of the Uppermost Allochthon where high strains combined with rare way-up criteria made erection of a lithostratigraphy difficult (Fig.6). In addition to the unconformity the recurrence of ophiolitic rocks as the substrate established five nappes in the complex and enabled a stratigraphy to be erected for the cover series in each. Yet another important consequence of the unconformity is the stratigraphic constraint it imposes

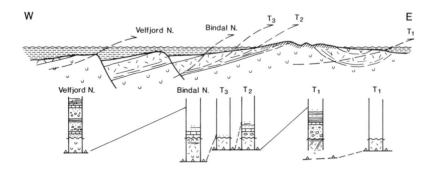

Fig. 6 Cartoon section depicting post-Finnmarkian erosion of the part of the Baltic Craton which was the provenance area of the Helgeland Nappe Complex. In the columns beneath the section the corrugated line denotes the unconformity developed on an ophiolitic substrate. Ornamentation: u - ultramafics, random dashes - gabbro, v - volcanic and hypabyssal segments, heavy oblique dashes - fossil soil, horizontal dashes - Upper Ordovician sequence. T1-3 are small nappes at base of the Helgeland Nappe Complex.

on the obduction of the Group I ophiolites (Sturt et al.,1984) throughout the
upper and uppermost nappes of coastal Norway. In the Trondheim Nappe
Complex the upper Middle Arenig age of the oldest dated sediments (Ryan et al.,
1982) in the Hovin Group, unconformably overlying the Storen Ophiolite together with the orogen fabric in the latter strongly indicates a Finnmarkian
obduction of this the uppermost nappe element in the Finnmarkian Belt.
Although the obduction surface is not seen the pre-unconformity emplacement
reinforces the view of the Finnmarkian as a thrust belt orogeny.

Elsewhere the unconformity is stratigraphically undatable with any certainty,
due to the absence of fossils, or it cannot be placed below Ashgillian in age.
In Bömlo-Karmöy where an upper Ordovician fauna has been obtained from limestone overlying thick fluvial conglomerates (Vikavågen Fm., Thon,1984) at the
base of the cover series it is clear that the obduction of the ophiolite is much
older. The conglomerate truncates the major ophiolite fragment and the
acid plutons of the WKIC which intrude it. As indicated earlier, radiometric
ages from one of these (Sturt & Thon,1978a) gave a Middle Ordovician age
(Ca.460 m.a.). This complex with its numerous xenoliths of cratonic gneiss
and ophiolite also post-dates the polyphasal orogenic fabric in the volcanic
layer of the ophiolite pseudostratigraphy and the conformably overlying
pelagic sediments of the Torvastad Gp. The ophiolite, therefore, was
obducted onto the Baltic craton and tectonised in the process prior to the
Middle Ordovician.

Form and Spatial Relationships of the Finnmarkian Orogenic Belt

Where it was first identified the Finnmarkian belt of northernmost Norway
is a sequence of extensive thin and flat-lying allochthonous nappes each comprising Caledonian metasediments (i.e. late Riphean - Upper Cambrian) and
an older Precambrian substrate. The repetition of the same stratigraphy
throughout the nappe stack, together with wide lateral persistence implies
that the provenance was an extensive stable margin to the Baltic Craton.
The continental character of the substrate in each of these nappes reveals
an encratonic setting for the whole sedimentary prism.

In the Vaddas Nappe the lower part of the sheet is similar in form to the
underlying Finnmarkian nappes but in its upper part Upper Ordovician sequences truncate large folds in the West Finnmarkian sequence. The polymict
conglomerates at or close to the base of the cover series contain a suite
of rocks representative of the W.Finnmark succession.

From Lyngen southwards there is a growing body of evidence to indicate
that the uppermost nappes in the Finnmarkian tectonostratigraphy were
ophiolitic in content. Over a distance of 1500 km one or other horizon of
the ophiolite occurs as the substrate or lower segment to most of the nappes
of the Upper and Uppermost Allochthon. These ophiolitic rocks reappear in
the clast population of the conglomerates in the lower part of the cover series
and many exhibit clear evidence of a pre-unconformity tectonic fabric.
Whether the ophiolite formed one major nappe or a nappe complex is not known
as a result of later dissection in the Scandian Orogeny and redistribution into
many of the nappes of the Upper and Uppermost Allochthon. For example,
in the five nappes so far identified, in the Helgeland Nappe Complex ophiolitic
rocks form the substrate beneath an Ordovician-Silurian cover. Scandian

reworking has largely obliterated original contact relationships between the
obducted ophiolite and the original cratonic substrate, although a few areas
permit speculation on this substrate, although the contact surfaces are now
Scandian thrust planes. For example, in the Trondheim Nappe Complex the
Storen ophiolite overlies the higher grade metasediments of the Gula Nappe;
in the Major Bergen Arc the Gullfjell Ophiolite overlies the Samnanger Complex
(Furnes et al.,1982). In each of these cases the substrate has been identified
as or compared with known Finnmarkian sequences.

It has veen demonstrated that Finnmarkian events have affected rocks
of the Middle to Uppermost Allochthon, along the entire length of the belt.
At present there is no evidence of Finnmarkian effects beneath the Middle
Allochthon in Central Scandinavia, although in Finnmark the first cleavage
in the autochthon has yielded an age of ca. 490 Ma. (Sturt et al.,1978).

This raises important questions concerning the setting and nature of this
orogenic activity, i.e. in addition to polyphasal deformation and metamorphism
did it involve nappe formation and if so, how far did the nappes travel cratonwards?

A non-balanced palinspastic restoration of the nappes of the Mid to
Uppermost Allochthon results in a broad apron of Ordovician-Silurian rocks
which unconformably overlie a nappe complex, with its front sited somewhere
to the west of the present coastline (Fig.7). Evidence of Finnmarkian thrusting
and nappe emplacement has already been presented and the case argued. It
is proposed that the nappe stack was sited to the west of the present coastline
save in Finnmark where it may have reached relatively close to its present
site (Fig.7). Orogenic events began in the more distal parts of the marginal prism
during the Upper Cambrian (530-550 Ma.) while undisturbed sedimentation
continued in East Finnmark (Sturt et al., 1978). It was not until the Lower
Arenig (490-500 Ma.), and the Second deformation phase in the allochthon.
that the nappes had penetrated sufficiently cratonwards to affect the foreland
of East Finnmark.

Scandian thrusts may have reactivated suitable orientated sections of
Finnmarkian thrusts, but it is believed in the main they were new structures
which sliced up the lower nappes together with their younger covers. It should
be possible to confirm this by investigation of the radiometric age of mylonitic
horizons within Scandian Nappes. The restriction of Group I Ophiolites to
the Upper and Uppermost Allochthons implies a provenance on the western
side of the Finnmarkian Orogen. How far east these Ophiolite Nappes extended
is not known but deep post-orogenic erosion probably removed them and restricted preservation to the west.

Sense of Lower Ordovician Plate Motion

No record of early Ordovician plate margins can be identified in the geological
record of Scandinavia and even the sense of plate motions is not known with any
certainty, although the growing body of data permits speculation. Earlier published models for Finnmarkian evolution assumed an eastward subduction beneath
the Baltic Craton (Ramsay,1973; Robins & Gardner, 1975) but the thinking behind
this was largely a backward extrapolation of later Scandian motions. On closer
scrutiny of the evidence however, this proposition is difficult to substantiate.
The change from a passive plate margin to a digesting one, following the onset
of ocean closure in late Cambrian times, must have effected dramatic changes

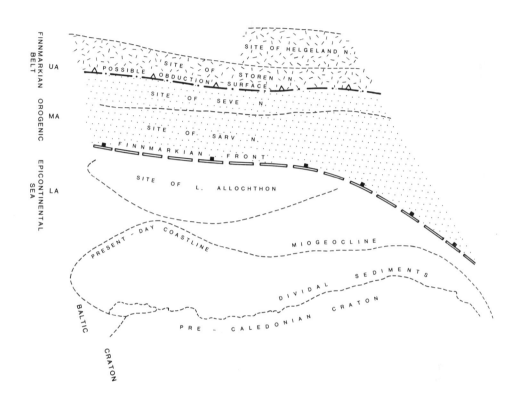

Fig. 7 Sketch map showing suggested provenance areas for the Scandian nappe complexes.

in geomorphology reflecting the new structural and volcano-sedimentological regime. In the stratigraphic record of the Tremadoc-Lower Arenig in Scandinavia however, no trace of such events can be identified. Sedimentation in the distal parts of the prism, as reflected in the sequences involved in orogenesis, ceased somewhere in the Mid-upper Cambrian, persisting into Tremadocian in the autochthon of East Finnmark.

There are unfortunately very few stratigraphic controls in the allochthon and in the Finnmark-Troms the only horizon that can be confidently dated is that of the tillite and this only from the Laksefjord Nappe. Previously it was considered that the Falkenes Marble of Söröy, could be dated as Lower Middle Cambrium on the basis of Archaeocyathids (Holland & Sturt, 1970), but the identification of archaeocyathids has subsequently been shown to be erroneous (Debrenne, 1984). In Central Norway control is provided by the Dictyonema Shale (Tremadoc) in the subunconformity sequence of the Trondheim Nappe (Grenne and Lagerblad, 1984). Radiometric ages of $530^{\pm}$ from anatactic products of the synorogenic Hasvik gabbro (Pringle and Sturt, 1969) dates the D1/2 interval and provides an upper time limit for the early stages of Finnmarkian orogenesis.

In none of the Riphean-Cambrian sequences is there any real hint or premonition of subduction nor any of the characteristic volcanogenic-plutonic products associated with it. There are no Tremadoc-Arenig volcanogenic developments, arc-related sediments or syn- to late orogenic acid plutons. However, in the Bömlo district of S.W. Norway a quartz keratophyre of the Geitung Group, a bimodal volcanic suite overlying the northward extension of the Karmöy Ophiolite, has yielded an age of $535^{\pm}45$ m.a. (Furnes et al., 1983). Similarly ensimatic arc volcanics of the Fundsjö Group in the eastern Trondheim area, underlying phyllites correlated with the Tremadocian Dictyonema Shale (Grenne & Lagerblad, 1984). This is indicative of the early stages in the formation of an island arc structure on top of the ocanic crust with possible westerly subduction polarity. The arguments for westerly subduction along the whole margin of the craton are somewhat negative, their greatest strength being the absence of criteria supporting sub-cratonic subduction. Positive support is suggested by the large-scale obduction of an extensive segment(s) of oceanic crust over the continental margin. It is contended that westerly subduction of Baltic Craton created Finnmarkian orogenesis, a feature supported by the similarity in ages of the known arc volcanism in Bömlo and Fundsjö and the metamorphism in the West Finnmark Nappes.

Perhaps the strongest argument in support of the above proposition is provided by a comparison of the pre- and post-Finnmarkian histories of the Baltic plate margin. The Riphean to Cambrian period is characterised by widespread lithofacial simplicity and uniformity and an absence of significant volcanic-plutonic activity. In the Ordovician-Silurian, on the other hand, from the earliest post-Finnmarkian records there is dramatic spatial and temporal variability of sedimentary environments i.e. continental, carbonate platform and deep basinal regimes together with abundant Island and Back Arc volcanic/plutonic developments. The post-Arenig history of the belt, therefore, is replete with the diagnostic features attributable to a margin developed in response to eastward subduction beneath the Baltic Craton.

A major tectonothermal disturbance broadly contemporaneous with the Finnmarkian is the Grampian Orogeny, affecting the late Riphean-Cambrian sequences of the Moine and Dalradian supergroups in the northern part of the

British Isles. Although developed on opposite sides of the Iapetus Ocean there are broad similarities in the evolutionary stages leading up to the orogeny. Cratonic rifting with thick fluvial and shallow marine deposits in the Riphean-Vendian, culminated in plate separation, ocean development and a widespread marine transgression in the L.Cambrian. By M.Cambrian major sedimentary prisms had developed along both margins terminating in the development of thick turbidite sequences.

Orogenesis along the margins, heralding the onset of oceanic closure, must have commenced at broadly the same time in both sectors. Although on opposite sides of the Iapetus Ocean, the major lithospheric changes which heralded change in the onset of closure found their expression in orogenesis along the cratonic margins. A major difference between the two sectors was the sense of subduction. The late Precambrian-early Cambrian age of old granites in the Scottish Highlands such as Carn Chuinneag and Portsoy and the Midland Valley derived granite now found as pebbles in upper Ordovician conglomerates suggest that sub-cratonic subduction had commenced well before the onset of orogenesis. In contrast there is no evidence of a similar pattern in the more extensive Finnmarkian belt, rather the indications are of westerly subduction.

The pattern of structural development in both belts during the ensuing orogeny appears to be dissimilar. In Norway activity was dominated by thrusting and major nappe formation just as it is in the later Scandian Orogeny. In Scotland on the other hand, the role of thrusting is obscure, as most thinking on Grampian deformation has been dominated by large scale folding. In recent years, however, evidence of thrusting during the Grampian deformation has been accumulating although the large scale significance of this is a matter for future study.

The post-orogenic history of these belts also follows a quite different pattern. In the British Isles there is no trace of post-Arenig-Silurian sediments on a Grampian substrate save along the Highland Boundary and here the mutual relations are a matter of some debate. In Norway on the other hand, Ordovician-Silurian rocks blanketing an eroded Finnmarkian base, and the unconformable contact forms a prominent marker in the later tectonostratigraphy. To some extent this may reflect some marked difference in the physiographic expression of the respective belts in post-orogenic times.

The Problems of Faunal Provincialism

No discussion of the Scandinavian Caledonides, even when restricted to the Finnmarkian component, would be complete without some comment on the upper nappes of the Trondheim Nappe Complex, the Gula and Storen Nappes, and faunal provincialism. This pair of nappes. especially the upper Storen Nappe with its well-documented Ordovician-Silurian stratigraphy and faunal assemblages of apparent American affinity, have been the focus of much debate and at the heart of several contradictory models for belt evolution. It has been argued that the Trondheim Nappe Complex represents an exotic flake of the Laurentian plate emplaced over the Baltic Plate during the major collision of the Scandian Orogeny (Stephens & Gee,1984; Bruton & Bockelie,1980). If this is indeed the case, then it should display a stratigraphy and tectono-thermal history compatible with flanking segments along the eastern margin of the Laurentian Plate.

The Storen Nappe comprises a basal ophiolite complex of oceanic affinity unconformably overlain by the Middle Arenig-Silurian Hovin and Horg Groups. A direct stratigraphic linkage between the oceanic substrate and the shallow marine cover is represented by the conglomerates at or close to the base of the cover sequence, i.e. the Lille Fundsjö conglomerate or its equivalents. The conglomerate is a shallow water mass-flow deposit composed of a clast suite representative of the ophiolite pseudostratigraphy and ensimatic material. This is succeeded by a succession of shallow water sandstones, shales and limestones. The geometry of the mid-Ordovician sedimentary and volcanic facies, together with evidence of derivation, is consistent with a basinal origin sited on the western margin of the Baltic Craton (Ryan & Skevington, 1982; Roberts et al.,1984).

The ophiolite and the ophiolite-derived clasts in the cover series possess a pre-unconformity fabric and a distribution suggestive of major displacements of a slice of oceanic crust which brought the complex to a structural level where it experienced erosion. This tectonostratigraphic association of ophiolitic substrate and Ordovician-Silurian cover, including the tectonothermal patterns, is a common one, as detailed earlier, throughout the Upper and Uppermost Allochthons along much of the lengt of coastal Norway. The main variable is the age of the oldest cover rocks and likely age of the unconformity. Outside the Storen Nappe anad Bömlö the only ages which can be confirmed are upper Ordovician, e.g. Ashgillian, as recorded from Troms and Karmöy. Karmöy however, provides the only concrete evidence of the structural setting of the ophiolite prior to the deposition of the cover sequence. As detailed earlier the xenolith population of the post-ophiolite West Karmöy Igneous Complex proves an encratonic setting had been established before mid Ordovician times. None of the other Group I Ophiolites of Scandinavia reveal the substrate of the ophiolite conclusively, as the base is always a later Scandian thrust plane. The pre-unconformity orogenic fabric which is common to the Storen and all other Group I Ophiolites tempts the proposition that they share a common history in the early Ordovician (Finnmarkian) orogenesis and were perhaps derived from a single slab or a series of related slabs from the same general provenance. The present distribution in several upper nappes is a function of the Scandian thrusting.

In translating the Storen Nappe with its American faunas the ophiolitic substrate was also being translated, which with its Finnmarkian fabric must have been in a high level situation prior to the shallow water Hovin sedimentation. If the Storen Nappe is American in origin then the possibility that the whole Upper Allochthon of Scandinavia is also American must be entertained. In addition the Uppermost Allochthon with its several extensive nappes which derived from a location to the west of the Upper Allochthon must also be considered a fragment of the Laurentian Plate.

The balance of structural, petrological and tectonostratigraphic evidence however, favours a provenance for the Trondheim Nappe Complex on the eastern side of the Iapetus Ocean, so the paradox presented by the palaeontological evidence remains and perhaps the position is best summed up by Ryan et al. (1980, "That the Lower and Mid Ordovician geology and palaeontology of Sör Tröndelag demonstrates that faunal provincialism cannot be solely explained in terms of tectonic controls."

Discussion

The Finnmarkian Orogeny (530-490 m.a.) was the first of three orogenic cycles whose combined effects have structured the Caledonide Orogen in Scandinavia. In common with the later cycles this one comprised a sedimentation, tectonothermal and denudation stage. From its first identification in Finnmark(Sturt et al., 1967; Ramsay & Sturt,1977; Sturt et al.,1978),and later in Hordaland and Rogaland (Furnes et al.,1979; Sturt & Thon, 1978b) evidence for this orogeny is steadily accumulating in the rest of the belt, within what is now the Middle, Upper and Uppermost Allochthons. The Finnmarkian Orogeny developed a sequence of major and minor nappes moving eastwards to take up a position west of the present coastline. A major zone of mylonites on Seiland predates syn-D2 plutons of the Seiland Igneous Complex, while in a more ambiguous setting in the central part of the belt, mylonites of the Särv Nappe yielded Finnmarkian ages (485 m.a., Claesson,1980). In Finnmark the effects of this deformation penetrates further east than elsewhere in the belt, to the autochthon of East Finnmark where first cleavage in the Stappogeidde Shales yielded a radiometric age of 490 m.a. (Sturt et al.,1978), i.e. D2 Finnmarkian.

It is in Finnmark alone that individual and group geometry and morphology of these nappes can be defined with any degree of certainty (Ramsay et al.,1984). The six major nappes which constitute the Kalak Nappe Complex each comprises a basal segment of Precaledonian gneiss with an unconformable cover of late Riphean to Cambrian sediments. A distinctive Caledonian stratigraphy can be recognised from nappe to nappe and presents the picture of a very wide pre-orogenic continental margin with uniform sedimentary environment.

South of N.Troms Finnmark assemblages underlie thick Ordovician-Silurian sequences and are sliced up with them into a sequence of new nappes during later Scandian Orogenesis. In this later deformation, while evidence of penetrative polyphasal Finnmakian strain within the nappes is abundant, its nature is fragmented and does not allow any large scale reconstruction of the morphology of Finnmarkian Nappes.

The detailed evidence of Finnmarkian tectonothermal activity is both geochronological and geological. Radiometric ages from different structural situations within the allochthon and autochthon (Sturt et al., 1967,1978) fall in the range 530-490 m.a. This spread encompasses two major deformation phases D1 and D2 and the time span of diachronous orogenic progression cratonwards from the interior to the foreland.

Geological evidence revolves around the major first-order unconformity between the pre- and post-Finnmarkian stratigraphic sequences. Where palaeontological evidence is available in the cover sequences a minimum age of Ashgillian can be proposed for the unconformity but in the Trondheim area the age of the break can be extended back to the Upper Arenig-Lower Llanvirn (Ryan et al.,1980). Over its extensive outcrop the unconformity represents a surface of both continental and marine erosion. This surface truncates Finnmarkian orogenic structures in the substrate and following later orogenesis marks a distinct hiatus in the metamorphic and tectonic histories. Where tectonic convergence has rendered structural relations ambiguous polymict conglomerates at or close to the unconformity assume great importance. The clast lithology and internal fabric reveal the lithostratigraphy, metamorphic state and structural history of the substrate prior to erosion.

The unconformity between the Finnmarkian assemblages and the cover sequences forms an important marker horizon enabling regional correlations both within and between nappes. Thus after palinspastic reconstruction the extent of the former Finnmarkian belt and the depth of its dissection can be appreciated. In addition, the surface of unconformity within each nappe forms a datum plane or base to the cover sequence, allowing the sense of younging to be identified and laying the basis for a lithostratigraphy where palaeontology or other sedimentological criteria are missing.

Restoring the Scandian nappes reveals a provenance for the Middle, Upper and Uppermost Allochthons in a sequence of zones sited west of the present coastline. This was the site of Ordovician sedimentation but also the final resting place of Finnmarkian orogenic displacements. At present evidence of Finnmarkian deformation does not extend beneath the Middle Allochthon, save in Finnmark, where it affects the autochthon. The pattern of internal strain together with the large-scale pattern, amplified by dating mylonites from central Scandinavia and the proof of ophiolitic obduction all point to the Finnmarkian Orogen being a thrust nappe belt broadly similar to the later Scandian one. Crustal thickening promoted considerable isostatic uplift and deep erosion to expose metamorphic rocks at the Ordovician land surface. The pattern of sedimentary facies in space and time implies that the early Ordovician palaeogeography was somewhat unstable and subject to rapid and significant change (Fig.8). Primarily the Finnmarkian nappes. capped by obducted ophiolite formed on archipelago to the west of the Baltic craton.

The repeated occurrence of ophiolitic rocks as the substrate to Ordovician-Silurian cover sequences in the nappes of the Upper and Uppermost Allochthons, along the west coast of Norway points to their being the uppermost members in the Finnmarkian nappe pile. The emplacement of this oceanic crust by a process of obduction can be approximately dated from the relationships of the post-ophiolite granitoids of the West Karmöy Igneous Complex which has a Mid-Ordovician age (Sturt & Thon,1978b; Sturt et al.,1984, and Sturt 1984) have argued for a westward subduction polarity during Finnmarkian orogenesis and present a variety of evidence in favour of this proposition. This sense of plate motion is reinforced by the absence of criteria which would support the counter notion of eastward subduction, namely, significant island arc volcanism in the Cambrian sequence, acid plutonism in the cover plate and a wide longitudinal and transverse uniformity and simplicity of sedimentary facies.

Establishing polarity of subduction leads on to the question of the site of the digesting plate margin. This problem has been heightened to controversy by the arguments of faunal provincialism. A significant number of species from several orders suggest an affinity with the Laurentian plate, at least during the Lower Ordovician. This led to the proposition that the Trondheim Nappe Complex is a flake from the Laurentian plate which transgressed the plate boundary during the Scandian collision and as a result of subduction of the Baltic beneath the American plate. This model would appear to conflict with the one based on a wide range of geological evidence which calls for easterly subduction beneath the Baltic Craton. These arguments on plate motion, supported by varying amounts of data relate to the Scandian Orogeny, but is this relevant to the Finnmarkian Orogeny?

If the model based on faunal provincialism is accepted the substrate to the Ordovician sedimentation in the Storen Nappe, i.e. the Storen Ophiolite was initially somewhere on the eastern seaboard of the American continent. This in its turn suggests that the ophiolite must have been obducted to elevate the ocean floor sequences to the shallow marine setting of the U.Arenig - L.Llanvirn. The complex pre-unconformity tectonic fabric in the ophiolite reflects orogenic activity for which there is no convincing evidence in Greenland. If the size of this oceanic complex was much greater than that of the present Storen Ophiolite as suggested in the lithostratigraphy of the Uppermost Allochthon it becomes more difficult to isolate the fabric-producing episode from the history of Greenland.

On the other hand, the tectonostratigraphic relations between Ordovician cover and ophiolite substrate, together with the tectonic fabric in this substrate, have striking similarities with all other Group I Ophiolite occurrences in Norway. The balance of evidence can be used to build a model of Finnmarkian orogenic evolution in terms of westerly subduction of the Baltic plate beneath an arc-trench system located in the eastern Iapetus Ocean. This promoted the asymmetrical pattern of cratonwards directed nappes, culminating in the obduction of oceanic crust as a major flake or series of flakes.

The Scandinavian Caledonides are a montage of several orogenic cycles sited along the western edge of the Baltic Craton. Each orogeny effected a progressive eastwards translation and interleaving of older and younger sequences during the course of nappe stacking. The Finmarkian Orogeny being the first, was largely responsible for the intimate involvement of major Precaledonian basement horizons, and oceanic crust, in the tectonostratigraphic framework of the belt.

References

Akselsen,J.1981: Unpublished Cand.Real.Thesis,
University of Bergen.
Austrheim,H.; Amaliksen,K.G. and Nordås,J.1983: 'Evidence for an incipient
early Caledonian (Cambrian) orogenic phase in south-west
Norway.'
Geol.Mag.
Bruton,D. and Bockelie,J.F.,1980: 'Geology and palaeontology of the
Hölonda area, western Norway - a fragment of North America?'
In: Wones,D.R. (ed.) The Caledonides in the U.S.A. Virginia
Polytechnic Inst. and State Univ., Dept. of Geol.Sci.Mem.**2**,
pp 41-7.
Claesson,S.,1980: 'A Rb-Sr isotope study of granitoids and related mylonites
in the Tännäs Augen Gneiss Nappe, southern Swedish Caledonides'.
Geol.fören.Stockholm, Förh.102,pp 403-420.
Dallmeyer, R.D.S.; Gee,D.G. and Beckholmen,M.1983: $^{40}Ar/^{39}Ar$ evidence of
superposed metamorphism in the Central Scandinavian
Caledonides, Jämtland, Sweden'(Abs.).
Geol.Soc. Amer. SE section meeting.
Full 1983, p 553.
Debrenne ,F.,1984; 'Archaeocyatha from the Caledonian rocks of Söröy
North Norway - a doubtful record.
Norsk.geol.Tidsskr. **64**,pp 153-154.
Furnes,H.;Roberts,D.;Sturt,B.A.; Thon,A. and Gale,G.H. 1980:'Ophiolite
fragments in the Scandinavian Caledonides.'In: Panayiouton,A.
(ed.). Ophiolites. Proc.Int.Ophiolite Symp.Cyprus 1974 582-600.
Furnes,H.; Thon.A.; Nordås,J. and Garmann,L.B.1982: 'Geochemistry of
Caledonian metabasalts from some Norwegian ophiolite
fragments. Contrib.Mineral. Petrol.**79**.pp 295-307.
Grenne, T. and Lagerblad,B.1984: 'The Fundsjö Group, central Norway -
A Lower Palaeozoic island arc sequence: geochemistry and
regional implications.
In: Gee,D.G., and Sturt, B.A.(eds.). The Caledonide Orogen -
Scandinavia and Related Aeas ,pp 745-760.
John Wiley, New York.
Holland,C.H.; and Sturt,B.A. 1970: 'On the occurrence of archaeocyathids in
the Caledonian metamorphic rocks of Söröy, and their
stratigraphical significance.
Norsk.geol.Tidsskr.**50**,pp 345-55.
Minsaas,O.; and Sturt,B.A.1984:'The Ordovician clastic sequence immediately
overlying the Lyngen gabbro complex, and its environmental
significance.'
In: Gee,D.G. and Sturt,B.A. (eds.) The caledonide Orogen -
Scandinavia and Related Areas , pp 379-393.
John Wiley,New York.
Miyashiro,A; Aki,K. and Sengor,A.M.A. 1982: 'Orogeny '
John Wiley,New York.
Oftedahl,C.1980: 'Geology of Norway'
Norges geol. Unders. **356**, pp 3-114.

Pringle,I.R. and Sturt,B.A.1969: 'The age of the peak of the Caledonian
Orogeny in west Finnmark, north Norway.
Norsk geol. Tidsskr.**49**,pp 435-436.
Quernardel,J.M. and Zwaan.K.B.1980:'Evolution palaeogeodynamique
d'un segment des Caledonides Scandinaves septentrionales.'
26.Int.Geol.Cong.Abstr.,1.p.379. Paris.
Ramsay,D.M. 1971: 'Stratigraphy of Söröy.
Norges geol.Unders.**269**, 314-317.
" 1973:'Possible existence of a stillborn marginal ocean in the
Caledonian orogenic belt of northwest Norway.'
Nature Phys.Sci.**245**,107-109.
" and Sturt,B.A.,1976:'The syn-metamorphic emplacement of the
Mageröy Nappe.
Norsk geol.Tidsk.,**56**,291-307.
" 1977: 'A sub-Caledonian unconformity within the Finnmarkian
nappe sequence and its regional significance'.
Norges geol.Unders.**334**,107-106.
" and Andersen,T.B.1979:'The sub-Caledonian Unconformity on
Hjelmsöy - New Evidence of Primary Basement/Cover
Relations in the Finnmarkian Nappe Sequence.'
Norges geol.Unders..**351**,1-12.
" Zwaan,K.B. and Roberts,D.1984:'Caledonides of
northern Norway.'In. Gee,D.G. and Sturt,B.A. (eds.)
The Caledonide Orogen - Scandinavia and Related Areas.
pp 163-184. JOhn Wiley,New York.
Roberts,D.;Sturt,B.A. and Furnes,H.1984:'Volcanite assemblages and
environments in the Scandinavian Caledonides and the
sequential development history of the mountain belt.'
In:Gee,D.G. and Sturt,B.A.(eds.).The Caledonide Orogen -
Scandinavia and Related Areas,pp 919-929.
John Wiley,New York.
Robins,B. and Gardner,P.M.1975:'The magmatic evolution of the Seiland
Province, and the Calaedonian plate boundaries in northern
Norway.
Earth planet Sci.Lett.**26**, pp 167-178.
Ryan,P.D.;Williams,D.M. and Skevington,D. 1980: 'A revised interpretation of
the Ordovician stratigraphy of Sör-Tröndelag, and its im-
plications for the evolution of the Scandinavian Caledonides.'
In.Wones,D.R. (ed.). The Caledonides in the U.S.A.
Virginian Polytechnic Inst. and State Univ. Dept.of
Geol.Sci.Mem.**2**, pp . 99-106.
Stephens M.B. and Gee,D.G.1984:'A plate tectonic model for Caledonian
orogenesis in the central Scandinavian Caledonides'.
In. Gee,D.G. and Sturt,B.A.(eds.) The Caledonide Orogen -
Scandinavia and Related Areas. pp 953-978.
JOhn Wiley, New York.
Sturt,B.A.1984:'The accretion of Ophiolitic Terranes in the Scandinavian
Caledonides.'
Geol.en.Mijenbouw.**63**,pp 201-212.
" ,Miller,J.A. and Fitch,F.J.1967:'The age of alkaline rocks from
West Finnmark,Northern Norway, and their bearing on the

dating of the Caledonian Orogeny.'
Norsk geol.Tidssk.**47**,pp 255-273.

Sturt,B.A. & Thon,A.,1976:'A comment on The age of the orogenic deformation in the Swedish Caledonides'.
Amer.J.Sci. **276**,385-390.

" and Thon,A.1978a:'An ophiolite complex of probable early Caledonian age discovered on Karmöy.'
Nature,Lond.,**275**,pp.538-9.

" 1978b:'A major early Caledonian igneous complex and a profound unconformity in the Lower Palaeozoic sequence of Karmöy, southwest Norway.'
Norsk geol.Tidssk.**58**,pp 224-

" Pringle,I.R. and Ramsay,D.M.1978:'The Finnmarkian Phase of the Caledonian Orogeny.'
J.geol.Soc.London. **135**,pp 597-610.

Sturt,B.A.;Thon,A. and Furnes,H.1979:'The Karmöy Ophiolite, southwest Norway.'
Geology,**7**,pp 316-320.

" ;Roberts,D. and Furnes,H.1984:'A conspectus of Scandinavian Ophiolites.'
In.Gass,I.G.;Lippard,S.J. and Shelton,A.W.(eds.).
Ophiolites and Oceanic Lithosphere. Geol.Soc.Lond. Sp. Pub.**13**,pp 381-391.

Thon,A.1984: 'The Gullfjellet Ophiolite Complex and the structural evolution of the Major Bergen Arc, west Norwegian Caledonides'.
In: Gee,D.G. and Sturt,B.A. (eds.).
The Caledonide orogen -Scandinavia and Related Areas,
pp 671-677. John Wiley,New York.

Worthing,M.,1971: Unpublished Ph.D.Thesis.
Univ.London.

THE MOINE THRUST ZONE: A COMPARISON WITH APPALACHIAN FAULTS AND THE STRUCTURE OF OROGENIC BELTS

R. D. Hatcher, Jr.
Dept. of Geology
University of South Carolina
Columbia, SC 29208
USA

ABSTRACT. The Moine Thrust transported rocks of the Moine Series over a thin foreland sequence overlying continental Lewisian basement (1.75 Ga.). Mylonitic rocks were generated below or within the brittle-ductile transition zone and later (following uplift?) transported northwestward along the brittle Moine Thrust. Deformation in the underlying foreland sequence took place in classic thin-skinned style, with some additional penetrative strain, producing typical detachment-ramp and duplex structures resulting in a southeastward-thickening wedge in weak and strong lower Paleozoic sedimentary units. Within part of the Moine Thrust zone, intermediate level thrusts, e.g., the Arnabol, transported Lewisian basement into the thrust pile along low angle thrusts, which exhibit thin-skinned behaviour as well.
 The Moine Thrust complex may be compared to large Appalachian faults, e.g., the Brevard, Towaliga, and possibly the Lake Char, which exhibit comparable histories of early deformation producing mylonites followed by final emplacement along a brittle fault. The foreland-intermediate level-crystalline mylonite thrust transition illustrated by the Moine Thrust may also be used as a smaller scale model of the foreland to metamorphic core transitions that occur on a much larger scale in other orogenic belts, e.g., the Alps, southern and central Appalachians and Canadian Cordillera.

"... in the district between Eriboll and Assynt the whole Silurian succession from the basal breccia to the lowest limestone occurs repeatedly above the first great thrust plane, separated by wedges of highly sheared gneiss." B. N. Peach and J. Horne (1885)

INTRODUCTION

The Moine Thrust (Fig. 1) is considered by many to be the classic thrust fault. It was one of the first discovered and carefully documented owing to the work of Peach, Horne, Clough and others of Her Majesty's Geological Survey in the late 19th Century, followed by the work of many others which continues through the present.

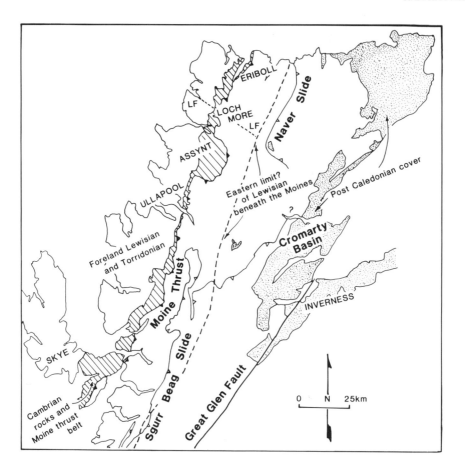

Figure 1. Moine Thrust zone in northwest Scotland (after Butler and Coward, 1984).

One of the interesting things about the work of Peach and Horne is that it was carried out objectively, particularly since this time produced abundant discoveries relating the nature and transport of rocks in thrust sheets (Bailey, 1935a).

Studies in the Moine Thrust may be classified into several categories. The classical studies of Peach and Horne (1885; Peach and others, 1907) on the geometry of the thrust were accompanied by the first recognition by Lapworth (1885) of the significance and the nature of mylonites. Lapworth even recognized the ductile character of mylonitic deformation even though his emphasis upon the brittle character of deformation and the association with faults as brittle

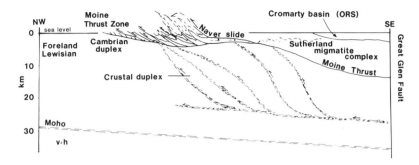

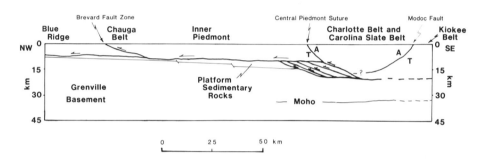

Figure 2. Crustal section (upper) through the Moine Thrust zone after Butler and Coward (1984) showing the deep structure from seismic reflection data beneath the thrust zone. Lower section is drawn through the southern Appalachians based upon surface, potential field and seismic reflection data.

structures has tended to cloud that observation. Bailey (1935b) undertook a study of the Glencoul thrust in the Assynt district in order to study the geometry of the thrust in that area.

Modern studies may be divided into three categories, including the studies of mylonite by Christie (1960) and Johnson (1967), studies of geometry of the thrusts in the Moine complex by Coward (1980, 1982), Elliott and Johnson (1980), Butler (1982), Boyer and Elliott (1982), Butler and Coward (1984), and Soper and Barber (1982). In addition, studies of deformation and strain in the rocks along the Moine Thrust zone are numerous, including studies by Johnson (1960), Chrisite (1963), Coward and Kim (1981), Butler (1982) and Fischer and Coward

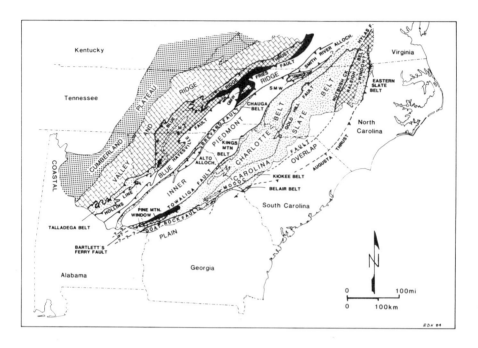

Figure 3. Map of the southern Appalachians showing the major subdivisions and the major faults. GMW - Grandfather Mountain window. SMW - Sauratown Mountains anticlinorium. Grenville basement is patterned black.

Figure 4. Schematic section through the southern Appalachians in Georgia showing the times of relative movement and reactivation (1 oldest, 4 youngest) of major faults. BR - Blue Ridge fault. HL - Hollins Line fault. AE - Allatoona-Enitachopco fault. B - Brevard fault. T - Towaliga fault. GR - Goat Rock fault. Heavy stipple is Avalonian rocks, light stipple is quartzite, the cross-hatch pattern is allochthonous Grenville basement.

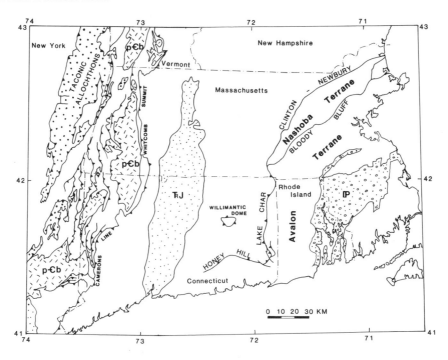

Figure 5. Map of southern New England showing the distribution of major faults, Grenville basement (pЄb), Pennsylvanian sediments of the Narragansett Basin (P) and Triassic-Jurassic sediments (TJ).

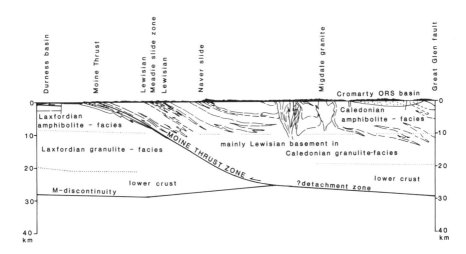

Figure 6. Crustal section through the Moine Thrust zone from Soper and Barber (1982).

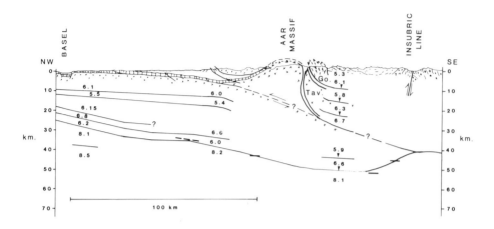

Figure 7. Crustal section through the Alps after Hsu (1979).

1982). Careful early mapping of the Moine Thrust zone, particularly in the Assynt region, together with its excellent exposure and accessibility, probably account for the large numbers of studies which have been conducted on this structure. Also, considering that it is the classic thrust fault instills in many geologists the desire to conduct studies related to the numerous problems which remain within the structure.

Research by the writer in the Appalachians related to thrust sheets and basement/cover interaction has been sponsored by National Science Foundation grants GA-20321, EAR-7615564, EAR-7826316, EAR-7911802, EAR-8018507, EAR-810852, EAR-8206944 and EAR-8212875. This support is gratefully acknowledged.

PROPERTIES OF THE MOINE THRUST ZONE

The Moine Thrust zone consists of a complex of northwest-directed thrusts which become younger westward (Butler and Coward, 1984). These thrusts apparently propagated from a relatively deep higher temperature zone transporting rocks which were deformed ductilely over cooler and more brittle rocks (Fig. 2). Thrusts propagated into a thin Cambro-Ordovician stratigraphy (380 m) which rests above the late Precambrian Torridonian succession. The rocks involved in the thrust zone are as young as the Ordovician Durness Limestone and as old as the 2.6 Ga Lewisian basement rocks. Basement is not involved in thrusting in the Moine Thrust zone along its northwestern edge, whereas in the transition zone from the frontal thrusts into the more internal Moine

Thrust, basement is involved in a number of the thrusts, including the Arnabol, Kinlochewe, Glencoul, Kishorn and Tarskavaig thrusts, and in the Moine Thrust zone as well (Soper and Barber, 1982).

Another property exhibited by rocks of the Moine Thrust zone is the penetrative strain present in the rocks of the foreland succession in the footwall sequence. Particularly notable is the strain which has been studied in the Cambrian Piperock quartzite underlying the Moine Thrust (Coward and Kim, 1981). Strain here is quite variable depending upon position in the pile from almost no strain in the more external units to considerable flattening strain in the more internal thrust sheets (Coward and Kim, 1981).

Much of the movement of the Moine Thrust can be related to the Grampian event which apparently began in the Cambrian and continued into the Ordovician (Butler and Coward, 1984). However, since the Ordovician Durness limestone is involved here, the more brittle events associated with the Moine Thrust deformation are probably Ordovician. Also, the Loch Borralan intrusive is involved in Moine thrusting. Its Pb-Pb age is 430 Ma. (van Breemen and others, 1979); consequently, movement on the Moine Thrust must be post 430 Ma. If the concept of the northwestward propagating character of the thrust mass is correct, Ordovician deformational events could have occurred while some of the foreland units were being deposited and shortly afterward were overthrust by the Moine Thrust zone. The ductile events therefore could have occurred earlier during the Cambrian.

COMPARISON TO LARGE APPALACHIAN FAULTS

The purpose of this section is to draw comaprisons between the Moine Thrust zone and some of the large faults which occur in the Appalachians. Several contain similar structures and lithologic types which would suggest that comparisons may be in order.

The Brevard fault zone extends from northern North Carolina to Alabama and contains a succession of stratigraphic units all of which are metamorphosed and mylonitized to varying degrees (Fig. 3). It has been suggested that ductile deformation in the Brevard zone occurred at least during two major events, perhaps separated by as much as 100 Ma. and then later succeeded by an event of brittle deformation (Hatcher, 1978). The Brevard fault zone, like the Moine, has been suggested (Odom and Fulalgar, 1973; Rankin, 1975) to be a suture. Yet, many who have conducted field studies along it have suggested that the Brevard cannot be a suture between two large continental blocks, and evidence continues to accumulate that there is no suture here. Ductile mylonites may have begun forming as early as 480 Ma. ago (Sinha and Glover, 1978). Mylonites also equilibrated Rb and Sr isotopes about 360 Ma ago (Odom and Fullagar, 1973). A series of events affected this fault zone which may be temporally unrelated to each other but resulted in some degree from stratigraphic localization of faulting. Further stratigraphic control of the later brittle deformational events may be demonstrated by study of the rocks into which the fault zone was emplaced. The brittle event also served to remove material from the footwall and carry it into the fault zone where it is exposed at the present level of erosion. This material is representative of basement beneath the Blue Ridge-Inner Piedmont thrust sheet as well as the foreland sequence of carbonates and possibly shales. The foreland exotic rocks are correlative with rocks of the exposed Valley and Ridge province of the Appalachians to the west.

The principal difference between the Brevard fault zone and the Moine Thrust is that the Brevard is not developed along the edge of the old foreland, nor does it juxtapose an offshore sequence against the foreland assemblage. The Brevard fault zone is developed within the Composite Blue Ridge-Inner Piedmont thrust sheet and has been interpreted as a backlimb thrust in the sense of a fault propagating back of the leading outcrop trace of a thrust into or through the thrust sheet (Hatcher, 1971).

The Towaliga-Goat Rock-Barletts Ferry fault system of the central and southern Piedmont of Alabama and Georgia is an assemblage of mylonitic rocks which is the result of series of events occurring in early to late Paleozoic time (Fig. 3). They may be part of the large complex of thrusts and strike-slip faults along which the entire southern Appalachian orogen was transported during the Paleozoic (Fig. 4). However, along the Towaliga fault, a brittle overprint occurs along the northwest boundary of the mylonite zone.

The principal similarity between the Towaliga fault zone and the Moine Thrust zone is that both are large mylonite complexes. However, the Towaliga is not developed adjacent to the foreland nor does it overthrust the foreland succession. It may be principally a strike-slip fault and is probably a major suture along most of its extent, except towards the southwest in western central Georgia and Alabama.

The Lake Char fault of southeastern New England (Fig. 5) is an extensive low angle thrust that connects into the steeply-dipping Bloody Bluff-Clinton-Newberrry fault system to the north (Dixon and Ludgren, 1968; Ludgren and Ebblin, 1973). The Lake Char is largely a west-dipping, low angle fault system along which a window may be developed in the Willamantic Dome (wintsch, 1979). This fault system separates Avalon terrane rocks from the Nesboba block in southeastern New England (Hall and Robinson, 1982). Goldstein (1982) suggested from study of mesoscopic structures that the Lake Char fault was emplaced from the east despite its dip towards the west. Comparisons to the Moine Thrust zone are difficult to make, except for the occurrence of extensive mylonites along both faults. The Lake Char is not a foreland related fault, unless it is the master sole fault beneath the entire orogen to the west. Rb-Sr studies indicate its time of movement may be as young as Permian (Alleghanian) (O'Hara and Gromet, 1983).

COMPARISONS TO STRUCTURE OF OROGENIC BELTS

The Moine Thrust zone may be readily considered as a model for larger scale development of the structure of orogenic belts. The propagation of the Moine Thrust from great depths, possibly as deep as along the Moho (Soper and Barber, 1982; Butler and Coward, 1984) indicate that this is a very large and fundamental fault in the British Caledonides, despite the fact that most do not consider it a suture (Figs. 2 and 6). Similar things have been said about the Brevard fault zone, although it is not considered to be a fault that originated from the Moho depths. I would suggest that the Moine Thrust developed not at Moho depths but at the depths of the ductile-brittle transition zone at its time of formation. The actual depth would be determined by the thermal conditions in the crust at the time it formed.

The crustal structure of the Alps may be compared with the Moine Thrust. Hsu (1979) suggested that the late stage (NeoAlpine) development of the Alps is related to a thrust that propagated from the ductile- brittle transition zone, or possibly the Moho, producing the Jura Mountains structures while further deforming and shortening a portion of the crust (Fig. 7). This structure formed long after Africa was sutured to Europe along the Insubric line.

Details of structure within the Alps differ considerably from those in other orogenic terranes. For example, there is no clearly definable attached foreland fold and thrust belt in the Alps. The Jura Mountains are separated from the Alps by the Molasse Basin. The Helvetic Alps contain properties of both foreland and more internal zones.

The transition zone from the external foreland to the internal metamorphic core of the Moine thrust zone is comparable with ease to the large thrusts which transport metamorphosed rocks over the rocks of the transition zone or the foreland in the Alps, the Appalachians, and the Canadian Cordillera.

Direct comparisons of the Appalachians and the Canadian Cordillera may be made using the Moine Thrust zone as a model. The imbricate zone beneath the large thrust complex and developed in the Cambro-Ordovician stratigraphy beneath the Moine Thrust is analogous to the Valley and Ridge and Canadian Rocky Mountains foreland fold and thrust belts. The transition zone containing the Arnabol, and other thrusts which transport basement are comparable to the zone where the external basement massifs occur in the Alps (Trumpy, 1960), Appalachians (Hatcher, 1983) and in the Canadian Cordillera (Campbell, 1973).

CONCLUSIONS

1. Studies along the Moine Thrust zone may be grouped into several categories: classical, geometric, deformational and strain, and studies of mylonites.
2. Mylonites along several Appalachian faults, the Brevard, Towaliga, Lake Char-Honey Hill are similar in many ways to Moine mylonites, since all formed at considerable depths, and most are retrograde.
3. The Moine Thrust zone may be considered a model for orogenic terranes illustrating the transition from an unmetamorphosed outer foreland to the polyphase deformed internal zone of the orogen.

REFERENCES

Bailey, E.B., 1935a. 'Tectonic essays, mainly Alpine'. Oxford Clarendon Press, 200 p.

―――― 1935b. 'The Glencoul nappe and the Assynt culmination'. Geol. Mag., 72, 151-65.

Boyer, S. & Elliott, D. 1982. 'Thrust systems'. Am. Assoc. Petroleum Geologists Bull., 66, 1196-230.

Butler, R.W.H. 1982. 'A structural analysis of the Moine Thrust zone between Loch Eriboll and Foinaven, northwest Scotland'. J. Struct. Geol., 4, 19-29.

Butler, R.W.H & Coward, M.P. 1984. 'Geological constraints, structural evolution, and deep geology of the northwest Scottish Caledonides'. Tectonics, 3, 347-65.

Campbell, R.B. 1973. 'Structural cross-section and tectonic model of the southeastern Canadian Cordillera'. Canadian J. Earth Sciences, 10, 1607-20.

Christie, J.M. 1960. 'Mylonitic rocks of the Moine Thrust zone in the Assynt region, northwest Scotland'. Trans. Geol. Soc. Edinburgh 18, 79-93.

―――― 1963. 'The Moine Thrust zone in the Assynt region, northwest Scotland'. Univ. Calif. Pub. in Geol. Sci., 40, 345-440.

Coward, M.P. 1980. 'The Caledonian Thrust and shear zones of northwest Scotland'. J. Struct. Geol., 2, 11-7.

―――― 1982. 'Surge zones in the Moine Thrust zone of northwest Scotland'. J. Struct. Geol., 4, 247-56.

Coward, M.P. & Kim, J.H. 1981. 'Strain within thrust sheets'. In: McClay, K.R. & Price, N.J. (eds.), Thrust and nappe tectonics Spec. Publ. Geol. Soc. London, 9, 245-92.

Dixon, H.R. & Lundgren, L.W., Jr. 1968. 'The structure and eastern Connecticut'. In: Zen, E-an, White, W.S., Hadley, J.B., and Thompson, J.B., Jr. (eds.), Studies of Appalachian geology: Northern and Maritime. New York, Wiley-Intersci., p. 219-29.

Elliott, D. & Johnson, M.R.W. 1980. 'Structural evolution in the northern part of the Moine thrust belt, northwest Scotland'. Trans. R. Soc. Edinburgh, 71, 69-96.

Fischer, M.W. & Coward, M.P. 1982. 'Strain and folds within thrust sheets: An analysis of the Heilam sheet, northwest Scotland'. Tectonophysics, 88, 291-312.

Goldstein, A.G. 1982. 'Geometry and kinematics of ductile faulting in a portion of the Lake Char mylonite zones, Massachusetts and Connecticut'. Am. Jour. Sci., 282, 1378-405.

Hall, L.M. & Robinson, P. 1982. 'Stratigraphic-tectonic subdivisions of southern New England'. In: St. Julien, P. & Beland, J. (eds.), Major structural zones of the northern Appalachians. Geol. Assoc. Canada Spec. Paper, 24, 15-41.

Hatcher, R.D., Jr. 1971. 'Structural, petrologic and stratigraphic evidence favouring a thrust solution to the Brevard problem'. Am. Jour. Sci. 270, 177-202.

―――― 1978. 'Tectonics of the western Piedmont and blue Ridge, southern Appalachians: Review and speculation'. Am. Jour. Sci., 278, 276-304.

_____ 1983. 'Basement massifs in the Appalachians: Their role in deformation during the Appalachian orogenies'. Geol. Jour., 18, 255-65.
Hsu, K.J. 1979. 'Thin-skinned plate tectonics during neoAlpine orogenesis'. Am. Jour. Sci., 279, 353-66.
Johnson, M.R.W. 1960. 'The structural history of the Moine thrust zone at Lochcarron, Wester Ross'. Trans. R. Soc. Edinburgh, 64, 139-68.
_____ 1967. 'Mylonite zones and mylonite banding'. Nature, 213, 246-7.
Odom, A.L. & Fullagar, P.D. 1973. 'Geochronologic and tectonic relationships between the Inner Piedmont, Brevard zone, and Blue Ridge belts, North Carolina'. Am. Jour. Sci., 273-A, 133-49.
O'Hara, K.D. & Gromet, P.L. 1983. 'Textural and Rb-Sr isotopic evidence for late Paleozoic mylonitization within the Honey Hill fault zone, southeastern Connecticut'. Am. Jour. Sci., 283, 762-99.
Lapworth, C. 1885. 'The Highland controversy in British geology: Its causes, course and consequences'. Nature, 32, 558-9.
Lundgren, L.W., Jr. & Ebblin, C. 1972. 'Honey Hill fault in eastern Connecticut: Regional relations'. Geol. Soc. Am. Bull., 83, 2773-94.
Peach, B.N., & Horne, J. 1885. 'The geology of Loch Eribol with special references to the Highland controversy'. Nature, 32, 558.
Peach, B.N., Horne, J., Gunn, W., Clough, C.T., Hinxman, L.W. & Teall, J.J.H. 1907. 'The geological structure of the northwest Highlands of Scotland'. Mem. Geol. Surv. UK.
Sinha, A.K. & Glover, L., III 1978. 'U/Pb systematics of zircons during dynamic metamorphism'. Contr. Mineral. Petrol., 66, 306-10.
Soper, N.J. & Barber, A.J. 1982. 'A model for the deep structure of the Moine thrust zone'. J. Geol. Soc. London, 139, 127-38.
Trumpy, R. 1960. 'Paleotectonic evolution of the central and western Alps'. Geol. Soc. Am. Bull., 71, 843-908.
Van Breeman, O., Aftalion, M., & Johnson, M.R.W. 1979. 'Age of the Loch Borrolan complex, Assynt and late movements in the Moine Thrust'. J. Geol. Soc. London, 136, 489-96.
Wintsch, R.P. 1979. 'The Willimantic fault: A ductile fault in eastern Connecticut'. Am. Jour. Sci., 279, 367-93.

THE MOINE THRUST STRUCTURES

M. P. COWARD
IMPERIAL COLLEGE OF SCIENCE & TECHNOLOGY
DEPARTMENT OF GEOLOGY
ROYAL SCHOOL OF MINES
PRINCE CONSORT ROAD
LONDON SW7 2BP

ABSTRACT. The Moine thrust zone carries Proterozoic (Moine) metasediments of the Caledonian belt over the foreland sequence of Archaean to Proterozoic (Lewisian) gneiss, late Proterozoic (Torridonian) sandstones and the unconformable sequence of Cambro-Ordovician shelf quartzites and limestones. The thrust structures occur on all scales from minor duplex zones a few centimetres across, to large thrust slabs on a kilometre scale. The thrust sequence is generally piggy-back, that is, the easternmost (highest level) thrust formed first. The thrust direction was to N295 $\mp$ 10.
  The lower thrust sheets have produced folds by the stacking of imbricate slices, though in Eriboll on the north coast, buckle folds are common in the hangingwall and footwalls of the lower minor thrusts. The higher thrust sheets carry buckle folds, which are sometimes large scale structures and generally oblique to the thrust transport direction. This suggests that the buckle folds formed by differential movement, the northern part of the thrust zone moving farthest to the WNW. Textural studies suggest that this extra movement in the north occurred at slower rates under more ductile conditions.
  The amounts of displacement, estimated from offsets of Lewisian structures, are of 35 km for the Glencoul sheet, 45 km for the Kishorn sheet. Using balanced cross sections in the region between Foinaven and Assynt, shortening values range from 35 km+ to 55 km+. The Foinaven imbricates are important in that they do not affect the lowest Cambrian sediments or their basement, that is, the basement rocks must continue back some 55 km beneath the Moines, the distance equivalent to the restored length of the middle to upper Cambrian sediments. This implies that any crustal ramp to the Moine thrust zone must lie more than 55 km E of the present outcrop of the thrust. However, off the north coast of Scotland, deep seismic reflection profiles show moderately dipping events much further to the northwest and if these represent the crustal scale ramp, this ramp must be offset by a major tear or transfer fault along the north Scottish coast.
  In the southern part of Assynt, in the central part of the thrust

zone, there are several extensional fault systems and also a late
(extensional?) fault at the local base of the Moines. This Moine
fault cuts across earlier thrust and extensional faults. The
extensional movements suggest a gravity spreading mechanism for some
Caledonide thrusting.

The timing of thrusting can be bracketed by the 430 my age for
the Borrolan igneous intrusion, which predates most major thrust
movements and by the 415 my age for the Ross of Mull granite which is
later than the thrusting. It is difficult to find the driving
mechanism for the Moine thrust at this time.

1. INTRODUCTION

The structure of the Moine thrust zone (Fig.1) aroused considerable
controversy throughout the 19th century, when the geologists were
split between "fixists" and "transporters". Fixists believed that
rocks stayed more or less in place and that situations where
metamorphic rocks overlay Palaeozoic sediments were due to metamorphic
inversions. The "transporters" believed the metamorphic rocks were
emplaced as large slabs. In 1884 the problem was resolved when the
Scottish Geological Survey began their major work in the NW Highlands
and soon realised the importance of thrust tectonics. It was in
1884 that Geikie defined the term "thrust" using Highland examples,
the major survey work being reported by Peach et al. (1888, 1907).
This work delineated the main thrust sheets and described the local
foreland stratigraphy consisting of Lewisian basement, overlain by
Torridonian and Phanerozoic sediments. However since this original
survey work, there have been few publications on Caledonide thrust
geology, except for several papers on aspects of mylonite formation
(Christie 1960, Johnson 1967), until the recent applications of
concepts developed in the thin-skinned thrust regions of the
Rocky Mountains and Appalachians (Barton 1978, Elliott & Johnson 1980).
This paper will summarise some of these applications of thin-skinned
tectonics to the Moine thrust zone. It will also consider the nature
of the thrust structures at depth and the driving forces associated
with thrust movement.

2. STRATIGRAPHY

2.1. Below the Moine thrust

The Lewisian gneisses of the foreland range in age from Archaean to
Proterozoic (Sutton & Watson 1951, Giletti et al. 1961, Moorbath et al.
1969) with important deformation and metamorphic events at 2600 my
(the Scourian) and 1600-1800 my (the Laxfordian). Phases of basic
intrusive activity occurred before the Scourian deformation and also
between 2400 to 1800 my, the latter phase forming the prominent
Scourie-dyke set (Sutton & Watson 1951). The gneisses include
metasediments of pre-Scourian age in the Outer Hebrides and northern

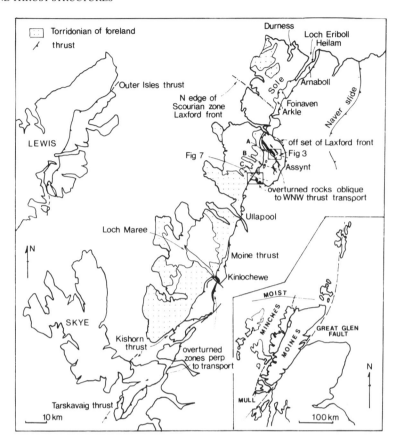

Figure 1. Location map for the Moine thrust zone. Section lines A, B, C refer to Fig.2. The black B indicates the distribution of overturned rocks on the hangingwall of the Glencoul, Ben More and Kishorn thrusts in the Assynt and Kishorn region.

parts of the Scottish mainland and also Proterozoic sediments, the Loch Maree Group (Bowes 1969).

The nature of the Scourian tectonics is difficult to define, but the later Laxfordian events involved deformation on large-scale shears and thrusts, thickening the crust, interslicing Loch Maree sediments with Scourian basement on the mainland and uplifting granulite facies lower crustal rocks in parts of the Outer Hebrides, (Coward 1984). The overthrust direction was to the NW and produced locally thick sheets of intensely sheared out gneiss. These sheets were later refolded by several generations of more upright structures, presumably continuations of the same deformation. The Lewisian gneisses show no evidence of Grenville deformation or metamorphism; the best correlations are with the Nagssugtoqidian of Greenland (Bridgewater et al. 1973, Escher et al. 1976).

The overlying Torridonian rocks form a thick sequence of conglomerates, fanglomerates, feldspathic sandstones and shales, which infill a highly disected topography at their base, but pass up into several kilometres of generally monotonous sediments (Stewart 1969, 1975). The main sedimentary sequence has been dated at around 800 my (Moorbath 1969) but on the west coast of Scotland, an older Stoer group lies with slight unconformity beneath the main Torridonian sequence and is considered to be about 1000 my old (Moorbath 1969).

The sediments appear to have been deposited from large alluvial fans whose apeces lie near the west side of the Minch (Williams 1969). However in Skye, in the southern part of the Moine thrust zone, the lower part of the Torridonian sequence involves shallow marine sediments with well bedded flags.

The Torridonian sediments were folded gently before Cambrian sedimentation; the differences in dip between Torridonian and Cambrian rocks are often of about 15° and trace out large scale, gently southward plunging folds, with a wavelength of about 45 km (Elliott & Johnson 1980, Soper & Barber 1979).

The Cambrian unconformity is remarkably planar and the overlying sequence has minor quartz rich conglomerates at its base, passing up into relatively pure, well sorted quartz rich grits and sandstones with rare feldspar clasts. This 75 to 100 m thick sequence of Basal quartzite is followed by about 75 m of pure quartz sandstone with worm burrows (genus Skolithos), the Pipe Rock. This unit is divisible into several zones based on the appearance of the pipes and these zones maintain their identity and thickness for long distances not only along the length of the thrust zone but also across strike, across the major thrust sheets. The Pipe Rock is overlain by about 20 m of dolomitic shales and sands, the Fucoid Beds, named after the abundance of burrows sub-parallel to bedding and resembling seaweed to early geologists. This is followed by about 10 m of impure, often dolomitic quartzites, the Serpulite Grit, with small shell fossils. The Fucoid Beds and Serpulite Grit make up the An-t'Sron Formation of Swett (1969). The uppermost sediments are a series of limestones and dolomites, the Durness Limestone. This sequence is over 1 km thick at Durness on the north coast and ranges in age from Cambrian to Lower Ordovician. The top is never seen, having been scraped off by the thrusts.

2.2. Above the Moine thrust

There is a pronounced change in stratigraphy across this thrust. The Moines form a monotonous series of metamorphised sandstones and shales, with few marker horizons. Thus original thickness is unknown and indeed the stratigraphic correlations between different parts of the Moines are under dispute (Harris et al. 1978, Piasecki 1980). They have suffered some effects of Precambrian deformation and metamorphism (Grenville and possibly late Precambrian-Morarian) giving pre-Caledonian ages of about 1100 my to 550 my Brook et al. 1977, Brewer et al. 1979). Further discussion of the Moines is given by Fettes & Harris (this volume); here it is suffice to say that no Moines and no effects of Grenville deformation have been detected west of the Moine Thrust. Presumably these rocks have been transported a large distance. Along the Moine thrust zone the Moines have been intensively deformed producing a complex mylonite zone, up to 70 m in thickness, with greenschist facies mineralogy.

2.3. Intrusives of the Moine thrust zone

These include a suite of felsites, porphrites and lamprophyre sills in the Assynt area and also large alkaline intrusives at Loch Borrolan and Loch Ailsh. They are considered to have intruded the sediments during the early effects of thrusting (Sabine 1953, Wooley 1971) and a Rb/Sr date of 430± 4 my on the Borrolan complex (van Breemen et al. 1979), dates some Moine thrust movements.

The intrusives also assist the correlation of different thrust sheets. The feldsparphyric Canisp porphyry is restricted to the foreland, while a set of groruderites (aegirine-felsites) are restricted to the Ben More and Glencoul sheets in Assynt, defining an Upper Assynt sheet (Sabine 1953).

3. THE THRUST STRUCTURES

3.1. Below the Moine thrust

Thrusting occurs on all scales from minor imbricates which produce duplex zones a few cm thick, to large thrust sheets, over 1 km thick. Several duplex zones have been described from Eriboll and Assynt (Coward 1982, Coward & Kim 1981, Butler & Coward 1984). Back-steepened duplexes and antiformal stacks occur, where early thrusts have been reorientated by later movements. These indicate a general piggy-back thrust sequence, where the eastern-most thrust formed first, the thrusts developing towards the WNW.

The thrust transport direction may be determined from sheared pipes and from the range in orientation of oblique to lateral folds and thrust ramps. These indicate a fairly consistent WNW overthrust direction, though locally there may be fold axial and thrust traces

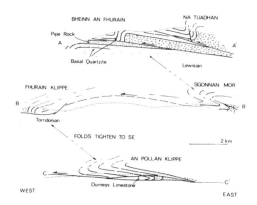

Figure 2. Cross sections through the oblique folds of central and southern Assynt, drawn parallel to the thrust direction. The section lines are shown in Fig. 1.

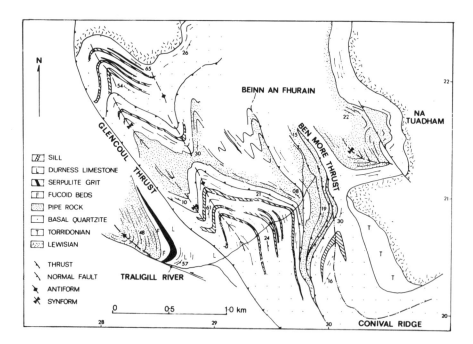

Figure 3. Map of the oblique fold structures of Na Tuadham and Beinn An Fhurain.

oblique to this trend. In these areas the strains are often oblique, as shown by the elliptical sections made by the pipes on the bedding planes (Coward & Kim 1981). Such regions of oblique structure are interpreted as due to the differential movement of the thrust sheet Coward & Kim 1981, Fisher & Coward 1982).

The major thrust sheets of Assynt may be divided not only on the presence or absence of groruderite intrusives but also on the structural style. The Assynt sheet, which contains the groruderites, is also characterized by major folds of Cambrian and Torridonian rocks, on both hanging and footwalls of subsidiary thrusts. These are shown on a series of sections in Fig.2 and are well exposed on Na Tuadham (Fig.3) and Sgonnan Mor. They are locally recumbent and sometimes tight to isoclinal, with a cleavage preserved in Torridonian sands and grits. On the north face of Conival and on Sgonnan Mor, the basal Torridonian conglomerates are deformed, the pebbles flattened in the cleavage and elongated in an ESE direction. There are close folds in the Pipe-Rock above the Glencoul thrust in the northern part of the Assynt sheet (Fig.3) and these folds often carry a weak pressure solution cleavage.

The Assynt sheet is cut off by the Moine thrust, which is a late, low angle structure in southern Assynt (Butler & Coward 1984), but similar large recumbent folds occur below the Moine thrust in the Kinlochewe and Kishorn thrust sheets in the southern part of the Moine thrust zone. In the south-east Skye, the fold structures are large; above the Kishorn thrust the overturned limb of a major syncline, the Lochalsh fold, can be traced about 10 km across strike.

North of Assynt, folds occur in Pipe Rock and Basal Quartzite and deform the Cambrian/Lewisian contacts above the Arnaboll thrust at Eriboll (Rathbone et al. 1983). All the above-described structures seem to involve a buckling component and bedding parallel shortening strains in Torridonian and Cambrian sediments, along with the development of small shear zones in the underlying Lewisian gneiss (Rathbone et al. 1983). The structures presumably formed by sticking of the thrust sheet with the development of shortening strains and eventually major buckle folds at the tip zone. These folds were then truncated as the thrust broke through the deformed zone behind the old tip line (Coward & Potts, in press).

In the central part of the Moine thrust zone, the majority of the folds in the Assynt and Kinlochewe sheets are oblique to the transport direction (Fig.1), suggesting a wide zone of sinistral differential movement. The northern part of the Assynt sheet must have moved several kilometres further than the southern part. This zone of differential movement should be traceable back into the Moines, assuming a piggy-back sequence, either reorientating Moine structures or producing new oblique folds. Alternatively, the differential movement zone would have to pass this extra movement on to the higher thrust sheets south of Assynt. There is a swing in trend of Moine structures east of Assynt but it is difficult to correlate structures across the Moine thrust, because of locally poor outcrop.

The lower thrust structures in and to the south of Assynt produce folds more by the stacking of imbricate slices rather than by buckling. This is especially evident in the Kinlochewe to Kishorn areas where the beds have very variable dips, but small-scale folds are uncommon and cleavage rare. However on the north coast, beneath the Arnaboll sheet, the lower thrusts produce well-developed folds by buckling, sometimes with a good cleavage and layer parallel shortening strains as shown by deformed pipes (Fisher & Coward 1982). This suggests a variation in deformation conditions, during the later thrust movements, with an increase in rock ductility from south to north. This may reflect a variation in thickness of thrust overburden.

The amount of displacement may be estimated from the offset of earlier structures, which can be correlated across the thrust zone. One set of correlations has been made between the "Laxford front" on the foreland and in the thrust belt north of Assynt. This front forms the northern limit of Scourian granulites in the Lewisian, where Scourie-dykes are generally discordant to gneissic banding. To the north the dykes are smeared into parallelism with the banding by Laxfordian deformation, the granulites are retrogressed to amphibolites and the rocks are locally swamped with Laxfordian granites. Similar granite sheets occur in a basement slice which constitutes the Loch More klippe, below the Moine thrust. From this correlation, Elliott & Johnson (1980) estimate a displacement of 44 km on the imbricates below the klippe. The Laxford front also occurs in the Glencoul sheet, allowing an estimate of 35 km of thrust displacement (Coward et al. 1980). Similar estimates of about 44 km displacement may be made from the offset of Lewisian amphibolites between the foreland and Kishorn thrust sheet in the southern part of the thrust zone.

Another method of estimating displacement involves the construction of balanced and restored cross sections (Dahlstrom 1970, Elliott & Johnson 1980) and its application to the Moine thrust zone has been described by Butler & Coward (1984). In the region between Eriboll and Foinaven, Pipe Rock, with only thin slices of Basal Quartzite and An t'Sron formation, form a complex condominium above the Sole thrust and below the Moine thrust. This imbricate zone restores to a length of about 54 km.

This displacement estimate is not excessive. Elliott & Johnson (1980) estimate a slip of 77 km on all the thrusts in the Loch More area. Similar displacement estimates of 35 km have been obtained from balanced sections at Foinaven and Arkle (Parish, pers. comm. 1984). 15 km of displacement have been estimated from structures above the Glencoul thrust in the Loch More area (Butler 1984), so that the total displacement is of $\sim$ 35 km for the Glencoul sheet, plus 15 km, giving $\sim$ 50 km. In north Assynt and south Assynt, my unpublished shortening estimates, based on section construction, are of 47 km and 45 km respectively.

The south-Eriboll section is important in that it involves mainly Pipe Rock and not the lower part of the Basal Quartzite, nor the Lewisian basement and hence these rocks must continue back 54 km to the ESE of their present outcrop. Thus the Moine thrust, which overlies and pre-dates these Pipe Rock imbricates, cannot cut up from lower crustal levels within 54 km of its present trace. The

Moine thrust zone must be a thin-skinned structure, scraping off the Cambrian sediments above the Sole thrust, to pile them up south of Eriboll.

In Assynt however, the thrusts involve Lewisian rocks, the Lewisian sheets having been transported some 35 km (Coward et al. 1980). Thus branches of the Sole thrust must cut down through the Cambrian by lateral ramps to bring up a thin (1 to 2 km thick) slice of Lewisian from east of Assynt (Fig.4).

West of the Moine thrust zone the foreland appears unaffected by Caledonian effects but in the Outer Hebrides there is a large eastward dipping thrust zone, the Outer Isles thrust, involving phyllonitic mylonites and thick developments of pseudotachylite (Francis & Sibson 1973, Sibson 1977). Francis & Sibson (1973) suggest a thrust displacement of about 10 km from offsets of Lewisian folds and metamorphic zones in the southern Hebrides. Presumably these displacements link with those of the Moine thrust by decoupling in the lower crust or on the Moho (Butler & Coward 1984).

3.2. Mylonites of the Moine thrust zone.

The mylonites were first described by Lapworth (1885) from the Eriboll district and later by Peach et al. (1907), Christie (1960), White (1979) and Law et al. (1984). The most westerly exposed thrusts in the mainland, involve An t'Sron and Durness formations, generally producing clean faults with very little alteration or mylonite production. Most of the strain in these thrust sheets is associated with flexure during ramp climb or folding (Fisher & Coward 1982). Quartz rich mylonites are locally developed in bedding-parallel shear zones within the Pipe Rock of the Ben Heilam area, Eriboll, but are most prominent in the higher thrust sheets, such as the Upper Arnaboll sheet at Eriboll and the higher parts of the Ben More sheet at Ben Arnaboll and a thick zone of intensely mylonite rock occurs in the Upper Arnaboll sheet.

The quartz mylonites show original detrital grains which have been broken into elongate ribbons with elongations of about 20 : 1, with the production of small recrystallised sub-grains at their margins. In the intense mylonites, the original and ribbon grains have been obliterated to leave a mass of fine recrystallised quartz $(5-14 / 1m)$. Strong crystallographic preferred orientations are found in these mylonites. The most intensely deformed rocks are encountered at the bases of thrust sheets (Law et al. 1984, in press). They are recrystallised L-S tectonites with asymmetrical quartz c-axis fabrics, the sense of asymmetry being consistent with WNW deformation, from the development of symmetric c-axis fabrics and from conjugate sets of small-scale shear bands, symmetric to the mylonite fabric (Law et al. 1984).

Such fabrics have been found from several localities in the Eriboll district and in mylonites beneath the Moine thrust zone in the Glencoul area of north Assynt and Loch Ailsh region of south Assynt. The kinematic interpretation involves co-axial thinning and concomitant extension parallel to the thrust transport direction,

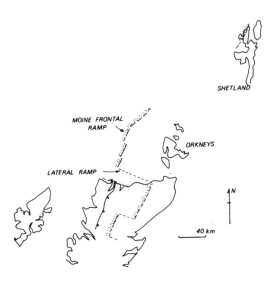

Figure 4. Simplified map showing where the Moine thrust cuts across Basal quartzites, that is, where it climbs the ramp between Lewisian basement and Cambrian cover.

accompanied by synchronous, similarly orientated shearing within
the lower levels of the thrust sheet (Law et al. 1984). This shearing
carried the mylonites on to less deformed rocks and also maintained
compatibility between the thinned and extended rocks above and the
unaffected sequence below. The extension was presumably related to
gravitational spreading or flow of the Moines above the thrust sheet,
as will be discussed later.

3.3. Thrust structures within the Moines

On the N coast of Scotland, several slices of Lewisian rocks occur
within the Moines, sometimes in fold cores, but often associated with
zones of intense deformation (R. Holdsworth, pers. comm. 1983, 1984).
These have been interpreted as a fold and thrust zone of Lewisian and
Moine, forming an imbricate stack carried on the hanging wall of the
Moine thrust (Holdsworth pers. comm. 1983, Butler & Coward 1984).
The eastern boundary of this imbricate zone is marked by a major
zone of deformation, the Naver slide, to the east of which, the style
and possibly the age of the deformation and metamorphism are different.
   In the central and southern Moines, an importance shear, the
Sgurr Beag slide, carries Moines of what is called the Glenfinnan
Division over western Moines of the Morar Division (Tanner 1971,
Rathbone & Harris 1979). This Sgurr Beag slide may be the southern
equivalent of the Naver slide. It has an apparent thrust sense of
displacement and can be traced for at least 25 km across strike around
later upright to west-verging folds (Lambert et al. 1979), with little
change in stratigraphic level. The slide is marked by a thick zone of
flaggy, intensely deformed rocks, in which all trace of sedimentary
structures has been removed (Rathbone & Harris 1979, Rathbone et al.
1983). It probably represents the higher grade but earlier
equivalent of the Moine thrust, preserved as a large folded flat.
Brewer et al. (1979) date the main deformation and metamorphism
on the slide at 467 my, though the nearby Moines cooled down to
413± 17 my, from Rb/Sr isochron data (Brewer et al. 1979). The later
upright folds, which deform the slide, may have developed above lower
level shear zones, such as those forming the Moine thrust
(Rathbone et al. 1983), suggesting a piggy-back development of thrusts
in the Moines. Thus, the Sgurr Beag slide was active first at about
460-470 my, followed by movement on the Moine thrust folding the
Sgurr Beag slide and producing the imbricate stack of Moines and
Lewisian rocks on the north coast. The Moines and Moine mylonites
were carried over 50 km on to foreland rocks, developing large
thrust sheets in the footwall to the Moine thrust. Much of the
deformation beneath the Moine thrust took place after 430 my,
indicating a long time span ($\sim$470 my to $\sim$430 my) for thrust
activity in northwest Scotland.

4. THE MOINE THRUST AT DEPTH AS SEEN ON SEISMIC PROFILES

The deep seismic reflection survey, the Moine and Outer Isles Seismic Traverse (MOIST) was shot north of the Scottish coast and aimed to detect the continuation of the major Caledonian faults at depth (Smythe et al. 1982, Brewer & Smythe 1984). The most prominent reflector is the Moho which lies at a depth of about 28 km and appears horizontal throughout, with hardly any changes in depth beneath the Caledonides.

Several crustal reflectors appear on the profile to the north and to the east of the offshore projection of the Moine thrust. They are interpreted as thrusts by Brewer & Smythe (1984) involving imbricated slices of Cambrian rocks with Moines or just Moines alone. Butler & Coward (1984) suggested another possibility, that the reflectors might be major shears in the basement rocks beneath the Moine thrust. However, from close inspection of shallow seismic data, these reflectors appear to be thrusts with thrust-type cut-offs. They probably represent intensely faulted and sheared Moines. The reflectors dip at about 30° to the ESE and thus represent a crustal-scale ramp (Soper & Barber 1982), where intensely deformed Moines were carried to higher levels. A probable continuation of this crustal ramp occurs in northwest Shetland where, at North Roe, Flinn et al. (1979) describe the edge of Caledonian deformation. Here, Lewisian rocks are intensely sheared and imbricated with previously intensely deformed Moines. The sheared rocks dip moderately to the east and the lineation generally suggests a WNW transport direction.

The crustal reflectors of the MOIST line flatten at about 18-20 km depth, suggesting this is the decoupling level from which the thrusts climbed. The mid-crustal reflectors are overlain by dipping shallow reflectors, representing post-Caledonian sediments deposited in a series of half-graben. These are bounded by normal faults which follow the earlier Caledonian structures at depth. On the shallow seismic data, the normal faults may be interpreted as growth faults and the main thickness of sediments can be correlated with Old Red Sandstone deposits on land. These normal faults continue to the northwest of the mid-crustal structures, as shown in Fig.5. They define the limits of the West Orkney basin, and must have displaced the mid-crustal ramps. Thus, the offset of these mid-crustal structures seen offshore must be over 60 km, if they are correlated with the Moine ramp onshore. The north Scottish coast must be marked by a tear fault or lateral ramp of Caledonian age.

To the west of the Moine thrust on the MOIST line, an eastward dipping reflector forms the western margin of a large half-graben but has been correlated with the Outer Isles thrust on Lewis (Smythe et al. 1982). If this correlation is correct, then the Outer Hebrides thrust must have been reactived offshore by post-Caledonian normal faults. There is evidence on land for some normal fault movement (White 1984, pers. comm.). Faults associated with the northern extension of the Minch basin appear to join the Outer Isles fault on the MOIST line and the fault bounding the western margin of the main

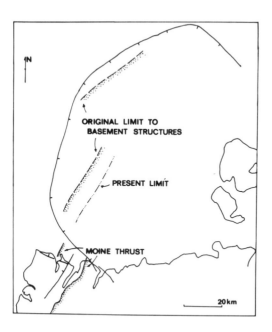

Figure 5. The shape of the bounding fault to the West Orkney basin (ticked line) and the original and present northwest limit of mid-crustal structures, compared to the position of the Moine thrust on land.

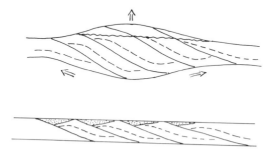

Figure 6. Model for the production of half-grabens by the reactivation of earlier crustal scale thrusts by extensional movements during isostactic uplift, after cessation of the compressional phase.

Minch graben may also join with the Outer Isles thrust east of the
Hebrides. West of the Outer Isles fault a prominent reflector
cross cuts the Moho and is termed the Flannan thrust by Brewer &
Smythe 1984). The true nature of this fault is unknown. However a
sedimentary basin near the edge of the continental shelf, near the
projection of the Flannan fault, may have formed during normal fault
movement on the Flannan structure.

Thus there is some evidence for reactivation of early thrusts to
generate post-Caledonian basins, particularly in the case of the
Outer Isles faults and West Orkney basin. Similar basins have been
detected on other deep seismic lines west of Britain (Brewer et al.
1983). Their possible origin is outlined in Fig. 6, which suggests
that the basins form by reactivation of crustal scale thrust ramps due
to isostatic uplift and spreading after the termination of a
compressional phase.

5. LOW ANGLE EXTENSIONAL FAULTS AND THE THRUST DRIVING MECHANISMS

In the southern part of Assynt there are several low angle faults
which have extensional geometry, cutting down across the bedding in
the movement direction. They are associated with the development of
smallscale extensional imbricate faults. South of Knockan (Fig.7),
thrust faults in limestones are cut by a set of extensional faults
with downthrow to the WNW. These normal faults are listric in shape
and curve into the Sole thrust, which had previously acted as a floor
fault for the local imbricates. They cause up to 30% extension of
the overlying sheet. All these thrust and extensional structures are
cut by the local Moine thrust, which joins the Sole about 1 km
southwest of Knockan village (Coward 1983, Butler & Coward 1984).
Similar extensional faults occur throughout the Assynt district
(Coward 1982, 1983) and can be found as far south as Ullapool, while
the late movements on the Moine thrust can be traced from southern
Assynt to south of Ullapool. Note that in northern Assynt however,
the Moine thrust is an early structure, folded and breached by
underlying thrusts.

Elsewhere in the thrust zone there are large thrusts, such as the
Glencoul and Kinlochewe thrusts, whichlocally cut down across bedding
in the transport direction, thinning or omitting Cambrian strata.
In 1982 I suggested these formed by extensional flow (Coward 1982).
However Butler (1984) has argued that they formed by compressional
flow and that the bedding had been previously tilted by the isostatic
effects of the overriding thrust sheets. According to Butler (1984),
the thrusts climbed section relative to datum but the beds had been
tilted steeper than the failure angle. This suggests that the maximum
compressive stress lay with a high angle relative to bedding.

Such a model is supported by observations of the thrusts which
carry large basement sheets; the Glencoul thrust slices across all
structures in the Lewisian rocks of its hanging wall and yet only
climbs at an angle of 2° relative to bedding in the overlying
Cambrian. Thus the orientation of the critical failure plane for the

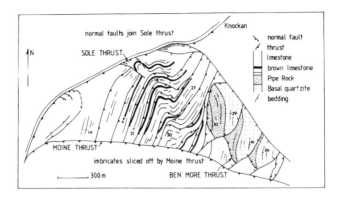

Figure 7. Map of the region south of Knockan in southern Assynt (see Fig. 1 for location), showing the imbricated limestones cut by normal faults and the Moine thrust.

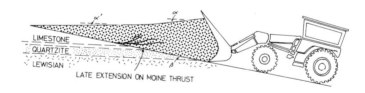

Figure 8. Schematic diagram of a thrust wedge to show how the surface shape of the wedge varies with basal shear strength, possibly explaining the local extensional structures as the thrust wedge climbs from strong basement to weaker sediments.

Lewisian was very close to the regional bedding orientation in Cambrian rocks. Generally the stresses were such that failure took place parallel to bedding in the Cambrian, but where the beds were tilted, or the orientation of the maximum compressive stress increased relative to datum, then the faults cut down across bedding. Thrust ramps developed at local stress perturbations or sticking points on the failure plane.

A maximum compressive stress at a high angle to datum suggests the development of a thrust wedge (Fig.8), with a surface slope dipping towards the foreland, contributing to the gravitational component of thrust movement (Davis et al. 1983). Such a wedge would be formed by ductile strain of the overriding mass and by the accretion of new material to the footwall. As described by Chapple (1978) and Davis et al. (1983), it is the development of this wedge which allows large displacements of the thrust zone.

The shape of the wedge depends on the strength of the overthrusting material and the basal friction, assuming a simple Coulomb failure criterion (Davis et al. 1983). Thus faults developing through Lewisian gneiss would require a greater critical taper angle for the thrust wedge than faults in more fissile shales. The Scottish thrusts have climbed from Lewisian basement to more fissile Fucoid Beds and Durness Limestone and thus the wedge shape, critical for movement in basement gneisses, would be unstable above the Cambrian sediments (Fig.8). Assuming that the overriding sheet was able to fail easily under tensile conditions, extension would occur, decreasing the wedge taper and producing localised normal faults as at Knockan. The late Moine "thrust" south of Assynt may be a large-scale example of this extensional flow. The early thrusts within the Moines would have contributed to the build-up of the wedge, while the extensional flow structures shown by the mylonites along the Moine thrust zone may be related to its collapse.

A possibly related problem is the origin of the 10° to 14° southeast dip of foreland and thrust zone. Soper & Barber (1982) suggest it is an isostatic effect due to overlying mass of Moines but Elliott & Johnson (1980) prefer it to be a post-Caledonian structure. Butler & Coward (1984) point out that the Moho, from LISPB (1984) and electrical profiles (Hutton et al. 1980) has a similar dip suggesting crustal tilting. Presumably there have been some post Caledonian isostatic effects and many of the late normal faults which cut the whole thrust zone, may be related to this uplift. Beneath the West Orkney basin, which possibly formed by more complete post-Caledonian isostatic rebound, the MOIST profile shows the Moho to be flat.

6. THE MOINE THRUST AND THE CALEDONIDES

Thrusting occurred before the development of Old Red Sandstone basins and after the intrusion of the Loch Borrolan igneous complex which is cut by most of the thrust generations in Assynt and has been dated at about 430 my (van Breemen et al. 1979). The Ross of Mull granite, dated at about 415 my (Pankhurst 1982) is not deformed and yet lies close to the probable trace of the Moine thrust zone. Thus Moine thrust movements probably took place in the period 430 to 415 my, that is in mid Silurian times. There is no obvious plate collision zone in Scotland coeval with this deformation nor with the slightly earlier deformation in the Moines (Coward 1983). In the Silurian the Grampian deformation had finished and the area now southeast of the Midland Valley was the site of an ocean trench (Leggett et al. 1979). The source of the compression, which produced the Moine thrust crustal ramp and then carried uplifted Moines over 50 km to the WNW, is unknown and may well have been removed by late to post-Caledonian movement on some major faults, such as the Great Glen fault, juxtaposing regions affected by different ages of Caledonian crustal compression.

7. ACKNOWLEDGEMENTS

Field work in the Moine thrust zone has been funded by NERC grant GR3/4100 and has involved Rob Butler, Sue Bowler, Steve Matthews, Richard Morgan, Henrique Dayan, J.H. Kim, Graham Potts, Marcus Parish, Phil Nell, Rick Law and Rob Knipe.

REFERENCES

Bamford, D., 1979. Seismic constraints on the deep geology of the Caledonides of northern Britain. In: A.L. Harris, C.H. Holland & B.E. Leake, The Caledonides of the British Isles Reviewed, Spec. Publ. Geol. Soc. London, 8, pp 93-96.

Barton, C.M. 1978. An Appalachian View of the Moine Thrust. Scot. J. Geol. 14, pp. 247-57.

Bowes, D. 1969. The Lewisian of Northwest Highlands of Scotland. In: Kay, M. (ed). North Atlantic Geology and Continental Drift. A Symposium. Mem. Amer. Assoc. Petrol. Geol. 12, pp 575-94.

Brewer, J.A., Matthews, D.H., Warner, M.R., Hall, J., Smythe, D.K. & Whittington, R.J. 1983. BIRPS deep seismic reflection studies of the British Caledonides - the WINCH profile.

Brewer, J.A. & Smythe, D.K., 1984. MOIST and the continuity of crustal reflector geometry along the Caledonian Appalachian orogen, J. Geol. Soc. London, 141, pp 105-120.

Brewer, M.S., Brook, M & Powell, D. 1979. Dating of the tectonometamorphic history of the southwestern Moine, Scotland. In : Harris, A.L., Holland, C.H. & Leake, B.E. (eds.). The Caledonides of the British Isles - Reviewed. Spec. Publ. geol. Soc. London, 8, pp 129-37.

Bridgewater, D., Watson, J.V. & Windley, B.F. 1973. The Archaean craton of the North Atlantic region. Phil. Trans. R. Soc. London. A273, pp 493-512.

Brook, M., Brewer, M. & Powell, D. 1977. Grenville events in Moine rocks of the Northern Highlands, Scotland, J. Geol. Soc. London, 133, pp 489-96

Butler, R.W.H. 1984. Structural evolution of the Moine thrust belt between Loch More and Glendhu, Sutherland. Scott. J. Geol. 20, pp 161 - 180.

Butler, R.W.H. & Coward, M.P. 1984. Geological constraints, structural evolution, and deep geology of the Northwest Scottish Caledonides. Tectonics, 3, pp 347 - 365.

Chapple, W.M. 1978. Mechanics of thin-skinned fold and thrust belts. Geol. Soc. Amer. Bull., 89, pp 1189-1198.

Christie, J.M. 1960. Mylonitic rocks of the Moine Thrust Zone in the Assynt region. North-West Scotland. Trans. Edinb. Geol. Soc. 18, pp 79-93.

Coward, M.P. 1982. Surge zones in the Moine thrust zone of NW Scotland, J. Struct. Geol. 4, pp 247-256.

Coward, M.P. 1983. The thrust and shear zones of the Moine thrust zone and the NW Scottish Caledonides. J. geol. Soc. London 140, pp 795-811.

Coward, M.P. 1984. Major shear zones in the Precambrian crust; examples from NW Scotland and southern Africa and their significance. In: Kroner, A. & Greiling, R. (eds). Pre-Cambrian tectonics illustrated, E. Schweizerbartische Verlagsbuch handlung Stuttgart, pp 207-235.

Coward, M.P., Kim, J.H. & Parke, J. 1980. The Lewisian structures within the Moine thrust zone. Proc. Geol. Ass. 91, pp 327-337.

Coward, M.P. & Kim, J.H. 1981. Strain within thrust sheets. In: McClay, K.R. & Price, N.J. (eds). Thrust and Nappe Tectonics, Spec. Pubb. geol. Soc. London, 9, pp 275-292.

Coward, M.P. & Potts, G. in press. Fold nappes, examples from the Moine thrust zone. Geol. J.

Dahlstrom, C.D.A. 1970. Structural geology of the eastern margin of the Canadian Rocky Mountains. Bull. Can. Petrol. Geol. 18, pp 332-406.

Davis, D., Suppe, J. & Dahlen, F.A. 1983. Mechanics of fold and thrust belts and accretionary wedges. J. Geophys. Res., 88 pp 1153-72.

Elliott, D., & Johnson, M.R.W. 1980. Structural evolution of the northern part of the Moine thrust belt, NW Scotland, Trans R. Soc. Edinburgh, 71, pp 69-96.

Escher, A., Jack, S. & Watterson, J. 1976, Tectonics of the North Atlantic Proterozoic dyke swarm. Phil. Trans. R. Soc. London, A280, pp 529-540.

Fischer, M.J. & Coward, M. P. 1982. Strains and folds within thrust sheets: an analysis of the Heilam sheet, NW Scotland. Tectonophysics, 88, pp 291-312.

Flinn, D., Frank, P.L., Brook, M. & Pringle, I.R. 1979. Basement-cover relations in Shetland. In: Harris, A.L. Holland, C.H. & Leake, B.E. (eds). The Caledonides of the British Isles - Reviewed, Spec. Publ. Geol. Soc. London 8, pp 109-116.

Francis, P.W. & Sibson, R.H. 1973. The Outer Hebrides Thrust. In: Park, G. & Tarney, J. (eds). The early Precambrian of Scotland and related rocks of Greenland. Univ. Keele, pp 95-104.

Giletti, B., Moorbath, S. & Lambert, R. St.J. 1961. A geochronological study of the metamorphic complexes of the Scottish Highlands. Q.J. geol. Soc. London. 117, pp 233-64.

Harris, A.L., Johnson, M.R.W. & Powell, D. 1978. The Orthotectonic Caledonides. (Moines and Dalradians) of Scotland. Pap. Geol. Surv. Can. 78, pp 79-85.

Hutton, V.R.S., Ingham, M.R. & Mbipom, E.W. 1980. An electrical model of the crust and upper mantle in Scotland, Nature, 289, pp 30-33.

Johnson, M.R.W. 1967. Mylonite zones and mylonite banding. Nature, 213, pp 246-247.

Lambert, R. St. J., Winchester, J.A. & Holland, J.G. 1979. Time, space and intensity relationships of the Precambrian and lower Palaeozoic metamorphism of the Scottish Highlands. In: Harris, A.L., Holland, C.H., & Leake, B.E. (eds). The Caledonides of the British Isles - Reviewed. Spec. Publ. geol. Soc. London, 8, pp.363-7.

Lapworth, C. 1885. The Highland controversy in British Geology. Nature, 32, pp 558-9.

Law, R.D., Knipe, R.J. & Dayan, H. 1984. Strain path partitioning within thrust sheets: microstructural and petrofabric evidence from the Moine Thrust zone at Loch Eriboll, northwest Scotland. J. Struct. Geol. 6, pp 477-497.

Law, R.D., Casey, M., Knipe, R.J. in press. The kinematic and tectonic significance of microstructure and crystallographic fabrics within quartz mylonites from the Assynt and Eriboll region of the Moine thrust zone, North West Scotland. Phil. Trans. R. Soc. Edinburgh. Earth Sci.

Leggett, J.K., McKerrow, W.S., Morris, J.H., Oliver, G.J.H. & Phillips, W.E.A. 1979. The north-western margin of the Iapetus Ocean. In: Harris, A.L., Holland, C.H. & Leake, B.E. The Caledonides of the British Isles - Reviewed. Spec. Publ. geol. Soc. London, 8, pp 499-511.

Moorbath, S. 1969. Evidence for the age of deposition of the Torridonian sediments of northwest Scotland. Scott. J. Geol. 5, pp 154-170.

Moorbath, S., Welke, H., & Gale, N.H. 1969. The significance of lead isotope studies in an ancient, high grade metamorphic basement complexes, as exemplified by the Lewisian rocks of north-west Scotland. Earth planet. Sci. Lett. 6, pp 245-56.

Pankhurst, R.J. 1982. Geochronological tables for British igneous rocks. In: Sutherland, D.S. (ed). Igneous rocks of the British Isles, John Wiley & SOns, pp 575-81.

Peach, B.N., Horne, J., Gunn, W. Clough, C.T., Hinxman, L.W. & Cadell, J.M. 1888. Report on recent work of the Geological Survey in the northwest Highlands of Scotland. Q. J. geol. Soc. London 44, pp.378-441.

Peach, B.N., Horne, J., Gunn, W., Clough, C.T., Hinxman, L.W. & Teall, J.J.H. 1907. The geological structure of the Northwest Highlands of Scotland, Mem. Geol. Surv. U.K.

Piasecki, M.A.J. 1980. New light on the Moine rocks of the Central Highlands of Scotland. J. geol. Soc. London, 137, pp 41-59.

Rathbone, P.A. & Harris, A.L. 1979. Basement-cover relationships at Lewisian inliers in the Moine rocks. In: Harris, A.L., Holland, C.H. & Leake, B.E. (eds). The Caledonides of the British Isles - Reviewed. Spec. Publ. geol. London, 8, pp.101-7.

Rathbone, P.A., Coward, M.P. & Harris, A.L., 1983. Cover and basement: A contrast in style and fabrics. In: Harris, L.D. & Williams, H. (eds). Tectonics and Geophysics of Mountain Chains, Mem. Geol. Soc. Am., 158, pp 213-223.

Sabine, P.A. 1953. The petrography and geological significance of the post-Cambrian minor intrusions of Assynt and the adjoining districts of N.W. Scotland. Q.J. geol. Soc. London, 109, pp 137-71.

Sibson, R.H. 1977. Fault rocks and fault mechanisms. J. geol. Soc. London, 134, pp. 191-214.

Smythe, D.K. Dobinson, A., McQuillin, R., Brewer, J.A., Matthews, D.H., Blundel, D.J. & Kelk, B. 1982. Deep structure of the Scottish Caledonides revealed by the MOIST reflection profile, Nature, 299, pp 338-340.

Soper, N.J. & Barber, A.J. 1979. Proterozoic folds on the Northwest Caledonian Foreland. Scott. J. geol. 15, pp 1-11.

Soper, N.J. & Barber, A.J. 1982. A model for the deep structure of the Moine thrust zone. J. geol. Soc. London, 139, pp 127-138.

Stewart, A.D. 1969. Torridonian rocks of Scotland reviewed. Mem. Am. Assoc. Petrol. Geol. 12, pp 595-608.

Stewart, A.D. 1975. Torridonian rocks of western Scotland. In: Harris, A.L. et al. (eds). A correlation of the Precambrian rocks in the British Isles. Spec. Rep. geol. Soc. London. 6 pp 43-52.

Sutton, J. & Watson, J.V. The pre-Torridonian metamorphic history of the Loch Torridon and Scourie areas in the north-west Highlands of Scotland and its bearing on the chronological classification of the Lewisian. Q. J. geol. Soc. London. 106 pp 241-307.

Swett, K. 1969. Interpretation of depositional and diagenetic history of Cambro-Ordovician succession of northwest Scotland. In: Kay, M. (ed) North Atlantic Geology and Continental Drift. Mem. 12, Amer. Assoc. Petrol. Geologists, pp 630-646.

Tanner, P.W.G. 1971. The Sgurr Beag slide, a major tectonic break within the Moinian of the Western Highlands of Scotland. Q.J. geol. Soc. London, 126, pp 435-63.

Van Breemen, O., Aftalion, M. & Johnston, M.R.W. 1979. Age of the Loch Borralan complex, Assynt and late movements on the Moine thrust. J. geol. Soc. London, 136, pp 489-96.

White, S.H. 1979. Grain size and sub-grain size variation across a mylonite zone. Contrib. Mineral. Petrol. 70, pp 193-202.

Williams, G.E. 1969. Petrography and origin of pebbles from Torridonian strata (late precambrian) northwest Scotland.

Mem. Am. Assoc. Petrol. Geol. **12**, pp 609–29.

Wooley, A.R. 1971. The structural relationships of the Loch Borrolan complex, Scotland. Geol. J. **7**, pp 171–82.

SOME ASPECTS OF GEOPHYSICS IN THE CALEDONIDES OF THE UK

Richard T Haworth
British Geological Survey
Keyworth, Nottingham, UK
NG12 5GG

INTRODUCTION

Geological similarities between the Caledonides of the United Kingdom and the Appalachians have long been recognized and form the basis of many attempts to reconstruct a single, continuous orogenic belt incorporating them. Surface continuity is examined in numerous ways (e.g. structure, metomorphism, palaeontology) but extrapolation between exposures is only possible with the assistance of geophysical methods. The author has made several attempts to use geophysical data to correlate gross structural zones in the Canadian Appalachians with their postulated counterparts in the British Isles (Haworth, 1980; Haworth and Jacobi, 1983; Jacobi and Kristoffersen, 1981; Lefort, 1980). Participation in the field symposium reported in this volume provided an opportunity to inspect the geology of the two areas and comment upon their correlation in light of the similarities and differences in their geophysical expression. This paper attempts to summarize the geophysical character of each of the areas visited by the field symposium to identify the key geological questions to whose answer geophysics might contribute, and to identify the problems that might only be answered by geophysical work. The attempt will necessarily be incomplete because of the author's limited contact with detailed interpretation of geophysical data in the United Kingdom and the fact that the field programme, although comprehensive, could not be all inclusive. Nevertheless it may provide a guide to those less familiar with geophysics or its application in the UK and may provide a basis for defining those activities that might contribute significantly to a regional geophysics programme for the United Kingdom.

Canadian Appalachian: British Caledonides correlation

The five major terranes of the Canadian Appalachians have a reasonably definitive geophysical expression by which their continuity to the continental margin can be readily defined (Haworth and Jacobi, 1983; Jacobi and Kristoffersen, 1981; Haworth, 1981). The clearest of these boundaries is the northern margin of the Avalon terrane which

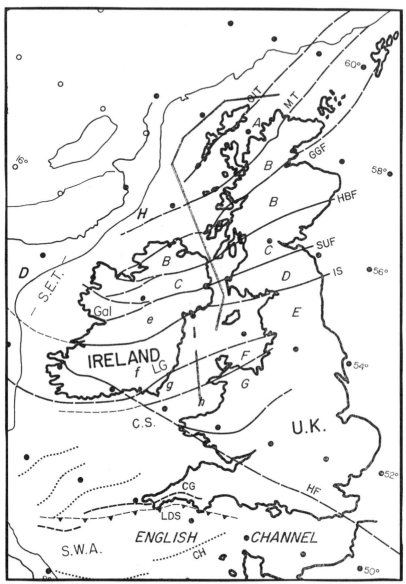

1. Geological zonation map of the British Isles with upper case letters (A through G) denoting the zones of Dewey (1974) and the lower case letters (e through g) denoting those of Phillips et al., (1976). The bold line between UK and Ireland is the approximate location of the WINCH profile (Brewer et al., 1983). OIT = Outer Isles Thrust; MT = Moine Thrust; GGF = Great Glen Fault; HBF = Highland Boundary Fault; SUF = Southern Uplands Fault; IS = Iapetus Suture. From Haworth and Jacobi (1983), by kind permission of the Geological Society of America.

internally is segmented into zones of structural "highs" and "lows" which have correlative high and low gravity and magnetic anomalies. Its boundary the Gander terrane, which, because of its granite character, exhibits predominantly negative gravity anomalies, is marked in Newfoundland by the Dover Fault, which extends across the continental shelf to the western end of the Charlie Fracture Zone (Haworth, 1977). However, the Delta terrane (Haworth, 1980) is interposed between the Gander and Avalon terranes half way across the shelf and the British landfall of the Dover Fault equivalent might be separating different terranes from those at its Canadian landfall. The Dunnage terrane, representing the remnants of the crust of the ocean that closed in Newfoundland during the Ordovician, has high gravity anomalies with lineated magnetic anomalies that represent slivers of that oceanic crust in local synforms superimposed on a generally southeastward dipping sheet. This geophysical character is in marked contrast with the adjoining Humber terrane which has chaotic magnetic anomalies and low gravity anomalies associated with the exposure of Precambrian (Grenville) Basement. The Meguma terrane is not present in Newfoundland, but its boundary with the Avalon terrane crosses the Scotian Shelf where it is marked by a major positive gravity and magnetic anomaly, the Collector Anomaly (Haworth and Lefort, 1979). Extension of that anomaly across the southern Grand Banks to the Newfoundland Seamounts, and correlation of the latter with seamounts west of Lisbon on most pre-drift reconstructions of the Atlantic, indicate that its correlative would not be anticipated on the continental shelf adjacent to the British Isles.

Differences in regional geophysical character may only be demonstrated by comprehensive data compilations on the same scale and projection. These have been accomplished for both gravity and magnetic data at scales of 1:1 million and 1:5 million within IGCP Project 27 (Haworth and Jacobi, 1983). It is not claimed that the base for these compilations is an entirely valid reconstruction for the time immediately prior to the opening of the North Atlantic, but it is believed to be sufficiently accurate to provide the first iteration in correlating structures across the North Atlantic from which second order adjustment (e.g. Max and Lefort, 1984) may proceed.

Whereas the geological terranes of Newfoundland have boundaries that are marked by large magnetic anomalies, the amplitudes of anomalies in the British Isles are generally reduced to less than half and there are few major lineated anomalies. The most obvious of these on a regional scale is associated with the Great Glen Fault, with which is also associated a major gravity gradient. High gravity anomalies are confined to the thrust zones of the North Western Highlands of Scotland and the Outer Isles, and the Southwest Approaches. Only in the latter area are these anomalies so lineated as to be possibly correlative with the lineated anomalies of the Avalon terrane. This difference in geophysical character must surely be diagonostic of a geological difference that is more fundamental than the greater thickness of cover sequences found in the British Isles.

There has been considerable speculation based primarily on palaeomagnetic results, regarding transcurrent movement roughly along the axis of the orogen (Kent and Opdyke, 1978; Irving, 1979). Roy et al (1983) warned of the dangers of such speculation because of its possible foundation on an incorrectly resolved repetitive history of magnetization. Briden et al (1985) further substantiate this viewpoint and imply that such interpreted transcurrent movement is of a considerably smaller scale than initially proposed (2000 km in Van der Voo and Scotese, 1981). Any transcurrent movement or separate movement of an Armorican block (Van der Voo, 1979) is likely to have occurred during Devonian or Carboniferous time so that trans-Atlantic correlation of early Palaeozoic features on a Mesozoic reconstruction is fraught with danger. Readers should also clearly recognize the data sets on which the reconstruction is based. Although the reconstructions of Scotese et al (Scotese et al, 1979) are commonly claimed to be "palaeomagnetic", it is clear from their description that palaeomagnetic results form only a (small?) subset of the data (much of which is interpretational e.g. palaeobathymetric data) on which the reconstructions are based.

Few reconstructions of the North Atlantic would not require an offset between the Canadian Appalachians and the British Caledonides which Haworth et al (1985) claim to be the largest in the entire orogen. Phillips et al (1976) note that transcurrent movement between terranes with such a serrated edge would produce significant structural contrast along the length of the orogen. Differences in geophysical contrast might not therefore be so surprising.

Regional Geophysical Data in the British Isles

The recent BIRPS (British Institutions Reflection Profiling Syndicate) programme has provided an excellent deep (15 second) profile across the width of the orogen between Ireland and Great Britain (Brewer et al, 1983). It was intended that this deep profile complement the shallow detail presumably resident in the comprehensive network of reflection profiles collected by industry (but not yet publicly available) during geophysical exploration for hydrocarbons on the continental shelf. Limited comparison between the two data sets indicates little loss of detail in the shallow section of the deep profile (Matthews, D., personal communication).

The BIRPS data do not resolve the Highland Boundary Fault and the Southern Uplands Fault (Brewer et al, 1983) either because there is insufficient impedance contrast across the faults or they are too steep. The Moine Thrust can be interpreted from reprocessed data (Brewer and Smythe, 1984) but it is still not clear. Good images are however seen of the Outer Isles Thrust, the Great Glen Fault, the south Irish Sea lineament and, possibly, the Iapetus suture. The latter is seen as a dipping reflector between 3 and 8 seconds in depth, projecting to the surface at the northern edge of the Solway (Firth) Basin, and having few deep reflectors to the north.

Correlation of these data with that from North America have been attempted (Cook et al, 1985) with limited success but with a clear recognition of several general characteristics:

- (a) a remarkably transparent upper crust with few reflections other than over sedimentary basins,
- (b) a concentration of reflections in the lower crust leading to speculation as to its physical state (e.g. microfactures, filled with water?),

and (c) the Moho at relatively constant depth, but remarkably variable in character.

These characteristics also seem valid on land (Whittaker and Chadwick, 1983) so that initial concerns that the loss of upper crustal reflectors might be an artifact of working at sea rather than land seem to be ill founded.

The presence of reflectors in the lower crust has stimulated considerable speculation as to the physical state of that zone. Correlation of a high conductivity layer with the depths of those reflectors (Hutton and Haak, 1984) suggests that pore fluids might be the cause (Vine and Toussaint-Jackson, 1984) but insufficient modelling of the seismic reflection data in conjunction with the conductivity data has been done to confirm this. Partial melting and/or high temperatures at lower crustal depths are discounted in view of the generally low heat flow throughout the UK (Vine and Toussiant-Jackson, 1984). However, Matthews (1984) hypothesizes that the fracturing of the lower crust might be due to crustal stretching.

Seismic refraction coverage of the United Kingdom seems quite comprehensive (J Hall, personal communication) but the data sets are inhomogeneous. The LISP data set (Bamford et al, 1978) is the one most commonly used as the basis for regional studies because of its extent throughout Great Britain. Lines transverse to LISP (J Hall, personal communication) complement that earlier work both in detail and areal extent. Unfortunately the southern leg of LISP has until recently (Unpublished MSc thesis, University of Birmingham) received only scant attention (Blundell, 1981).

Interpretation of seismic refraction and magneto-telluric data depends on a comprehensive knowledge of the physical properties of rocks. Although a good start has been made on determining these properties (Hall, 1984), a much greater effort is needed, particularly in measurement of such properties at high temperature and pressure.

Regional interpretations of magnetotelluric data are based on long time series observations. Hutton and co-workers (Hutton et al, 1980) have provided this for Scotland in particular, but additional information is now available for northern England (Sule and Hutton, 1984). It is anticipated that the depth dependence of this method involving physical properties different from those primarily affecting seismic methods make the two methods complementary. With the recent success of seismic reflection profiling in examining the structure of orogenic belts, the impetus for more magnetotelluric work will grow.

The Heat Flow map of the United Kingdom is dominated by high values associated with Devonian age granites. Current observations are dominated by those made as part of Geothermal Energy programmes contracted by the UK Department of Energy centred on "hot dry rock"granites in S.W. England, the Lake District and the Grampian Highlands and "low enthalpy" aquifers in the sedimentary basins of eastern and southern England. Although this detail tends to emphasize the anomalous areas, extrapolation throughout the United Kingdom based on regional geological and geophysical data (primarily gravity) permits consideration of regional variations of heat production, conductivity and heat flow (Burley et al, 1984).

Hydro-fracturing at "hot dry rock" geothermal sites and concerns for environmental stability (of reservoirs, pipelines and drilling platforms) have also contributed to the nationwide monitoring of seismicity (Turbitt, 1984). In such a low seismicity area as the UK, comprehensive and uniform distribution of monitoring stations is a critical factor in determining the principal seismotectonic zones. Whereas the Great Glen Fault has been inferred from historical records as a prime slip zone, the advent of instrumental monitoring has not specifically justified this conclusion, and recent extension of the monitoring network offshore has demonstrated that the greatest activity is associated with the central graben of the northern North Sea (possibly due to glacial unloading, (Browitt, 1984).

In general, therefore, the geographical coverage of geophysical data, collected in recent years on a more localized scale for specific projects, has now reached the stage where a regional crustal interpretation of UK structure might proceed on a relatively firm foundation. Considerable compilation of data will be required before this is possible, but without it the inconsistencies and data gaps will not be recognized. Such compilations as have been made to date are of variable density of coverage and quality as to permit only a superficial analysis of the principal structures examined by the field symposium, but far more detailed interpretation would be possible with existing data. As a result, there may be more projects proposed than questions answered for each of the areas visited which will be discussed in turn in the following sections.

SOUTH WALES

The Welsh Basin contains a thick sequence of Ordovician and Silurian marine clastic rocks. The nature and distribution of these sedimentary rocks together with the nature of the volcanic rocks at its northern and southwestern margin indicates that they evolved in an ensialic trough. The southeastern margin of the basin is provided by the Welsh borderland/central England "platform" whose intrusive history indicates that southeastward subduction might have taken place beneath it in the late Precambrian-early Palaeozoic (Thorpe et al, 1984). Exposure of this basement is extremely limited but geological-geophysical correlation would suggest that it is responsible for the high magnetic anomalies throughout southern and southeastern Wales (Arthur, personal communication). Such a correlation would be compatible with the suggestion (Rast et al, 1976) that the southern UK "platform" is correlative with the Avalon terrane of the Appalachians. This suggestion is also supported by their relative location on pre-drift reconstrutions of the North Atlantic and the postulated continuity of geophysical anomalies between the Grand Banks of Newfoundland and the South West Approaches (Haworth and Jacobi, 1983). Further similarity is provided by the unconformity between the Precambrian and Cambrian rocks in both areas (King, 1980), although in Newfoundland the Ordovician (and Siluirian?) sedimentary rocks conformably overlie the Cambrian rocks whereas an unconformity separates the two at least locally in southwest Wales.

Although small scale and regional maps of the magnetic field (Haworth and Jacobi, 1983; Hall and Dagley, 1970) indicate that the magnetic high associated with the basement rocks is lineated, larger scale maps (Anon, 1965, 1972), indicate that the high is a zone, of individual segments that are offset from one another. Arthur (personal communication) believes that the faults separating segments are en-echelon and trend northeastward, but this author believes that the faults might in fact be dextral and trend north to northwestward. The latter fault trend is compatible with that within Cambrian sediments of the Avalon terrane of Newfoundland and is perhaps justified by the, albeit small scale, faulting of the lowermost Cambrian and Precambrian rocks at Porth Clais.

The northern part of the Welsh Basin is clearly shown by the BIRPS profile ending in Cardigan Bay (Brewer et al, 1983). Beneath that basin one reflector dips to the northwest, but its significance and relation to the southern margin of the Basin will only be apparent when the results of SWAT (South West Approaches Traverse) are published by BIRPS. Correlating the reflectors might be made even more difficult because the profiles cross the Precambrian margin and the Hercynian Front at a location in which they intersect.

Onshore seismic refraction data are only available from a segment of the LISP experiment for which only preliminary results were available (Blundell, 1981) until very recently (Unpublished MSc thesis, University of Birmingham). More refraction and reflection across the southern margin of the Basin would provide the necessary control for the interpretation of gravity and magnetic data which as in Newfoundland are valuable in deciphering the structural pattern of the Precambrian (Avalonian?) basement.

NORTH WALES AND ANGLESEY

A major positive magnetic anomaly lies on the northern side of the Welsh Basin coincident with lower Palaeozoic rocks. Magnetic anomalies increase rapidly to the north when crossing the Bala Fault, the highest anomalies being coincident with the Harlech Dome where lowest Cambrian rocks are exposed. A borehole on the dome penetrated 500 metres into volcanic rocks which Howells (M.F., personal communication, 1984) does not believe represents the basement to the Dome. Howells and his colleagues are currently reviewing the regional geology of Northwest Wales incorporating the geophysical data available and their conclusions will not be pre-empted here. Suffice it to say that the major positive gravity and magnetic anomalies of the area appear to be associated with its basement and volcanic rocks.

Individual fault blocks can be recognized within the extreme northwest of the area. Major changes in both gravity and magnetic anomaly coincide with the Menai Strait and the Berw Fault which runs across southeast Anglesey parallel to the Strait. In the northwestern half of the island the magnetic lineations run northwestward implying another significant change in structure. Southwest of the Lleyn Penninsula, on which lies the epicentre of a relatively major earthquake which occurred in 1984 (Turbitt et al, 1984), a north northeast trending segment of the WINCH profile showed a conspicuous reflector (the South Irish Sea lineament) which dips NNW at 25° to 15 to 18 km depth (Brewer et al, 1983). The focus of the earthquake at a depth of 18 km cannot be associated specifically with any of the faults passing through the area because their dip is not well known and seismic reflection profiles to link the fault imaged by the WINCH profile with the mainland are not available. However it is reasonable to hypothesize that the significance of the WINCH reflector (it is probably the most prominent of all dipping events imaged by BIRPS) coupled with the major change in potential field implies a most significant crustal boundary in the vicinity of the Menain Strait. Current (late 1984) monitoring of the aftershock sequences of the earthquake on Lleyn Penninsula has not yet indicated a focal plane that might add weight to its possible association with this crustal boundary.

It has been suggested elsewhere (Haworth and Jacobi, 1983) that this major boundary might be equivalent to the northwestern edge of the Avalon terrane in eastern Canada. Although structures in NW Anglesey are orthogonal to structures in SE Anglesey just as they are between the Delta terrane and Avalon terrane at the Canadian Margin, this may be purely fortuitous. Such orthogonality of structure is not seen elsewhere along the boundary which must pass through or close to southeast Ireland. Nevertheless, the hypothesis is not contradicted by this new information and might be worthy of detailed modelling.

LAKE DISTRICT

The primary field stop in the Lake District was in the vicinity of the Shap Granite, which has a Siluro-Devonian age (Wadge et al, 1978). The granite lies within a prominent negative gravity anomaly which encompasses all the granites of the Lake District. Lee (in press) has conducted a detailed interpretation of the gravity field and concludes that all the granites are interconnected at a depth of less than 4 km beneath the surface sediments and that they dominate the crust for an area of over 2000 sq km to a depth of about 9 km. This would represent a tremendous volume of granite, but Wones (D.R., personal communication 1984) suggests that granite was derived from an oxidized granulite source and may therefore have come from subducted Borrowdale Volcanics having been subducted.

The granites of the Lake District have been subject to detailed investigation for their geothermal potential. The Shap and Skiddaw granites appear to be the most radiothermal of the intrusions. Descriptions of the subsurface structure, heat flow and heat production throughout the area will be published soon (Downing & Gray, in press; Lee, in press).

Interpretation of geophysical data in the Lake District has been necessarily dominated by consideration of the granites. Now that their distribution has been well defined, surface correction should be possible to permit the examination of other data sets, such as seismic refraction. Until deep seismic information is obtained there will be little hope of testing the various hypotheses for the evolution of the Caledonides.

SOUTHERN UPLANDS

The sedimentary rocks of the Southern Uplands are generally thought to have been deposited within an accretionary prism. Leggett et al (1983) believe that the subduction zone associated with that prism dipped northwestwards with eventual under thrusting of the continental crust of northern England. The details of the model seem however still to be in dispute. Some models invoke detachment of the surface structure by major thrusts. Other models imply lack of continuity of structure at depth across the Southern Uplands fault because of large strike slip movement, with the accretionary prism having developed south of Newfoundland or Rockall before translation to its current location. In such circumstances, geophysics has a major role to play in determining the deep structure and therefore the setting in which the sedimentary prism developed. The reader is reminded that such large scale translation along what was a highly irregular margin would have resulted in major deformation of the original structures.

The Southern Uplands have been divided into distinct stratigraphic zones (Leggett et al, 1979) separated by faults whose significance and sense of movement is still under investigation. Seismic refraction data (Hall et al, 1983) show that some of them undoubtedly have major significance. Hall et al (1983) have examined the results of two orthogonal refraction lines in detail and shown that the basement south of the Southern Uplands Fault can be divided into blocks of high 6.0 km/s) and low (5.6 km/s) velocity. There is also a good correlation between areas of positive gravity and magnetic anomalies which may also correlate with the high velocity blocks in much the same way as exists, although in a different setting within the Avalon terrane (Haworth and Lefort, 1979). Within the Southern Uplands the northern and southern edges of the northern high velocity block roughly correspond to the northern edge of zone 4 (the Kingledores Fault) and the southern edge of zone 8 (the Hawick Line) respectively. The northern edge of the southern high velocity block lies almost at the southern edge of the Southern Uplands. The upper edge of the high velocity material is at the surface immediately north of the Southern Uplands Fault but lies at a depth of 5 km south of the Fault. The LISP profile is the most definitive indication of structure at greater depth (Bamford et al, 1978; Hutton et al, 1980). Major changes in structure occur at both the Highland Boundary Fault and Southern Uplands Fault such that north and south of them respectively the crustal section is quite different, with a significant increase in thickness of the lower Palaeozoic sediments (within the prism) to a depth of 15 km.

Initial analysis of magnetotelluric data along a profile close to that followed by LISP, indicated that the relatively low conductivity sediments of the sedimentary prism in the Southern Uplands, was underlain by a high conductivity zone to a depth of 90 km (Hutton et al, 1980). These data have now been supplemented on a local scale to indicate that the conductive layer begins at a depth of 17 km (Sule and Hutton, 1984), and it is apparent that further work could contribute significantly to elucidating the zonation of shallow structure discussed earlier in this section.

McKerrow (personal communication) has used the LISP data to support the model for northwestward subduction beneath the Southern Uplands (Leggett et al, 1979), whereas Hall et al (1983) sees some support for Bluck's model which involves northward translation of the entire accretionary prism. Critical examination of such hypotheses will be a prime role for geophysics.

The Southern Uplands Fault and/or the Ballantrae complex (Bluck, 1978) are marked by a prominent magnetic anomaly which can be easily traced across the North Channel to Belfast Lough. Its correlation with a specific feature of the gravity field is less obvious. High resolution magnetic field and VLF data have been correlated with the geology of the Ballantrae complex (Carruthers,   ), but without succeeding in establishing regional continuity of the highly faulted and segmented outcrop. A future project should concentrate on establishing whether the magnetic anomaly along the Southern Uplands

Fault is a consequence of that fault or of the mafic and ultramafic rocks (exposed at Ballantrae) lying close to the fault.

HIGHLAND BOUNDARY COMPLEX

The Highland Boundary Fault is marked by a narrow magnetic anomaly roughly parallel to the magnetic anomaly that correlates with the Southern Uplands Fault. Between the two, high magnetic anomalies having a more easterly trend, are presumably associated with the results of Carboniferous tectonism in the Midland Valley. The gravity field exhibits a major change north of the Fault, but the most intense lows are associated with granite outcrop, so it is anticipated that, as in the Lake District, most of the outcrop is connected at depth. Whereas it is also anticipated that the granites do not extend south of the Fault, the change in gravity itself need not be diagnostic of the Fault's location or attitude.

Henderson & Robertson,(1982) note a major change in sedimentary structure approximately 10 km north of the Highland Boundary Fault. This seems to coincide with a northwards increase in the thickness of the upper crustal layer shown by the LISPB profile (Bamford et al, 1978). The velocity of the upper crustal layer also changes in the vicinity of the Highland Boundary Fault but its exact location cannot be defined by the LISP refraction data. A thin high conductivity layer also spans the surface location of the Fault (Hutton et al, 1980), but within the Midland Valley the conductivity structure is uncertain. These observations indicate that a comprehensive investigation of the shallow conductivity and velocity structure of the region adjacent to the Highland Boundary Fault might indicate the reason for the change from steeply dipping Dalradian metasediments immediately north of the Fault, to more gently dipping metasediments 10 km farther north. There seems little evidence from the WINCH profile (Brewer et al, 1983) for this change in structure, although the Highland Boundary Fault itself might be recognizable as a reflector

THE GREAT GLEN FAULT

The area between the Highland Boundary Fault and the Great Glen Fault is dominated by negative Bouguer gravity anomalies, the most intense of which are correlated with the granites of the Grampian Highlands. These negative anomalies are abruptly truncated to the SW along a line correlative with a band of magnetic anomalies due to a NW trending Tertiary dyke swarm (Hipkin & Hussain, 1983). The fundamental importance of this line, which also marks significant geological and geochemical contrasts, deserves much more investigation. Isostatic anomalies calculated by Hipkin and Hussain (1983) follow this NW lineation and appear to cross-cut the Great Glen Fault, the Highland Boundary Fault and possibly the Southern Uplands Fault.

Northeast of this boundary, the Great Glen Fault is marked by significant geophysical anomalies. There is in general a marked northwestward increase in gravity across the Fault. It is also the centre of an approximately 30 km wide zone of positive magnetic anomalies which extends into the Moray Firth and can be followed northwards to provide evidence for extension of the Great Glen Fault to the Shetlands.

The Fault is believed to have a large component of lateral slip. Paleomagnetic results have been interpreted to suggest that up to 2000 km of displacement might be accommodated by the Fault (Van der Voo & Scotese, 1981) but these interpretations have been faced with increasing scepticism (Briden et al, 1984). There seems little hope for providing independent evidence for the amount of displacement by matching similar geological or geochemical markers on opposite sides of the Fault.

Historically, the Great Glen Fault has been identified as a zone of higher than normal seismicity (Lilwall, 1976). The significant events reported by inhabitants of the Fault zone might however be more of an indication of population concentration over a long period than seismicity. Certainly, seismicity monitored instrumentally over the past few years (Turbitt, 1984) shows little support for the Fault being active.

The LISP seismic refraction profile (Bamford et al, 1978) indicated that the Caledonian metamorphic rocks were about 3 km thicker between the Great Glen Fault and the Highland Boundary Fault than they were to the north of the Fault. Although there is no direct corroboration of this by the WINCH seismic reflection profile (Brewer et al, 1983) the Fault zone does seem to correlate with an extremely steep northward dipping reflector that separates two zones having contrasting reflection characteristics. This correlation is somewhat tenuous because of the bifurcation of the Fault and its possible offset of the Moine Thrust, but the Fault must lie within 20 km of this marked change in crustal character.

Independent evidence for the nature of this crustal contrast might be anticipated from the interpretation of electrical conductivities from magneto-telluric and magnetic variation studies. However, the regional results of Hutton et al (1980) suggest a gradual decrease in conductivity of the lower crust north of the Highland Boundary Fault, the Great Glen Fault only being marked by a thicker zone of higher conductivity rocks which are near surface in the vicinity of the Fault.

In summary therefore the Great Glen Fault has a marked geophysical expression indicating contrasts across the Fault zone at least to the depth of the Moho. The nature of that contrast and the way in which it was effected are not, however, known.

2. Gravity anomaly map of the British Isles. From Haworth and Jacobi (1983) by kind permission of the Geological Society of America.

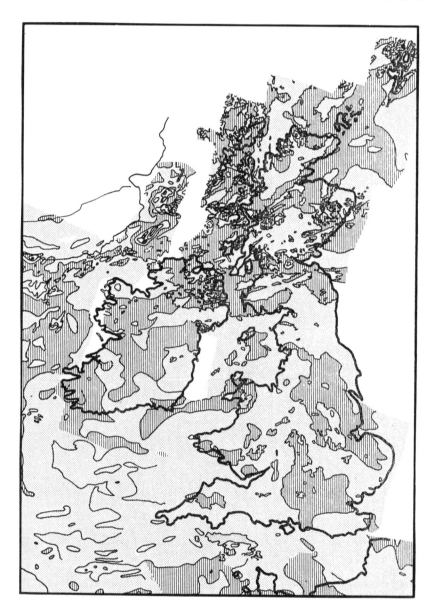

3. Magnetic anomaly map of the British Isles. From Haworth and Jacobi (1983), by kind permission of the Geological Society of America.

NORTHWEST HIGHLANDS OF SCOTLAND

The Moine Thrust and Outer Isles Thrust are the most significant of the structural boundaries in northwest Scotland and each has a significant geophysical expression.

The Outer Isles Thrust is coincident with a significant northwestward increase in the gravity field. That gravity gradient does not however have a clear extension north and south of the Outer Hebrides. Neither gravity nor magnetic data provides a clear indication of the location in which the Outer Isles Thrust might be expected to be seen on the WINCH profile (Brewer et al, 1983). However there is clear evidence on that seismic reflection profile for a reflector in the vicinity of the geometrical projection of the thrust. The Outer Isles Thrust is one of the few features on the BIRPS profiles which extends from the surface into the lower crust. It dips southeastward at 25° to a depth of 18 to 20 kilometres.

The Moine Thrust correlates with a much smaller gradient in the gravity field and the reflectors seen on the WINCH or MOIST seismic reflection profiles in the vicinity of the extension of the Thrust are less well correlated with it. Brewer and Smythe (1984) have interpreted the MOIST line as indicating that the Moine Thrust is offset eastwards by about 30 km, a suggestion supported by an offset, although of greater amplitude, in the gravity and magnetic anomaly trends almost coincident with the north coast of Scotland. The interpretation by Brewer and Smythe (1984) has provided some constraints for the thrust models for the Moines proposed by Coward (1980) and Soper and Barber (1982), but even more constraint might be available following thorough analysis of the gravity data. In particular, the methods applied by Royden and Hodges (1984) to the determination of the thermal uplift and erosion of thrust nappes in Scandinavia, and by Quinlan and Beaumont (1984) to the thrusting in the Appalachians might be able to discriminate between the thrust models proposed for the Moines.

FUTURE POTENTIAL

Much of this paper has served to indicate where further work needs to be done to elucidate the deep structure and tectonic development of the Caledonides of the UK. The gravity and magnetic maps of the UK provide a significant base from which to work. Digitization of the magnetic data and organisation of the digital gravity data is underway within the British Geological Survey to facilitate that analysis. The BIRPS programme has now completed a full cross section of the Caledonides, with additional sections to the northeast of Scotland and west of Ireland expected to be profiled before 1988. Recent seismic refraction experiments in the vicinity of the Iapetus suture have stimulated the anticipated acquisition of considerably more refraction data in the Irish Caledonides in 1985. Conductivity profiles will provide independent evidence for the nature of the lower crust suggested by these seismic experiments. Such work will, however, need the support of a major laboratory programme to investigate the physical properties of rocks at the tempreatures and pressures anticipated in the lower crust. Analysis of longer period magnetic variations will provide information on changes in conductivity beneath the crust. Such changes have been interpreted as indicating the remains of ancient subduction zones (e.g. Hutton et al, 1977; Wright & Cochrane, 1978). Sub-crustal topography can also be inferred from seismic travel-time anomalies (Stewart, 1978) and the possibilities of determining lithospheric structure beneath the Caledonides by similar means is currently under investigation (El-Haddadeh & Fairhead, 1984).

Our understanding of the overall plate tectonic setting of the British Caledonides will be considerably enhanced if each of these geophysical initiatives is pursued. Only when that is done will it be possible to put the detailed local work into the perspective it needs.

REFERENCES

ANON, 1965, 1972. Aeromagnetic Map of Great Britain, Sheet 1 (1972) Sheet 2 (1965) British Geological Survey

BAMFORD, D., NUNN, K., PRODEHL, C., & JACOB, B. 1978. LISPB-IV Crustal structure of Northern Britain. Geophysical Journal of the Royal Astronomical Society, 54, 43-60

BLUCK, B.J. 1978. Geology of a continental margin 1. The Ballantrae Complex in Crustal Evolution in Northwestern Britain and Adjacent Regions, (eds) Bowes, D.R., and Leake, B.E. Geological Journal Special Issue 10, 151-162

____ 1984. Precarbiniferous history of the Midland Valley of Scotland. Transactions of the Royal Society of Edinburgh: Earth Sciences, 65, 275-295

BLUNDELL, D.J. 1981. The nature of the continental crust beneath Britain. In: Petroleum Geology of the Continental Shelf of North West Europe (eds) Illing, L.V., and Hobson, G.D., 58-64.

BREWER, J.A., MATTHEWS, D.H., WARNER, M.R., HALL, J., SMYTHE, D.K. & WHITTINGTON, R.J. 1983. BIRPS deep seismic reflection studies of the British Caledonides. Nature, 305, 206-210

____ & SMYTHE, D.K. 1984. MOIST and the continuity of crustal reflector geometry along the Caledonian-Appalachian orogen. J. Geol. Soc. London, 141, 105-120

BRIDEN, J.C., KENT, D., LAPOINTE, P., LIVERMORE, R.A., ROY, J., SEGUIN, M.K., SMITH, A.G., VAN DER VOO, R., & WATTS D.R. 1985. Palaeomagnetic Constraints on the Evolution of the Caledonide-Appalachian Orogen. In: Evolution of the Caledonide-Appalachian Orogen (eds) Harris, A.L. & Fettes, D.

____ TURNELL, H.B., & WATTS, D.R. 1984. British Paleomagnetism, Iapetus Ocean, and the Great Glen Fault. Geology, 12, 428-431

BROWITT, C.W.A., TURBITT, T. & NEWMARK, R.H. 1984. Advances in the investigation of North Sea earthquakes. Proceedings of Oceanology International Conference. OI2.7/1 to 2.7/3

BURLEY, A.J., EDMUNDS, W.M. & GALE, I.N. 1984. Catalogue of geothermal data for the land area of the United Kingdom: Second Revisiion April 1984. Report Series on the Investigation of the Geothermal Potential of the UK, British Geological Survey

CARRUTHERS, R.M. 1978. Detailed airborne surveys, 1978. An assessment of geophysical data from the Girvan-Ballantrae district. Applied Geophysics Report 78. British Geological Survey

COOK, F., MATTHEWS, D.H., & JACOB, A.W.B. 1985. Crustal and Upper Mantle Structure of the Appalachian-Caledonide Orogen from Seismic Results. In: Evolution of the Caledonide-Appalachian Orogen (eds) Harris, A.L. & Fettes, D.

DOWNING, R.A. & GRAY, D.A. (in press) Geothermal resources of the UK. J. Geol. Soc.

EL-HADDADEH, B.R.H., & FAIRHEAD, J.D. 1984. Teleseismic delay time study of the UK. Geophysical Journal of the Royal Astronomical Society, 77, No.1, 299

JACOBI, R.D., & KRISTOFFERSEN, Y. 1981. Transatlantic correlations of geophysical anomalies on Newfoundland, British Isles, France and adjacent continental shelves. In: Canadian Society of Petroleum Geologists Memoir 7, 197-229 (eds) Kerr, J.W., & Fergusson, A.J.

KENT, D.V., & OPDYKE, N.D. 1978. Paleomagnetism of the Devonian Catskill Red Beds: Evidence for motion of the coastal New England-Canadian Maritime Region Relative to Cratonic North America. J. Geophys. Res, 83, 4441-4450

HALL, D.H. & DAGLEY, P. 1970. Regional magnetic anomalies - An analysis of the smoothed aeromagnetic map of Great Britain and Northern Ireland. Institute of Geological Sciences 70/10

HALL, J. 1984. Physical properties of the lower continental crust. Geophysics J R Astronomical Society, 301

HALL, J., POWELL, D.W., WARNER, M.R., EL-ISA, Z.H.M., ADESANYA, O., & BUCK, B.J. 1983. Seismological evidence for shallow crystalline basement in the southern uplands of Scotland. Nature, 305, 418-420

HAWORTH, R.T. 1977. The continental crust northeast of Newfoundland and its ancestral relationship to the Charlie Fracture zone. Nature, 266, 246-249

_____ 1980. Appalachian structural trends northeast of Newfoundland and their trans-Atlantic correlation. Tectonophysics, 64, 111-130

_____ 1981. Geophysical expression of Appalachian-Caledonide structures on the continental margins of the North Atlantic. In: Geology of the North Atlantic Borderlands, Canadian Society of Petroleum Geologists, Memoir 7, 429-446

_____ & LEFORT, J.P. 1979. Geophysical evidence for the extent of the Avalon zone in Atlantic Canada. Canadian Journal of Earth Sciences, 16, 552-567

HAWORTH, R.T., & JACOBI, R.D. 1983. Geophysical correlation between the geological zonation of Newfoundland and the British Isles. In:(eds) Hatcher, R.D. Jnr., Williams, H., & Zutz, I., Contributions to the tectonics and geophysics of mountain chains. Geological Society of America, Memoir 159

―――― HIPKIN, R., JACOBI, R.D., KANE, M., LEFORT, J.P., MAX, M.D., MILLER, H.G. & WOOLF, F. 1985. Geophysical framework and the Appalachian-Caledonide connection. In: Evolution of the Caledonide-Appalachian Orogen (eds) Harris, A.L. and Fettes, D.

HENDERSON, W.G., & ROBERTSON, A.H.F. 1982. The Highland Border rocks and their relation to marginal basin development in the Scottish Caledonides. Journal of the Geological Society, 139, Part 4, 433-450

HIPKIN, R.G., & HUSSAIN, A. 1983. Regional gravity anomalies, 1. Northern Britain. Rep. Inst. Geol. Sci. 82/10

HUTTON, R., & HAAK, V. 1984. Electrical conductivity in continental lithosphere. Geol. Soc. Newsletter, 13, No.5, 9

HUTTON, V.R.S., INGHAM, M.R., & MBIPOM, E.W. 1980. An electrical model of the crust and upper mantle in Scotland. Nature, 287, 30-33

―――― SIK, J.M. & GOUGH, D.I. 1977. Electrical conductivity and tectonics of Scotland. Nature, 266, 617-620

IRVING E, 1979. Paleopoles and paleolatitudes of North America and speculations about displaced terrains. Can. J. Earth Sci., 16, 669-694

KING, A.F. 1980. The birth of the Caledonides: Late Precambrian rocks of the Avalon Penninsula, Newfoundland, and their correlatives in the Appalachian orogen. In: The Caledonides in the USA (ed) Wone, D.R., Department of Geological Sciences, Virginia Polytechnic Institute and State University, Memoir 2, 3-8

LEE, M.K. (in press) A new gravity survey of the Lake District and three-dimensional model of the granite batholith. Geol. Soc.

LEFORT, J.P. 1980. Un 'fit' structural de l'Atlantique nord: Arguments geologiques pour correler les marqueurs geophysiques reconnus sur les deux marges. Marine Geology, 37, 355-369

LEGGETT, J.K., McKERROW, W.S., MORRIS, J.H., OLIVER, G.J.H., & PHILLIPS, W.E.A. 1979. The north-western margin of the Iapetus Ocean. In: The Caledonides of the British Isles, Geo. Soc. of London, 499-511

―――― McKERROW, W.S., & SOPER, N.J. 1983. A model for the crustal evolution of southern Scotland. Tectonics, 2, 187-210

LILWALL, R.C. 1976. Seismicity and seismic hazard in Britain. Seismol. Bull. Inst. Geol. Sci., 4

MATTHEWS, D.H. 1984. BIRPS above, gas below: UK deep reflections. Geol. Soc. Newsletter, 13, No.5, 11

MAX, M.D. & LEFORT, J.P. 1984. Does the Variscan Front in Ireland follow a dextral shear zone? In: Variscan Tectonics of the North Atlantic region. J. Geol. Soc. 177-183

PHILLIPS, W.E.A., STILLMAN, C.J., & MURPHY, T. 1976. A Caledonide plate tectonic model. J. Geol. Soc. London, 132, 579-609

RAST, N., O'BRIEN, B.H., & WARDLE, R.J. 1976. Relationships between the Precambrian and Lower Paleozoic rocks of the "Avalon Platform" in New Brunswick, the north-east Appalachians and the British Isles. Tectonophysics, 30, 315-338

ROY, J.L., TANCZYK, E., & LAPOINTE, P. 1983. The Paleomagnetic record of the Appalachians in Regional trends. In: The geology of the Appalachian-Caledonian-Hercynian-Mauritanide Orogen (ed) Schenk, P.E., D. Reidel Publishing Co. NATO ASI Series C: Mathematical and Physical Sciences, 116, 11-26

ROYDEN, L., & HODGES, K.V. 1984. A technique for analyzing the thermal and uplift histories of eroding orogenic belts: A Scandinavian example. J. Geophys. Res., B Paper, 4B0610, EOS, 65, No.21, 361-368

SCOTESE, C.R., BAMBACH, R.K., BARTON, C., VAN DER VOO, R., & ZIEGLER, A.M. 1979. Paleozoic Base Maps. Journal of Geology, 87, 217-277

STEWART, I.C.F. 1978. Teleseismic reflections and the Newfoundland lithosphere. Can. J. Earth Sci., 15, 175-180

SULE, P.O., & HUTTON, V.R.S. 1984. A tensorial magnetotelluric study in S E Scotland. Geophysics J R Astronomical Society, 77, 289-330

THORPE, R.S., BECKINSALE, R.D., PATCHETT, P.J., PIPER, J.D.A., DAVIES, G.R., & EVANS, J.A. 1984. Crustal growth and late Precambrian - early Palaeozoic plate tectonic evolution of England and Wales. Journal of the Geological Society of London, 141, 521-536

TURBITT, T., BARKER, E.J., BROWITT, C.W.A., HOWELLS, M., MARROW, P.C., MUSSON, R.M.W., NEWMARK, R.H., REDMAYNE, D.W., & WALKER, A.B. 1984. The Earthquake of 19 July 1984, Lleyn Peninsula, U.K. (Magnitude 5.4). Global Seismology Report 239, British Geological Survey

TURBITT, T. 1984. Catalogue of British Earthquakes recorded by the BGS seismograph Network 1979, 1980, 1981. Global Seismology Report 210 British Geological Survey

VAN DER VOO, R. 1979. Paleozoic assembly of Pangea: A new plate tectonic model for the Taconic, Caledonian and Hercynian Orogenies, EOS, 60, No.18, 241

―――― & SCOTESE, C. 1981. Paleomagnetic evidence for a large (2000 km) sinistral offset along the Great Glen fault during Carboniferous time. Geology, 9, 583-589

VINE, F.J. & TOUSSIANT-JACKSON, J.E., 1984. Implications of laboratory investigations of electrical conductivity. Geol. Soc. Newsletter, 13, No.5, 9

WADGE, A.J., GALE, N.H., BECKINSALE, R.D., & RUNDLE, C.C. 1978. A Rb-Sr isochron for the Shap Granite. Proceedings Yorkshire Geol. Soc., 42, 297-305

WHITTAKER, A., & CHADWICK, R.A. 1983. Deep seismic reflection profiling onshore United Kingdom. First Break, 9-13

WRIGHT, J.A., & COCHRANE, N.A. 1978. Geomagnetic sounding of an ancient plate margin in the Canadian Appalachians. Advances in Earth and Planetary Sciences, 9, Supplement Issue to Journal of Geomagnetism and Geoelectricity - Electromagnetic Induction in the Earth and Moon, (ed) Schmucker, U.

THE CALEDONIAN GEOLOGY OF THE SCOTTISH HIGHLANDS

D J Fettes
British Geological Survey
Murchison House
West Mains Road
Edinburgh, UK

A L Harris
Department of Geological Sciences
The University
Liverpool, UK

with a contribution on the development of the Tay Nappe by
L M Hall
Department of Geology and Geography
University of Massachussets
Amherst, Mass. 01003
USA

INTRODUCTION

The orthotectonic zone of the British Caledonides lies on the northern margin of the orogen and consists of a varied sequence of late Middle Proterozoic to Cambrian metasediments and meta-igneous rocks resting wholly on continental crust. In Scotland the zone is bordered by the Highland Boundary fault to the SE and by the Moine Thrust zone, which forms the Caledonian front, to the NW. To the SE of the Highland Boundary fault, in the Midland Valley, lies a thick sequence of dominantly Devonian and Carboniferous sediments and associated intrusives while to the NW of the Moine Thrust zone are late-Archaean to Middle Proterozoic Lewisian gneisses and granulites, which are overlain by clastic and carbonate sequences of varying age from late Middle Proterozoic to Lower Ordovician age.

Four major lithological groups are recognised within the Caledonian orthotectonic zone:
1) the Moine, characterised by a rather monotonous sequence of psammites and pelites,
2) the Dalradian, a lithologically more varied sequence than the Moine with notable carbonate, graphitic pelite and volcanic formations,
3) Lewisian gneisses that occur as fault slices within the Moine Thrust zone and as highly reworked inliers in the Moine rocks NW of the Great Glen fault,
4) a varied series of rocks occurring largely as fault-bounded slivers along the Highland Boundary fault and referred to as the Highland Boundary Complex or Highland Border Series.

The term Moine was originally applied to the rocks lying to the NW of the Great Glen fault but gradually came to include the lithologically similar 'Central Highland Granulites' lying SE of the

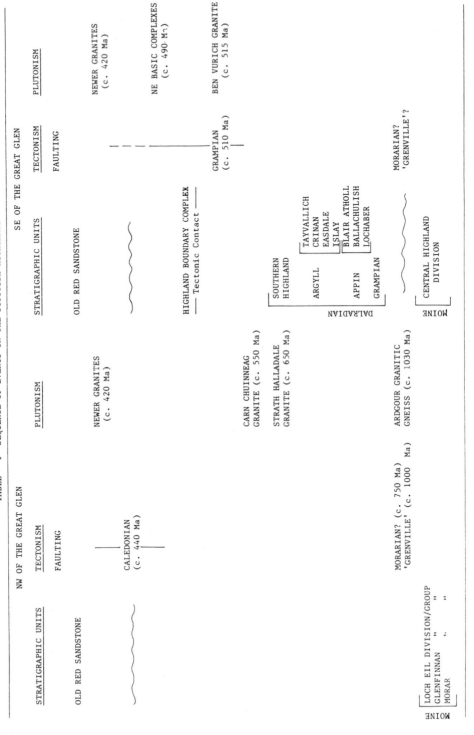

TABLE 1 SEQUENCE OF EVENTS IN THE SCOTTISH HIGHLANDS

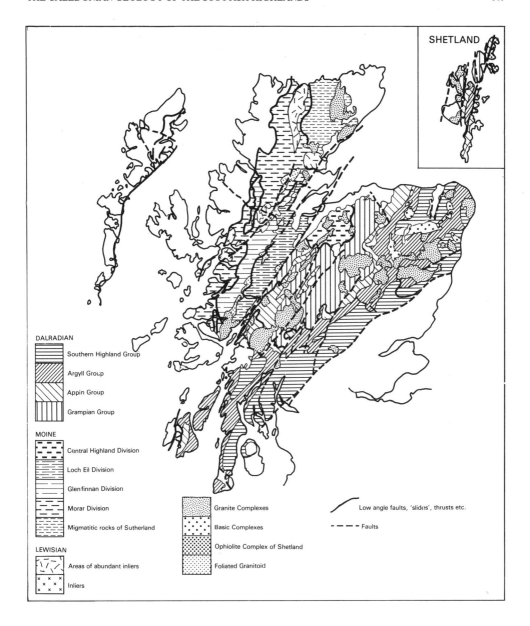

Fig. 1. Map of the Caledonides of the Scottish Highlands (pre- and post Caledonian rocks blank).

fault and occupying the ground between the fault and the Dalradian (see discussion in Johnstone, 1966, 1975). Although much of the boundary between the Moine 'Central Highland Granulites' and the Dalradian is marked by a major ductile shear (the Iltay Boundary slide) there are several localities (Bailey, 1934; Treagus, 1974; Harris and Pitcher, 1975) where the Moine can be seen to pass upwards into the Dalradian by normal sedimentary conformity.

Most geologists (e.g. Brook et al., 1977; Harris et al., 1978; Piasecki et al., 1981) now accept that the Moine can be divided into a cover-basement sequence; the 'younger' and 'older' Moine (Johnstone, 1975). The 'older' Moine rocks have an unconformable relationship on Lewisian (late-Archaean) basement and have themselves suffered polyphase deformation and metamorphism during the Grenvillian (c 1000-1100 Ma) and the Morarian (c 750 Ma) events. The 'younger' Moine rocks pass up into the Dalradian and have suffered only Cambro-Ordovician deformation and metamorphism. Piasecki et al. (1981) implied the existence of both 'older' and 'younger' Moine rocks NW of the Great Glen but Roberts and Harris (1983) have shown that, at least in the southern part of the area all the rocks belong to the 'older' or basement sequence. SE of the Great Glen, within the 'Central Highland Granulites'/Moine, cover-basement relationships have been described (Piasecki, 1980; Harris et al., 1981) although everywhere separated by a major ductile shear (the Grampian slide of Piasecki (1980)).

In order to clarify the stratigraphic nomenclature, Harris et al. (1978) suggested that the term Moine should be restricted to the 'older' or basement rocks and that the cover or 'younger' Moine sequence be referred to as the Grampian Group and included as the lowermost group within the Dalradian Supergroup. Piasecki (1980) working in the Central Highlands proposed the term Central Highland Division for the basement or 'older' Moine and Grampian Division for the cover or 'younger' Moine. Although Piasecki's (1980) term Grampian Division is broadly equivalent to the term Grampian Group there are significant differences as will be discussed below. In this paper the terms Grampian Group and Moine as defined by Harris et al. (1978) are used.

MOINE

Lithostratigraphy

Johnstone et al. (1969) subdivided the Moine of the N Highlands into the Morar, Glenfinnan and Locheil Divisions. SE of the Great Glen fault only the Central Highland Division (see above) can be even broadly related to the Moine of the N Highlands. The nature of the divisions recognised by Johnstone et al. (1969) are summarized in Table 2. Rocks of the Morar Division are bounded to the east by the Sgurr Beag slide, a major ductile thrust, first described by Tanner (1970) and Tanner et al. (1970). For reasons outlined later, the displacement on this ductile thrust is certainly of tens and may be

TABLE 2: Summary of lithological characters of Moine formations of western Inverness-shire (after Johnstone, Smith and Harris, 1969).

7. LOCH EIL PSAMMITE	Variably quartzose psammitic granulite (locally a micaceous 'salt and pepper' type) with very subordinate bands of pelitic and semipelitic schist. Calc-silicate ribs and lenticles present throughout and locally abundant. Locally abundant metabasic bands.
6. GLENFINNAN STRIPED	Banded siliceous psammite (locally quartzite) and pelitic gneisses; pods or lenses of calc-silicate granulite.
5. LOCHAILORT PELITE	Pelitic gneiss, with subordinate psammitic or semi-pelitic stripes. Metasedimentary amphibolite and calc-silicate lenses are usually present.
	Position of Sgurr Beag Slide (ductile thrust)
4. UPPER MORAR PSAMMITE	Dominantly psammitic granulite, often pebbly, with common semipelitic bands; calc-silicate ribs present throughout.
3. MORAR (Striped and Pelitic) SCHIST	Dominantly pelitic rocks locally divided into: a. Rhythmically striped and banded pelite, semi-pelitic schists and micaceous psammitic rocks with abundant calc-silicate ribs. b. Pelitic schists with some subordinate semi-pelite stripes. c. Laminated grey, semi-pelitic and micaceous granulite, lo-ally with thin siliceous and calc-silicate ribs rare except towards the top; heavy-mineral bands present, but most common near the base.
2. LOWER MORAR PSAMMITE	Micaceous and siliceous psammitic granulites locally pebbly; subordinate semi-pelitic rocks developed locally and more thickly towards the top; heavy-mineral bands present, but most common near the base.
1. BASAL PELITE	Dominantly pelitic and semi-pelitic schists, thinly banded with psammite. Local basal conglomerate

hundreds of kilometres and hence the relationships of the lithostratigraphic sequence established for the Morar Division to that of the now adjacent and overlying Glenfinnan Division has been largely obscured. It is possible, however, that Glenfinnan rocks are younger than Morar since in the Ross of Mull, older rocks of Morar aspect pass up without a major structural break into younger rocks of Glenfinnan aspect.

A lithostratigraphic sequence for the Glenfinnan Division has now been established by Dr A.M. Roberts, while Strachan (1982, 1985) and Stoker (1983) have subdivided the Locheil Division rocks, distinguishing units of differing psammitic lithology. Roberts and Harris (1983) have confirmed a clear lithostratigraphic passage upwards from the older Glenfinnan into the younger Locheil, a passage which approximately coincides with the Loch Quoich line (Clifford, 1957), which will be discussed in more detail later. Roberts et al. (1984) have shown that the N Highland steep belt (Leedal, 1952) which lies to the W of the Loch Quoich line, contains outliers of Locheil rocks while the flat belt to the east locally carries inliers of Glenfinnan rocks.

In the light of the relationships in the Ross of Mull and the improved knowledge of the Glenfinnan-Locheil relationships it is proposed that the Moine rocks are accorded the status of a supergroup, comparable with the Dalradian and that the Morar, Glenfinnan and Locheil units are accorded the status of groups. Within each group formations are distinguished, but it is proposed, in view of facies variations, that these formations, like those of the Dalradian (Harris and Pitcher, 1975), are accorded local status only. It is further recognised that present knowledge does not allow the relationship between the Morar and Glenfinnan groups to be established firmly; despite the evidence from the Ross of Mull there remains a strong possibility that the Morar Group rocks accumulated in one basin floored by Lewisian while the Glenfinnan and Locheil accumulated in another. Thus they may be to some extent laterally equivalent, each having a cover-relationship with Lewisian basement.

The lithostratigraphic formations established for the three Moine groups are based on a consideration of gross lithological characteristics - psammitic, pelitic and heterogenous striped psammite and pelite. Lithologically similar psammitic or pelitic formations may be distinguished, however, by the presence, absence or relative abundance of minor lithologies such as heavy-mineral laminae, calc-silicates (usually occurring as thin ribs) and amphibolites (Johnstone et al., 1969). These refinements of gross lithological characteristics taken together with sedimentation structures showing the original way-up of strata have enabled the lithostratigraphic sequence within the groups to be determined. They have been used to establish the Glenfinnan-Locheil junction and to show that the Locheil Group is younger than the Glenfinnan Group.

Relationships across the Great Glen fault between the Moine units of the N Highlands and the Central Highland Division of Piasecki and van Breemen (1979) and Piasecki (1980) are unknown, although there is a resemblance which may be more than superficial between rocks of the Central Highland Division and those of the Glenfinnan Group. The thick monotonous garnetiferous pelitic gneisses of the two units are similar and both carry calc-silicates and mafic garnetiferous amphibolites. Psammites in both units are gneissic and are locally calc-silicate- and amphibolite-bearing; however, the thick units of heterogeneously striped and banded siliceous psammites and garnetiferous pelitic gneisses characteristic of much of the Glenfinnan Group are lacking in the Central Highland Division. Way-up within the Central Highland Division has not been established.

If the Central Highland Division can be correlated with the 'Old' Moine of the N Highlands, as seems likely, an unconformity is implied between the Moine of the Central Highland Division and the Dalradian (including Grampian Group) which contains near its top, formations of undoubted lower Cambrian age. The Dalradian suffered only lower Palaeozoic orogenesis principally during the Grampian episode (Tremadoc-Arenig) whereas the N Highland Moine underwent a prolonged polyphase orogenic history extending from the c. 1000 Ma (? = Grenville) to the late Silurian. The unconformity referred to above has not been firmly established at outcrop but marked contrasts between mid-high amphibolite-facies Central Highland Division rocks and much lower grade Grampian Group rocks are recorded across zones of major displacement such as the Grampian Slide (ductile shear zone) by Piasecki and van Breemen (1979) and Piasecki (1980).

Orogenesis

Unequivocal stratigraphic evidence to constrain the timing of orogenesis in the Moine rocks is not forthcoming. The Moine rocks of the Morar Group are unconformable on Lewisian gneisses and amphibolites. At Glenelg (e.g. Clough in Peach et al., 1910), and at Talmine on the north coast (Holdsworth, pers. comm.) Moine basal conglomerates have been recorded and field evidence indicates that the Lewisian rocks suffered a more prolonged and intense history of metamorphism, deformation and igneous activity than the adjacent Moine. An original, little modified cover-basement relationship has been inferred for the Glenfinnan Group rocks which rest on Lewisian slices emplaced above the Sgurr Beag ductile thrust. This evidence suggests that the Moine rocks were deposited later than, and suffered orogenesis later than the Laxfordian orogeny which affected the Lewisian gneisses. The youngest limit for Moine orogenesis is stratigraphically constrained only by Devonian or late Silurian Old Red Sandstone sediments which rest with strong unconformity on the Moine both in the N Highlands and SE of the Great Glen fault, to the SW of Inverness and near Elgin. Within the Moine terrain a complex history of deformation, metamorphism and igneous activity intervened between c. 1100 Ma and c. 400 Ma. The detailed timing of the orogenic events which made up this history can be constrained only by radiometric evidence. On the basis of this it has been concluded that orogenic activity occurred c. 1100 Ma to 1000 Ma ago and between about 460 Ma and 420 Ma (Brook et al., 1977; Powell et al., 1983; Powell and Phillips, 1985). An episode of pegmatite emplacement occurred about

750 Ma ago (van Breemen et al., 1974; Powell et al., 1983) but this of itself can hardly be regarded as constituting important orogenesis although it has recently been interpreted as coinciding with a period of crustal extension during the initiation of the Dalradian basin in which sedimentation must have commenced at about this time (Soper and Anderton, 1984.). Evidence for Precambrian deformation and metamorphism is forthcoming from several igneous bodies. The deformed Carn Chuinneag granite (550 ± 10 Ma) clearly cross-cuts pre-intrusion tectonic structures and imposes thermal metamorphic assemblages on garnet-grade regional phases, but does not further constrain their age. Units of the Strath Halladale Granite (649 ± 30 Ma, M. Brook quoted in Pankhurst, 1982) in the far NE of Scotland are later than deformation and high-grade metamorphism in the Moine-like country rocks of the Sutherland migmatite complex (McCourt, 1980). The contrast between the state of the Carn Chuinneag granite and the undeformed Strath Halladale granite has been taken to indicate the existence of a Caledonian front of deformation between the two bodies. A similar limit must exist between the Sutherland complex and the zone of major Caledonian thrusting to the west where Holdsworth (pers. comm.) has recorded ductile thrusting and folding as far east as the Naver/Swordly (Caledonian) slides. Powell et al. (1983) have demonstrated that granitic pegmatites > 776 ± 15 Ma) in western Inverness-shire have been strongly deformed but that the tectonic episode which deformed them is the third episode to affect the country rocks in which the pegmatites were emplaced. It is concluded that these country rocks had suffered intense deformation on two occasions prior to the emplacement of the pegmatites. The radiometric age of 1004 Ma obtained by Brook et al. (1977) from Morar Group rocks and interpreted by them as the age of peak metamorphism has been otherwise interpreted as dating of diagenesis. The interpretation of Brook et al. (1977) is, however, strongly supported by the age of 1030 ± 45 Ma obtained for the Ardgour granitic gneiss (Brook et al., 1976). This orthogneiss the nature and relationships of which are described by Barr et al. (1985) has undergone the entire hstory of deformation, metamorphism and igneous intrusion experienced by the Moine metasediments of both Glenfinnan and Locheil groups in which the protolith of the gneiss was emplaced. The gneiss itself, however, lacks an aureole, unlike other deformed granites of the Scottish Highlands such as Carn Chuinneag. This feature has been interpreted by Barr et al. (1985) as indicating that the protolith of the gneiss was emplaced into rocks already undergoing high-grade metamorphism and deformation. Two early episodes of deformation have been recognised alike in Ardgour granitic gneiss and Moine metasediments of all three lithostratigraphic groups. In the Glenfinnan and Locheil Groups these have been clearly shown to predate the 456 ± 5 Ma Glen Dessary syenite and the strong tectonic reworking which postdated the syenite but which predates c. 430 Ma pegmatites (Roberts et al., 1984). The earlier of the two Precambrian deformation episodes is believed (Powell, 1974) to have been responsible for the emplacement of Lewisian slices in the core of the major E-facing and W-rooting isoclinal folds of Glenelg and Morar, the envelope of which comprises

Morar Group metasediments. The second episode is widely recognised in
both Morar and Glenfinnan units as transposing the MS1 or MP1
gneissosity in pelitic gneisses into the S2 foliation. In the Locheil
and Glenfinnan Groups it has been shown to fold bedding and S1
foliation in sediments and thin (pre-DI) metabasic bodies into folds
having strongly curvilinear hinges with which is associated a N-S
stretching fabric (Holdsworth and Roberts 1984).

Heterogeneous overprinting of the Precambrian structures,
fabrics and metamorphic mineral assemblages took place during the
lower Palaeozoic (Caledonian). Major structures of this episode
include the Moine thrust zone, the Knoydart slide (ductile thrust),
the Sgurr Beag slide (ductile thrust) and the Grampian slide (ductile
shear zone). Because of the large displacement on the Great Glen
fault the relationship of the Grampian slide to the others is unknown.
The Moine thrust zone is thought to constitute a series of
WNW-directed foreland propagating thrusts of which the highest
individual thrust - the Moine thrust itself - is the oldest and
highest of a duplex system. Possibly the Moine tract as a whole is a
major duplex with the Sgurr Beag, Knoydart and Moine thrusts being a
foreland propagating series of thrusts with the Sgurr Beag ductile
thrust being the oldest and highest in the N Highlands (cf Soper and
Barber, 1982; Coward, 1980, 1983; Watson, 1984). Rathbone *et al*.
(1983) have suggested that the Sgurr Beag thrust gives insight to the
nature of the Moine thrust if it were to be traced down-dip into the
middle layers of the crust where ductile rather than brittle processes
prevailed. The slide zone is repeated by NNE-folding at the present
erosion surface, and juxtaposes Glenfinnan Group rocks against the
Morar Group. Regionally the Glenfinnan overlies the Morar Group and
occurs in the core of a major SSW-plunging synform in Morar, and
elsewhere overlies the now steeply E-dipping slide zone. Outliers or
klippe of the Sgurr Beag nappe (Rathbone *et al*., 1983) occur to the
west of the main thrust zone - in Kintail and in Fannich Forest. In
many places to the N of Kinlochourn the slide zone is marked by a thin
persistent slice of Lewisian basement - the Scardroy sheet; a major
break is inferred at the base of the slice while a cover-basement
relationship is thought to be preserved, albeit modified, at the top
of the slice. Where Lewisian rocks are absent, to the S of
Kinlochourn, the main zone of displacement is taken at the junction
between Morar and Glenfinnan lithologies. Morar Group rocks belonging
to various formations are blastomylonitic for many tens of metres
below the Lewisian slice. The absence of the Lewisian slices to the S
of Kinlochourn is interpreted as coinciding with a southward lateral
ramp in the ductile thrust zone. Although exposed for several
kilometres across the strike as the result of subsequent folding, the
Glenfinnan Group rocks adjacent to the thrust are remarkably
consistent in metamorphic grade (mid-high amphibolite facies). This
suggests that the Sgurr Beag thrust made a remarkably shallow cut
through the crust and this taken in conjunction with the major
lithostratigraphic break across the thrust argues that displacements
across the thrust are very large, being many tens or even hundreds of
kilometres. That displacement was towards the WNW as is shown by the

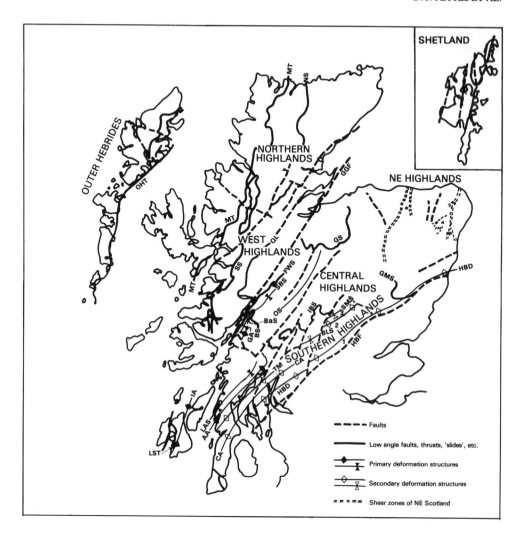

Fig. 2. Diagram illustrating structures referred to in text (after Ashcroft et al., 1984; Harris, 1983; Harte, 1979; Bradbury et al., 1979; Harte et al., 1984; Thomas, 1979; Roberts and Treagus, 1979; Piasecki et al., 1981). AA — Ardrishaig anticline; BAS — Ballachulish syncline; BLS — Ben Lawers synform; BS — Ballachulish slide; CA — Cowal anticline; FWS — Fort William slide; GA — Glen Creran anticline; GGF — Great Glen fault; GMS — Glen Mark slide; GS — Grampian slide; HDB — Highland Border downbend; HBF — Highland Boundary fault; IA — Islay anticline; IBS — Iltay Boundary slide; LAS — Loch Awe syncline; LST — Loch Skerrols thrust; MT — Moine thrust; NS — Naver slide; OHT — Outer Hebrides thrust; OS — Ossian steep belt; QL — Loch Quoich line; SBS — Stob Ban synform; SMS — Sron Mhor syncline; SS — Sgurr Beag slide; TM — Tarbert monoform. 1 — Highland Border steep belt; 2 — Flat belt; 3 — Loch Tummel steep belt.

orientation of LS shape fabrics related to the thrust. Powell et al., (1981) working in the SW part of the N Highlands have suggested, on the basis of calc-silicate mineral assemblages that relatively hot Glenfinnan Group rocks were emplaced above relatively cool Morar Group rocks and they have traced a history of thermal relaxation for the thrust zone. Barr (1983) has shown a contrast in Precambrian PT conditions across the Sgurr Beag thrust and suggests that the Morar Group carries garnet-to-kyanite grade assemblages consistent with 520-615°C and 5.5 - 7 kbar; to the east the Glenfinnan Group carries kyanite-to-sillimanite assemblages estimated at 590-640°C and 6-7 kbar. The age of the Sgurr Beag thrust has long been known to be Caledonian because in NE Scotland its characteristic SSE-plunging LS fabric also overprints and deforms the Carn Chuinneag (550 ± 10 Ma) granite. Kelley and Powell (1985) have recently shown that Sgurr Beag thrust fabrics predate those of the Moine thrust by some 25 Ma; this seems to confirm the foreland-propagating sequence because at least the final displacements on the Moine thrust postdate the Loch Borrolan (430 ± 4 Ma) syenite.

Before the Caledonian thrusting and the subsequent folding the regional attitude of the Moine rocks, their Precambrian structures and isograds were probably approximately horizontal and it is believed (Roberts and Harris, 1983) that they retain this Precambrian attitude to the east of the Loch Quoich line (Clifford, 1957) which marks the eastern limit of severe Caledonian reworking of the Precambrian crystalline rocks. Across the Loch Quoich line the N Highland steep belt, comprising largely Glenfinnan Group rocks, gives way eastwards to the flat belt comprising largely Locheil Group rocks. The steep belt is marked by upright folds on all scales which have markedly curvilinear hinges and a strong steep-to-vertical extension fabric. Fabrics including a well defined gneissosity in pelites are transposed into upright or very steep foliation accompanied by strong grain-size reduction. The upright folds and fabrics trend NNE-SSW and, where reworking strains are slight, produce remarkable interference structures with earlier intensely curvilinear (?Precambrian) isoclinal folds (Holdsworth and Roberts 1984). The folds and fabrics which comprise the steep belt are largely absent to the E of the L Quoich line and are constrained in age by the observation that they overprint the Glen Dessary syenite (456 ± 5 Ma) (Roberts et al., 1984). The syenite is folded into the core of an elogate periclinal synform and is foliated by the upright fabrics. It, itself, contains xenoliths of already folded and foliated Moine country rocks (Roberts et al., 1984). Progressive, incremental deformation during this episode of upright strain is indicated by the deformation of c. 440 Ma pegmatites and of microdiorites, members of which suites are considerably less deformed than the Moine country rocks (Talbot, 1984). It also constrains the bulk of the strain during reworking to the 456-440 Ma period - a period normally regarded as late Ordovician (McKerrow, pers. comm.). The NNE-SSW trending upright folds belong to the same set as those which folded the Sgurr Beag (and Knoydart) thrust(s), and hence the folding and possibly the syenite emplacement postdates all displacements in the Sgurr Beag thrust zone. In accordance with the Caledonian thrust duplex model for the N Highlands the folding of the Sgurr Beag thrust and the rocks to the east of and above it, are interpreted as the result of shortening in the hanging wall of the underlying and younger Moine thrust following 'sticking' of that thrust (Roberts et al., 1985). The termination of the folds and of reworking at the L Quoich line is interpreted as either marking the point to which a train of folds had worked back at the moment when 'unsticking' occurred or, as is more likely, the position in depth of a ramp of the Moine thrust (Roberts et al., 1985).

The late Ordovician age (456-440 Ma) of this deformation is at variance with the timing of orogenic events in the Dalradian to the SE of the Great Glen fault (qv) where the age of emplacement of granites and gabbros suggests early Ordovician orogenesis.

DALRADIAN

Lithology

The Dalradian Supergroup consists of four major lithostratigraphical Groups, from oldest to youngest, the Grampian Group, the Appin Group, the Argyll Group and the Southern Highland Group (Harris and Pitcher, 1975; Harris et al., 1978; Table 1, Fig 1). These groups comprise a varied succession of rocks with a cumulate thickness of c. 25 km, although vertical thickness of that magnitude never existed in any one area. The sequence youngs overall from NW to SE and shows a general change from a stable estuarine or intertidal environment (Hickman, 1975) to an unstable turbiditic environment.

Stratigraphic and palaeontological evidence for the age of the Dalradian rocks is imprecise. The base of the succession possibly post-dated the tectonothermal events affecting the Moine at c. 750 Ma. The tillites at the base of the Argyll Group may be equivalent to the Varanger tillite of Scandinavia which has been dated at 668 ± 23 Ma (Pringle, 1972) and the top of the succession is probably Lower to Middle Cambrian. Obviously the succession must span the Cambrian/Precambrian boundary and although there is little evidence for its precise position most workers place it in the lower part of the Argyll Group (Harris and Pitcher, 1975; Anderton et al., 1979; Harris et al., 1978). Anderton et al., (1979) have described how increasing tectonic instability led to major changes in the sedimentary environment following the deposition of the major quartzite formations which overlie the basal tillites of the Argyll Group. It is tempting to equate these quartzites with those marking the base of the Cambrian throughout the North Atlantic region (Swett, 1981), and consider the marked changes in the sedimentological record as part of the general transgression at the base of the Cambrian (Swett and Smit, 1972).

The Grampian Group although now regionally extensive is probably less than 1-2 km in overall thickness. It consists of quartzose and feldspathic sandstones, silts and muds characteristic of intertidal, coastal or estuarine environments (Hickman, 1975).

The base of the succeeding Appin Group (6-7 km thick) marks the first point in the Dalradian succession where subsidence proceeded faster than deposition (Hickman, 1975), leading to basin deepening and the deposition of muds and carbonates with local sand incursions. Although lateral facies change does occur the group is characterised by the general continuity of formations along strike with broad lithological associations traceable from Donegal in Ireland to the NE Highlands.

The Argyll Group has an overall cumulative thickness of c 12 km. The base is locally marked by tillite horizons. The group represents the beginnings of marked tectonic instability. Major growth faults with possible displacements of several kilometres were developed resulting in the deposition of thick wedges of clastic material (Anderton et al., 1979). Basins controlled by NW-SE and NE-SW trending faults have been recognised (Litherland, 1982; Anderton et al., 1979). Lateral facies variations are common and individual lithological units are generally much thicker than in the underlying groups. Harris et al., (1978) described a number of major basin-deepening and basin-shallowing sedimentary sequences from the Argyll Group, although overall the group reflects a transition to deeper water condition.

Anderton et al., (1979) noted that tectonic instability is characteristic of the group as a whole but becomes particularly noticeable above the basal quartzites and tillites. This coincides with the beginning of basic volcanicity which continued intermittantly thereafter, culminating in the Tayvallich lavas at the base of the Southern Highland Group. Graham and Bradbury (1981) noted that the episodic igneous activity corresponded to periods of basin deepening. They also showed that the chemistry of the lavas is similar to those formed at accreting plate margins and suggested the volcanics may represent embryonic oceanic crust.

The Southern Highland Group (3-4 km thick) comprises a series of grits, greywackes and muds indicative of proximal turbiditic fans. Lateral facies change is extreme and few formations can be traced for any distance (Harris and Fettes, 1972). These units are the products of highly unstable tectonic conditions.

The evolving nature of Dalradian sedimentation is therefore one of increasing instability with the relatively uniform shelf conditions breaking down into a series of fault-bounded basins as the crust became thinned and locally ruptured. It is probable that the main depocentre was moving south and east during this period (Hickman, 1975; Anderton et al., 1979). Palaeocurrent directions are varied but indicate considerable current movement along the basin. It is generally agreed that detritus was derived from a source to the NW, during the early and middle parts of the depositional history. However the proximal nature of the Southern Highland Group turbidites has been taken as evidence of continental detritus being derived from a landmass to the south.

Although its existence is inferred in the cover-basement arguments for the 'Moine', the unconformity at the base of the Grampian Group has not been satisfactorily demonstrated. Piasecki

(1980) has described a low angle fault, the Grampian slide, bringing together what he refers to as Grampian Division (cover) and Central Highland Division (basement) rocks. He argued (op. cit.) that this slide belongs to the second deformational event affecting the Grampian Division rocks and occurred during the Morarian (c. 750 Ma). Piasecki et al. (1981) however, agreed that the Grampian Division rocks pass without unconformity into the overlying Dalradian. They resolved the conflict in ages by suggesting that the Morarian tectonothermal event could be restricted to deep levels in the succession and might only be manifested by a sedimentary hiatus at higher levels such as at the Grampian Group and Appin Group boundary. The small overall thickness of the Grampian Group, however, probably argues against such a model.

The sedimentary transition of the Grampian Group into the Appin Group has long been recognised in the SW Highlands (eg. Bailey, 1934). In the Central Highlands, however, the Grampian Group is separated from the overlying Dalradian by a major southward dipping low angle fault, the Italy Boundary slide (IBS). Bailey (1934) had originally envisaged the Appin Group as a separate tectonostratigraphic unit (the Ballappel Complex) separated from the overlying Argyll and Southern Highland Groups (the Iltay Complex) by an extended IBS. This view, however, was largely abandoned when Rast and Litherland (1970) demonstrated a stratigraphic continuity between the Ballachalish and Blair Atholl formations of the Appin Group in the SW Highlands. In addition, recent work (Smith and Harris, 1976; Treagus and King, 1978; Upton, in press) has demonstrated the existence of Appin Group formations above the IBS in the Central Highlands.

Structure

The structure of the Dalradian rocks is dominated by a series of major, largely recumbent folds whose limbs are commonly modified or attenuated by low angle faults or 'slides'. The early interpretations of the structure owed much to the work of Clough and Bailey (see summary in Johnstone (1966, pp 14-17) who proposed a large fan like structure with an array of major folds facing to the northwest and southeast on either side of a central zone of divergence (Fig. 3). Bailey and his co-workers placed great faith in stratigraphic continuity and explained all discontinuities by low angle faults. Several workers have challenged the rigidity of this view and have preferred to attribute some of the structures to facies change (eg. Voll, 1964; Hickman, 1978). Despite these modifications, however, the current structural interpretations are still fundamentally similar to those of the early workers.

Recent research (e.g. Roberts and Treagus, 1977, 1979; Hickman, 1978; Thomas, 1979; Bradbury et al., 1979; Harte, 1979; Piasecki, 1980; Litherland, 1982; Haselock et al., 1982; Ashcroft et al., 1984; Harte et al., 1984; Mendum and Fettes, 1985) has largely concentrated on modifying the structural model and in elucidating the deformational history. Although the correlation of individual structural sequences within the Southern Highlands is reasonably well known correlation into the Central Highlands is complicated by poor exposure and a number of intervening granite complexes. In considering this recent work it is useful to divide the structural history into a primary and secondary deformational episodes. The former covers the early compressional phases and formation of the major folds; the latter covers the later compressional and uplift phases.

Primary deformation. This episode is locally polyphasal but can be resolved into two or three regional events (D1-D3 of Harris et al. (1976), Harte et al., (1984); D1-D2 of Roberts and Treagus (1979), Thomas (1979)). These events produced folds of major amplitude, the later phases resulting in large scale refolding and major reversals of facing in the early folds. The resultant fan-shaped array of recumbent folds or nappes is most simply seen in the high structural levels of the SW Highlands. In this region there is a simple tripartite structure with the NW-facing Islay anticline, the central Loch Awe syncline and the SE-facing Ardrishaig anticline (Figs. 2 and 3). The central syncline was originally traced into Perthshire where it was correlated with the Sron Mhor syncline (Sturt, 1961; Harris, 1963; Fig. 2), although it is now accepted that this is a later fold which locally reverses the facing of the early folds (Bradbury et al., 1979). Roberts and Treagus (1977, 1979) correlated the Loch Awe syncline with the Ballachulish syncline which can be traced around the second phase Stob Ban antiform in the Loch Leven area. Further east they (op. cit.) argued that the syncline tightens up and is cut out against the Ballachulish slide (Fig. 2). Similarly they correlated the Ardrishaig anticline with the Glen Creran anticline which is also cut out against the slide. These structural interpretations were contested by Hickman (1978) who re-interpreted the stratigraphic evidence for the Ballachulish syncline and slide in terms of facies change. Roberts and Treagus (1977) maintained that the folds interpreted to be cut out against the slide are also present below it as the tightly compressed Kinlochlaggan syncline and associated folds lying in the Ossian steep belt (Fig 2; Thomas, 1979). They (op. cit.) also correlatd the Iltay Boundary slide with the Ballachulish slide, across the Etive granite complex, and referred to it simply as the Boundary slide. They believed that movement on this boundary slide was composite, initiated as an early (D1) structure which was reactivated and propagated through the fold pile during later (their D2) deformation. Litherland (1982) did not accept many of the correlations of Roberts and Treagus (1977) in the Appin-Ballachulish area and argued for a simpler model with swings in the strike of early folds due to primary inflexions rather than later refolding. Thomas (1979) also argued against Roberts and Treagus' correlations and prefered to trace the Loch Awe syncline or 'zone of divergence' directly northwards into the Ossian steep belt (across the intervening granite mass), effectively turning the Ballachulish and Iltay Boundary slides northwards to flank the steep zone on the west and east sides respectively (Fig. 2). Van Breemen and Piasecki (1983) have traced the steep belt northeastwards into the Central Highland Division rocks.

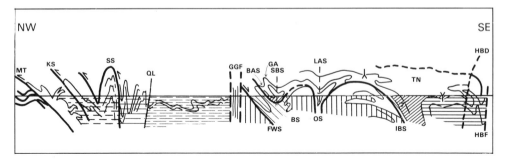

Fig 3. Composite section across the Scottish Highlands. Moine area in part after Powell (pers. comm.), Roberts and Harris (1983); Dalradian after Mendum and Fettes (1985) and Thomas (1979). The sections of Roberts and Treagus (1977, 1979) and Coward (1983) should also be considered. BS - Ballachulish slide; BAS - Ballachulish syncline; FWS - Fort William slide; GA - Glen Creran anticline; GGF - Great Glen fault; HBD - Highland Border downbend; HBF - Highland Boundary fault; IBS - Iltay Boundary slide; KS - Knoydart slide; LAS - Loch Awe syncline; MT - Moine thrust; OS - Ossian steep belt; QL - Loch Quoich line; SS - Sgurr Beag slide; TN - Tay nappe. Arrows (↙) indicate direction of younging of the stratigraphy.

The Southern Highlands are dominated by the southward facing Tay nappe complex. Although the Tay nappe was originally regarded as a single fold with a relatively uniform geometry along the Southern Highlands, recent work by Mendum and Fettes (1985) has indicated that at least in the central belt there are three major folds involved. Over much of the central part of the Southern Highlands only the lower inverted limb of the fold stack now remains giving rise to a 'zone of inversion'. Further northeast, Harte (1979) has demonstrated the existence of the right-way-up limb of a lower fold, the Tarfside nappe, exposed in a structural culmination and separated from the overlying Tay nappe complex by the Glen Mark slide.

In the NE Highlands the Dalradian succession is generally right-way-up. The structural relationship of rocks in this area to the Southern Highland structures remains uncertain. Read (1955) regarded the Tay nappe complex as plunging gently northwards from a culmination in the Southern Highlands such that the upper limb is exposed in the NE area. He (op. cit.) also argued for a major plane of discontinuity (the Boyne line) modifying the upper limb. More recently Ramsay and Sturt (1979) have argued for the NE area to be a major allochthonous block of pre-Dalradian rocks transported on a plane of discontinuity, different in detail but broadly similar to Read's "Boyne Line". However, Ashcroft et al., (1984) have argued against these models. They suggested that there is a good stratigraphic correlation between the NE area and the Southern Highlands and that the NE rocks are essentially autochthonous resting directly on their original basement. This they suggested implies that the Tay nappe complex reduces in amplitude and dies out into the NE area. Mendum (pers. comm.), however, has argued for the presence of some recumbent folding at least in the extreme east.

Secondary deformation. This episode (D4 of Harte et al., 1984; Harris et al., 1976; D3 of Roberts and Treagus, 1979) folded the nappe complex. It comprised a number of compressional and uplift phases and commonly produced belts of steep foliation. In the SW Highlands the Tay nappe complex is arched over the Cowal antiform (Fig. 2). The crest of this structure flattens out into the central belt so that the Tay nappe is disposed in a broad flat belt flanked to the northwest and southeast by steep zones (Fig. 2), the Tummel steep belt and Highland Border Steep Belt respectively. The large monoformal fold which rotated the rocks of the flat belt into the Highland Border steep belt is termed the Highland Border downbend (Harte et al., 1984). In the NE Highlands a series of shear zones may, in part, relate to the episode of secondary deformation (Ashcroft et al., 1984). In the Central Highlands this period of deformation was responsible for a number of major crossfolds (Thomas, 1979).

Deformational Models The nature of the deformation and correlation of the different deformational phases in the higher structural levels of the nappe pile (the Southern Highland area) is reasonably well known (see summary in Harte et al., 1984). Correlation into the deeper levels is, however, still uncertain. Much debate has centred on whether or not the nappe pile has a root zone and if so where it lies. Both Roberts and Treagus (1979) and Thomas (1979) favoured the existence of a root zone. Which, they argued, can be traced down through the pile from the Loch Awe syncline into the Ossian steep belt. This model has however been criticised by Litherland (1982). Also, Shackleton (1979), was sceptical of a fountain of nappes arising from a root zone and suggested that the opposed facing of the nappes may have been due to a tightly compressed refolded sequence of early folds. Shackleton (1979) also argued against the existence of a root zone preferring to relate the nappe formation to some form of gravity collapse structure. A further deficiency of the 'root zone' model would appear to lie in the geometry of the low angle faults lying on both sides of the proposed root zone. These features have the geometry of extensional faults (Bailey's 'lags') with younger rocks being emplaced on older (Thomas, 1979, figs. 4 and 6; Roberts and Treagus, 1977, fig. 5). Soper and Anderton (1984) have argued that the faults may have originated as growth faults during basin formation and have been reactivated during subsequent compressional phases (cf. Litherland, 1982). This model, however, largely based on the northward facing nappes of the Ballachulish area is less convincing when applied to the faults associated with the southward facing nappes. More recent models by Coward (1983) and Dewey and Shackleton (1984) envisaged a general northwestward directed movement with the southeastward facing nappes forming as facing reversals caused by a type of 'retrocharriage' or backfolding, possibly in response to gravitational collapse (Shackleton, 1979; Coward, 1983; cf. Bradbury et al., 1979).

Metamorphism

The metamorphism of the Dalradian rocks has recently been reviewed by Fettes et al. (1985) and by Harte (pers. comm.). The metamorphic grade generally rises from lower greenschist to upper amphibolite facies from the Highland Border towards the Central Highlands. The pressure at peak metamorphism increased from the NE to the SW Highlands (Fettes et al., 1976; Harte and Hudson, 1979). The majority of the rocks contain assemblages characteristic of Barrovian metamorphism, with the Buchan area of the NE Highlands exhibiting lower pressure assemblages. It is interesting to note that the geothermal gradients were greatest in the NE area where tectonic thickening was least, suggesting that the sedimentary pile had a blanketing effect on a heat source rather than engendering or promoting it (cf. Richardson and Powell, 1976). In recent years a considerable amount of new data have emerged (Harte, pers. comm.) on the PT conditions of metamorphism. This allied to detailed textural studies has revealed variations in the pressure conditions of metamorphism during the progressive phase of metamorphism. Harte (pers. comm.) in a review of metamorphism in the Scottish Highlands has recognised a number of areas or provinces with distinctive metamorphic styles.

In general terms the climax of metamorphism and main phase of porphyroblast growth occurred in the period between the primary and secondary deformations as described above at c 490 Ma.

Basic Intrusives of the NE Highlands

The NE Highlands contain a number of basic and ultrabasic complexes intruded after the climax of metamorphism and referred to as the Younger Basics, to distinguish them from basic rocks which predated the main deformation and metamorphism. The Younger Basic suite ranges from peridotites to ferro-gabbros, syenogabbros, and quartz-norites. During the shearing episode associated with the secondary episode of deformation the masses were locally deformed and rotated (e.g. Munro, 1984; Munro and Gallagher, 1984) in many cases being disassociated from their aureoles (Leslie, 1984). Ashcroft et al., (1984) remarked on the association of the basic masses with the regional shear zones. It is possible that early lineaments controlled the zones of basic intrusion and were subsequently reactivated as the shear zones seen at present. It is also possible that the presence of the large rigid basic plutons in the NE Highlands may have controlled the pattern of the secondary deformation in that area.

Timing of events

Recent reviews on the timing of events in the Dalradian rocks have been given by Fettes et al. (1985), Powell and Phillips (1985), Harte et al., (1984) and Harte (pers. comm.). In summary there are three main episodes, namely:
1)   The predominantly extensional phase from the Precambrian to the Lower/Middle Cambrian (630-530 Ma). This phase resulted in basin formation and probable thinning of the crust.
2)   The compressional phase from Middle Cambrian to Tremadoc (530-500 Ma). This resulted in tectonic thickening, nappe development and progressive metamorphism. The end of the phase was marked by the climax of metamorphism in the early Arenig followed closely by the intrusion of the basic masses (c. 490 Ma).
3)   The 'uplift' phase in the period Llandeilo to Caradoc (460-440 Ma). This phase was marked by the secondary deformation and retrogressive metamorphism. Dempster (1985) has shown that the uplift was episodic occurring at different rates in both space and time.

HIGHLAND BOUNDARY COMPLEX

The Highland Boundary Complex is a varied lithological assemblage lying in fault-bounded slivers adjacent to the Highland Boundary fault. The Complex comprises a mixture of spilitic lavas, serpentinites, black phyllites, cherts, limestones and greywackes. The exact nature of the Complex and its mode of emplacement has been the subject of considerable debate. Recent papers include work by Henderson and Robertson (1982), Robertson and Henderson (1984), Curry et al., (1984), and Harte et al., (1984). Most workers agree that the Complex is part of a dismembered oceanic crustal sequence. Robertson and Henderson (1984) argue that Dalradian components are largely absent and that the Complex must have formed in a basin relatively remote from its present position, adjacent to the Dalradian rocks.
        Curry et al., (1984) have presented extensive palaeontological data which indicate an Ordovician age for the assemblage with a maximum age of early Arenig and a minimum age of Caradoc. It follows that the rocks were deposited after the primary deformation of the Dalradian rocks with its associated nappe production but before the secondary uplift phase. At the present time the rocks of the Complex contain deformational fabrics which are broadly parallel to those in the neighbouring Dalradian. In addition, at a number of localities original contacts are apparently preserved (Henderson and Robertson, 1982; Harte et al., 1984). In the SW Highlands Henderson and Robertson (1982) have recorded thermal effects of Highland Boundary Complex rocks on the neighbouring Dalradian units, which they attribute to an obducting ophiolite suite.
        Much of the evidence relating to the mechanism of emplacement seems mutually contradictory. Harte et al. (1984) in a review of this evidence argued that a simple thrusting mechanism either before or after the formation of the Highland Border downbend (and the Highland

Border steep belt) is unacceptable as the sole means of emplacement although thrusting could form part of a polygenetic model. The consensus of opinion, at present, concludes that the most likely method of emplacement was one of lateral strike slip (see also discussion in Dewey and Shackleton, 1984; Harte et al., 1984; Curry et al., 1984).

GRANITE PLUTONISM

Granite and associated plutonism in the Scottish Highlands is generally divided into two groups, the 'Older Granites' which predate or are closely associated with the climax of metamorphism and the 'Newer Granites' which significantly postdate the main metamorphism. The Older Granites are confined to a few masses in the Moine and Dalradian assemblages. They are generally found as foliated granitic augen gneisses and have the chemical and isotopic signatures of crustal melts. The Newer Granites have been divided by Read (1961) into high level permitted granites and deeper level 'forceful' granites. More recently other workers have preferred to divide the Newer Granites on the basis of isotopic characteristics (eg. Pankhurst and Sutherland, 1982), into a late kinematic Middle Ordovician suite and a post kinematic Upper Silurian-Devonian suite. The late kinematic granites consist only of 'forceful' types. They are largely confined to the NE Highlands, although van Breemen and Bluck (1981) have argued for an originally more extensive suite, intruded into high crustal levels and largely eroded during Lower Devonian times.

Several workers (eg. Pankhurst and Sutherland (1982), Plant et al., (1980) van Breemen and Bluck (1981), have demonstrated the chemical, petrographic and isotopic differences between the late kinematic and the post kinematic granites. The former group are similar to the Older Granites and are characteristic of crustal or 'S' type granites. The post kinematic granites which contain both forceful and permitted types are characteristic of 'I' types.

The influence of lineaments on controlling granite plutonism has been discussed by Leake (1978) and the possible relationship to subduction models described by van Breemen and Bluck (1981).

GREAT GLEN FAULT

The significant transcurrent movement on the Great Glen fault in late-Caledonian to Middle Devonian times is accepted by most workers. The amount and direction of displacement is, however, unknown. Several workers (e.g. Fettes et al., 1985; Harte, pers. comm.) have commented on the differences in the timing and nature of the deformation and metamorphism across the fault. Also, Roberts and Harris (1983) and Harris et al., (1981) have highlighted the stratigraphic and deformational contrasts across the Great Glen. These lines of evidence argue for considerable displacements although conversely Smith and Watson (1983) have suggested maximum movements of c 200 km on the basis of gross similarities.

SUMMARY

The orthotectonic zone of the Scottish Caledonides comprises two major lithological assemblages, the Moine, a block of reworked basement and the Dalradian, a cover sequence of Late Precambrian to Middle Cambrian age. The Dalradian rocks lie wholly to the south and east of the Great Glen fault, the Moine occupies the ground to the north and west and extends, in part at least, under the Dalradian. The Moine rocks were extensively deformed and metamorphosed, locally to upper amphibolite facies, during the Precambrian. The lower portions of the Dalradian succession accumulated in shallow water under tectonically stable conditions. Increasing tectonic instability probably led to crustal thinning and fault-controlled basin development. The start of significant tectonism may well relate to the general transgression at the base of the Cambrian in the North Atlantic.

During the subsequent compressional episode of deformation both the Moine and the Dalradian rocks suffered polyphasal deformation and metamorphism. In the Dalradian rocks this resulted in the formation of a pile of largely recumbent fold nappes accompanied by prograde metamorphism (the Grampian Orogeny), which reached a peak in the Arenig. This was followed by a period of relative calm until renewed compression, uplift and retrograde metamorphism occurred in the Upper Ordovician. Geochronological evidence suggests that deformation and metamorphism in the Moine rocks was significantly later than in the Dalradian and was of varied intensity, with some areas suffering little Caledonian reworking.

The orthotectonic zone shows a marked asymmetry along its length. In the SW Highlands the Dalradian rocks appear to be thrust directly onto the foreland, the zone is narrow and it is probable that tectonic thickening of the Dalradian succession was greatest in this area. To the northeast along the orogen the zone opens out, the wedge of Moine rocks, and the pile of Dalradian nappes appears to diminish. It is in the NE Highlands where the tectonic thickening was least, that the temperatures (relative to pressure) of metamorphism were greatest and the majority of the granite plutons are concentrated.

Several tectonic models have been proposed for the Scottish Caledonides (see summary in Dewey and Shackleton, 1984). In recent years the concept of the Dalradian nappes relating to a deep seated root zone has probably been less in favour than the idea of the nappes forming as relatively high level features in response to some form of gravitational collapse. Coward (1983) related the deformation to underthrusting and related thrust-faults propagating northwestwards with time and culminating in the late movements on the Moine Thrust. Dewey and Shackleton (1984) envisage a two stage model with initial obduction of an opholite nappe and subsequent relaxation and backfolding. Any model must, however, account for the following features; the different timing of the major events between the Moine (Caledonian) and the Dalradian(Grampian); the SE-facing Dalradian nappes formed and emplaced before the Arenig; the apparent extensional geometry of the majority of Dalradian low angle faults as opposed to the thrust geometry of the Moine low angle faults. To date no model satisfactorily explains all of these points. Many of the early models were constrained by the assumption that the Scottish Caledonides had existed as a structural entity throughout orogenesis. However, the growing recognition that major strike-slip movements on the fault system particularly the Highland Border, Great Glen and Southern Upland Faults throws doubt on this assumption and suggests that the various structural units may thus represent different parts of the orogen brought together at a relatively late stage (Dewey and Shackleton, 1984; Coward, 1983).

Editors' Note   In the course of a year's study leave in the U.K. during 1982-83 Professor Leo Hall worked on the Dalradian rocks of the Southern Highlands of Scotland and developed a model for the evolution of Tay Nappe, based partly on his observations in Scotland and partly on his work in New England. The text outlining this model with which Drs Fettes and Harris are glad to be associated arrived too late for complete melding with the article by Fettes and Harris.

Hypothetical Model for the Development of the Tay Nappe

L M Hall

The backfolding phase involved the development of three folds with overfolding directed toward the southeast. The conceptual cross section that obtained after the backfolding developed is shown on Figure 4 where the folds are referred to from shallowest to deepest as the Ben Lui fold, Ben Lui' fold and Ben Lui'' fold. As a result of the backfolding the Tay nappe in the upper limb region of the Ben Lui fold was directed into a recumbent fold closing to the southeast. It was transported southeastward as a result of the development of many minor folds and associated slip cleavage by the process of simple shear as described by Harris et al. (1976). Notice that the initial lower inverted limb of the Tay nappe, indicated by the stipple pattern, is right-way-up on the upper limb of the Ben Lui fold as a result of the backfolding.

Folds with steep axial planes were developed during the latest of the main phases of deformation (Fig. 4). These folds fan through the cross section, dipping southeast in the northwest portion, vertically in the central region and northwest in the southeast portion. The Downbend fold was produced during this phase of folding and resulted in the present downward facing of the Tay nappe in the southeastern region of the Southern Highlands as originally described by Shackleton (1958). The topology of the cross section after latest phase deformation (Fig. 4) is very close to that of the diagrammatic cross section presented by Johnson et al. (1979, fig. 3) as would be

THE CALEDONIAN GEOLOGY OF THE SCOTTISH HIGHLANDS

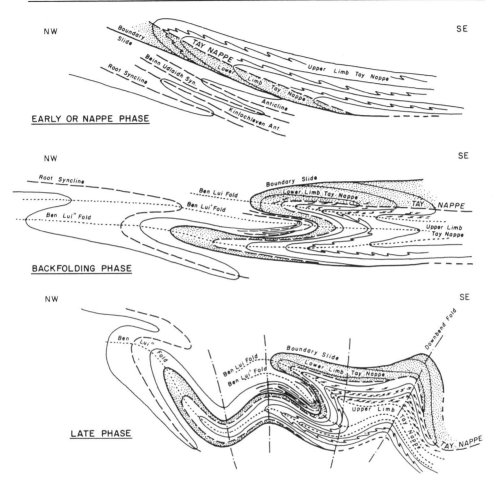

Figure 4.
Diagrammatic cross sections to illustrate a model for the structural development of the Tay Nappe and related features. Traces of the axial surfaces of the early phase folds are shown by long dashed lines, backfold phase folds by dotted lines and late phase folds by dot-dash lines. The initial lower limb of the Tay Nappe that resulted from the early phase of folding is shown in each cross section by the stipple pattern.

suspected since that cross section was drawn on in producing the model. Thus this cross section (Fig. 4) shows the essentials of the present structural configuration of the rocks.

This model envisaged the development of the present geometry of the Tay nappe as being due to three main phases of deformation, the early nappe phase, the backfolding phase and the late phase (Fig. 4). "Phase of deformation", as used here, is intended to identify a developmental aspect in the sequential history of deformation. Thus it does not necessarily mean a punctuated event because the deformational process may well have been essentially continuous.

The conceptual model proposed here is partially based on the topology of the structural geometry shown in the diagrammatic structure section of the Scottish orthotectonic zone presented by Johnson et al. (1979, fig. 3) and the block diagram of the Central Highlands presented by Thomas (1979, fig. 5). Numerous articles that discuss the Tay nappe, among which are those of Shackleton (1958), Roberts and Treagus (1979), Bradbury et al. (1979), as well as field observations were also very important in developing the model.

The early nappe phase of deformation involved the production of a series of recumbent folds overturned to the northwest and resulting from northwesterly directed tectonic transport. The names and axial surface traces of these folds are indicated on Figure 4 where the names have been chosen to conform with existing nomenclature. Among the most important feature to note here is the deepest recumbent syncline, which is referred to on the diagram as the "Root Syncline" because the stratigraphy roots into a relatively autochthonous section where it wraps around the hinge of this syncline. The Tay nappe is the tectonically highest recumbent fold shown in Figure 4 and its lower limb in its initial phase of development is shown by the stipple pattern in each diagram on Figure 4. This limb of the Tay nappe is truncated by the Italy Boundary slide, here referred to simply as the Boundary slide.

REFERENCES

ANDERTON, R., BRIDGES, P.H., LEEDER, M.R. & SELLWOOD, B.W. 1979. 'A dynamic stratigraphy of the British Isles'. London. (Geo. Allen & Unwin).

ASHCROFT, W.A., KNELLER, B.C., LESLIE, A.G. & MUNRO, M. 1984. 'Major shear zones and autochtonous Dalradian in the north-east Scottish Caledonides'. Nature, 310, 760-2.

BAILEY, E.B. 1934. 'West Highland Tectonics: Loch Leven to Glen Roy'. Q. J. Geol. Soc. London, 90, 462-525.

BARR, D. 1983. 'Genesis and structural relations of Moine migmatites'. PhD thesis (unpubl.), Univ. of Liverpool.

———, ROBERTS, A.M., HIGHTON, A.J. PARSON, L.M. & HARRIS, A.L.

1985. Structural setting and geochronological significance of the West Highland Granitic Gneiss, a deformed early granite within Proterozoic, Moine rocks of NW Scotland'. J. Geol. Soc. London, 142, 663-75.

BRADBURY, H.J., HARRIS, A.L. & SMITH, R.A. 1979. 'Geometry and emplacement of nappes in the Central Scottish Highlands'. IN: HARRIS, A.L., HOLLAND, C.M. & LEAKE, B.E. The Caledonides of the British Isles - reviewed. Spec. Publ. Geol. Soc. London, 8, 213-20.

BROOK, M., POWELL, D. & BREWER, M.S. 1976. 'Grenville age for rocks in the Moine of north-western Scotland'. Nature, 260, 515-7.

_____, _____ & _____ 1977. 'Grenville events in the Moine rocks of the Northern Highlands, Scotland'. J. Geol. Soc. London, 133, 489-96.

CLIFFORD, T. N. 1957. 'The stratigraphy and structure of part of the Kintail district of southern Ross-shire; its relation to the Northern Highlands'. Q. J. Geol. Soc. London, 113, 57-8.

COWARD, M.P. 1980. 'The Caledonian thrusts and shear zones of NW Scotland'. J. Struct. Geol., 2, 11-7.

_____ 1983. 'The thrust and shear zones of the Moine thrust zone and the NW Scottish Caledonides'. J. Geol. Soc. London, 140, 795-811.

CURRY, G.B., BLUCK, B.J., BURTON, C.J., INGHAM, J.K., SIVETER, D.J. & WILLIAMS, A. 1984. 'Age, evolution and tectonic history of the Highland Border Complex, Scotland'. Trans. R. Soc. Edinburgh, 75, 113-33.

DEMPSTER, T.J. 1985. 'Uplift patterns and orogenic evolution in the Scottish Dalradian'. J. Geol. Soc. London, 142, 111-28.

DEWEY, M.F. & SHACKLETON, R.M. 1984. 'A model for the evolution of the Grampian tract in the early Caledonides and Appalachians'. Nature, 312, 115-21.

ENGLAND, P.C. & RICHARDSON, S.W. 1977. 'The influence of erosion upon the mineral facies of rocks from different metamorphic environments'. J. Geol. Soc. London, 134, 201-13.

FETTES, D.J., GRAHAM, C.M., SASSI, E.P. & SCOLARI, F.P. 1976. 'The lateral spacing of potassic white micas and facies series variation across the Caledonides'. Scott. J. Geol., 12, 227-36.

———————, LONG, C.B., BEVINS, R.E., MAX, M.D., OLIVER, G.J.H., PRIMMER, T.J., THOMAS, L.J. & YARDLEY, B.W.D. 1985. 'Grade and time of metamorphism in the Caledonide Orogon of Britain and Ireland'. IN: HARRIS, A.L. (ed.) The Nature and Timing of Orogenic Activity in the Caledonian Rocks of the British Isles. Mem. Geol. Soc. London, 9, 41-53.

GRAHAM, C.M. & BRADBURY, H.J. 1981. 'Cambrian and late Precambrian basaltic igneous activity in the Scottish Dalradian: a review'. Geol. Mag., 118, 27-37.

HARRIS, A.L. 1963. Structural investigations in the Dalradian rocks between Pitlochry and Blair Atholl. Trans. Geol. Soc. Edinburgh, 19, 256-75.

——————— & FETTES, D.J. 1972. Stratigraphy and structure of Upper Dalradian rocks at the Highland Border. Scott. J. Geol., 8, 253-64.

——————— & PITCHER, W.S. 1975. 'The Dalradian Supergroup'. IN: HARRIS, A.L. et al. (eds.), A correlation of the Precambrian rocks in the British Isles. Spec. Rep. Geol. Soc. London, 6, 52-75.

———————, BRADBURY, H.J. & McGONIGAL, N.H. 1976. 'The evolution and transport of the Tay Nappe'. Scott. J. Geol., 12, 103-13.

———————, BALDWIN, C.T., BRADBURY, H.J., JOHNSON, H.D. & SMITH, R.A. 1978. 'Ensialic basin sedimentation: the Dalradian Supergroup'. IN: BOWES, D.R. & LEAKE, B.E. (eds.), Crustal evolution in northwestern Britain and adjacent regions. Spec. Issue Geol. Soc. London, 10, 115-38.

———————, PARSON, L.M., HIGHTON, A.J. & SMITH, D.I. 1981. 'New/Old Moine relationships between Fort Augustus and Inverness'. J. Struct. Geol., 3, 187-8.

HARTE, B. 1979. 'The Tarfside succession and the structure and stratigraphy of the eastern Scottish Dalradian rocks'. IN: HARRIS, A.L., HOLLAND, C.H. & LEAKE, B.E. (eds.), The Caledonides of the British Isles - reviewed. Spec. Publ. Geol. Soc. London, 8, 221-8.

——————— & HUDSON, N.F.C. 1979. 'Pelite facies series and temperatures and pressures of Dalradian metamorphism in E. Scotland'. IN: HARRIS, A.L., HOLLAND, C.H. & LEAKE, B.E. (eds.), The Caledonides of the British Isles - reviewed. Spec. Publ. Geol. Soc. London, 8, 323-37.

———————, BOOTH, J.E., DEMPSTER, T.J., FETTES, D.J., MENDUM, J.R. &

WATTS, D. 1984. 'Aspects of the post-depositional evolution of Dalradian and Highland Border Complex rocks in the Southern Highlands of Scotland'. Trans. R. Soc. Edinburgh : Earth Sciences, 75, 151-63.

HASELOCK, P.J., WINCHESTER, J.A. & WHITTLES, K.H. 1982. 'The stratigraphy and structure of the southern Monadhliath Mountains between Loch Killin and Upper Glen Roy'. Scott. J. Geol., 18, 275-90.

HENDERSON, W.G. & ROBERTSON, A.H.F. 1982. 'The Highland Border rocks and their relation to marginal basin development in the Scottish Caledonides'. J. Geol. Soc. London, 139, 433-50.

HICKMAN, A.H. 1975. 'The stratigraphy of late Precambrian metasediments between Glen Roy and Lismore'. Scott. J. Geol., 11, 117-42.

_____ 1978. 'Recumbent folds between Glen Roy and Lismore'. Scott. J. Geol., 14, 191-212.

HOLDSWORTH, R.E. & ROBERTS, A.M. 1984. 'A study of early curvilinear fold structures and strain in the Moines of the Glen Garry region, Inverness-shire'. J. Geol. Soc. London, 141, 327-38.

JOHNSON, M.R.W., SANDERSON, D.J. & SOPER, N.J. 1979. 'Deformation in the Caledonides of England, Ireland and Scotland'. IN: HARRIS, A.L., HOLLAND, C.H. & LEAKE, B.E. (eds.), The Caledonides of the British Isles - reviewed. Spec. Publ. Geol. Soc. London, 8, 165-86.

JOHNSTONE, G.S. 1966. 'THE GRAMPIAN HIGHLANDS'. 3rd Edition. British Regional Geology.

_____ 1975. 'The Moine Succession'. IN: HARRIS, A.L., et al. (eds.), A correlation of the Precambrian rocks in the British Isles. Spec. Rep. Geol. Soc. London, 6, 30-42.

_____, SMITH, D.I. & HARRIS, A.L. 1969. 'The Moinian Assemblage of Scotland'. IN: KAY, M. (ed.), North Atlantic - geology and continental drift. Mem. Am. Assoc. Petrol. Geol., 12, 159-80.

KELLEY, S.P. & POWELL, D. 1985. 'Relationships between marginal thrusting and movement on major, internal shear zones in the North Highland Caledonides, Scotland'. J. Struct. Geol., 7, 161-74.

LEAKE, B.E. 1978. 'Granite emplacement: the granites of Ireland and their origin'. IN: BOWES, D.R. & LEAKE, B.E. (eds.),

Crustal evolution in northwestern Britain and adjacent regions. Spec. Issue Geol. Soc. London, 10, 221-48.

LEEDAL, G.P. 1952. 'The Cluanie igneous intrusion, Inverness-shire and Ross-shire'. Q. J. Geol. Soc. London, 108, 35-63.

LESLIE, A.G. 1984. 'Field relations in the north-eastern part of the Insch mafic igneous mass, Aberdeenshire'. Scott. J. Geol., 20, 215-35.

LITHERLAND, M. 1982. 'The structure of the Loch Creran Dalradian and a new model for the SW Highlands'. Scott. J. Geol., 18, 205-25.

McCOURT, W.J. 1980. 'The Geology of the Strath Halladale-Altnabreac District'. EPNU 80-1 Rep. Inst. Geol. Sci.

MENDUM, J.R. & FETTES, D.J. 1985. 'The Tay nappe and associated folding in the Ben Ledi - Loch Lomond area'. Scott. J. Geol., 21, 41-56.

MUNRO, M. 1984. 'Cumulate relations in the 'Younger Basic' masses of the Huntly-Portsoy area, Grampian Region'. Scott. J. Geol., 20, 343-59.

_____ & GALLAGHER, J.W. 1984. 'Disruption of the 'Younger Basic' masses in the Huntly-Portsoy area, Grampian Region'. Scott. J. Geol., 20, 361-82.

PANKHURST, R.J. 1982. 'Geochronological tables from British igneous rocks.' IN: SUTHERLAND, D.S. (ed.), Igneous rocks of the British Isles. Wiley, 575-81.

_____ & SUTHERLAND, D.S. 1982. 'Caledonian granites and diorites of Scotland and Ireland.' IN: SUTHERLAND, D.S. (ed.), Igneous rocks of the British Isles. Wiley, 149-90.

PEACH, B.N. et al. 1910. 'The geology of Glenelg, Lochalsh and southeast part of Skye.' Mem. Geol. Surv. UK.

PIASECKI, M.A.J. 1980. 'New light on the Moine rocks of the Central Highlands of Scotland.' J. Geol. Soc. London, 137, 41-59.

_____ & van BREEMEN, O. 1979. 'The "Central Highland Granulites": cover-basement tectonics in the Moine'. IN: HARRIS, A.L., HOLLAND, C.H. & LEAKE, B.E. (eds.), The Caledonides of the British Isles - reviewed. Spec. Publ. Geol. Soc. London, 8, 139-44.

_____, _____ & WRIGHT, A.E. 1981. 'Late Precambrian Geology of Scotland, England and Wales. IN:

KERR, J.W. & FERGUSON, A.J. (eds.), Geology of the North Atlantic Borderlands. Mem. Can. Soc. Petrol. Geol., 7, 59-94.

PLANT, J., BROWN, G.C., SIMPSON, P.R. & SMITH, R.T. 1980. 'Signatures of metalliferous granites in the Scottish Caledonides.' Trans. Instn. Min. Metall., 89, B198-210.

POWELL, D. 1974. 'Stratigraphy and structure of the Western Moine and the problem of Moine orogenesis'. J. Geol. Soc. London, 130, 575-93.

_____, BAIRD, A.W., CHARNLEY, N.R. & JORDAN, P.J. 1981. 'The metamorphic environment of the Sgurr Beag Slide; a major crustal displacement zone in Protorozoic Moine rocks of Scotland'. J. Geol. Soc. London, 138, 661-73.

_____, BROOK, M. & BAIRD, A.W. 1983. 'Structural dating of a Precambrian pegmatite in Moine rocks of northern Scotland and its bearing on the status of the "Morarian orogeny".' J. Geol. Soc. London, 140, 813-24.

_____ & PHILLIPS, W.E.A. 1985. 'Time of deformation in the Caledonide Orogen of Britain and Ireland.' IN: HARRIS, A.L. (ed.), The nature and timing of orogenic activity in the Caledonian rocks of the British Isles. Mem. Geol. Soc. London, 9, 17-39.

PRINGLE, I.R. 1972. 'Rb-Sr age determinations on shales associated with the Varanger Ice Age.' Geol. Mag., 109, 465-72.

RAST, N. & LITHERLAND, M. 1970. 'The correlation of the Ballachulish and Perthshire (Iltay) Dalradian successions. Geol. Mag. 107, 259-72.

RATHBONE, P.A., COWARD, M.P. & HARRIS, A.L. 1983. 'Cover and basement: a contrast in style and fabrics.' Mem. Geol. Soc. Am., 158, 312-23.

RAMSAY, D.M. & STURT, B.A. 1979. 'The status of the Banff nappe.' IN: HARRIS, A.L., HOLLAND, C.H. & LEAKE, B.E. (eds.), The Caledonides of the British Isles - reviewed. Spec. Publ. Geol. Soc. London, 8, 145-51.

READ, H.H. 1955. 'The Banff Nappe: an interpretation of the structure of the Dalradian rocks of north-east Scotland'. Proc. Geol. Ass., 66, 1-29.

_____ 1961. 'Aspects of Caledonian magmatism in Britain'. Liv. & Manch. Geol. J., 2, 653-83.

RICHARDSON, S.W. & POWELL, R. 1976. 'Thermal causes of the Dalradian metamorphism in the central Highlands of Scotland'. Scott. J. Geol., 12, 237-65.

ROBERTS, A.M. & HARRIS, A.L. 1983. 'The Loch Quoich Line - a limit of early Palaeozoic crustal reworking in the Moine of the Northern Highlands of Scotland.' J. Geol. Soc. London, 140, 883-92.

————————, SMITH, D.I. & HARRIS, A.L. 1984. 'The structural setting and tectonic significance of the Glen Dessary syenite, Inverness-shire.' J. Geol. Soc. London, 141, 1033-42.

————————, ————————, ———————— & HOLDSWORTH, R.E. 1985. 'Discussion of the structural setting and tectonic significance of the Glen Dessary Syenite, Inverness-shire.' J. Geol. Soc. London, 142.

ROBERTS, J.L. & TREAGUS, J.E. 1977. 'Polyphase generation of nappe structures in the Da,lradian rocks of the southwest Highlands of Scotland.' Scott. J. Geol., 13, 185-8.

———————— & ———————— 1979. 'Stratigraphical and structural correlation between the Dalradian rocks of the SW and Central Highlands of Scotland.' IN: HARRIS, A.L., HOLLAND, C.H. & LEAKE, B.E. (eds.), The Caledonides of the British Isles - reviewed. Spec. Publ. Geol. Soc. London, 8, 199-204.

ROBERTSON, A.H.F. & HENDERSON, W.G. 1984. 'Geochemical evidence for the origins of igneous and sedimentary rocks of the Highland Border, Scotland.' Trans. R. Soc. Edinburgh, 79, 135-50.

RUSHTON, A.W.A. & PHILLIPS, W.E.A. 1973. 'A Protospongia from the Dalradian of Clare Island, Co. Mayo, Ireland.' Palaeontology, 16, 231-237.

SHACKLETON, R.M. 1958. 'Downward-facing structures of the Highland Border'. Q. J. Geol. Soc. London, 113, 361-92.

———————— 1979. 'The British Caledonides: comments and summary'. IN: The Caledonides of the British Isles - reviewed'. Spec. Publ. Geol. London, 8, 299-304.

SMITH, D.I. & WATSON, J.SMITH, 1983. 'Scale and timing of movements on the Great Glen fault, Scotland.' Geology, 11, 523-6.

SMITH, R.A. & HARRIS, A.L. 1976. 'The Ballachulish rocks of the Blair Atholl District.' Scott. J. Geol., 12, 153-7.

SOPER, N.J. & BARBER, A.J. 1982. 'A model for the deep structure of the Moine thrust-zone.' J. Geol. Soc. London, 134, 41-4.

_____ & ANDERTON, R. 1984. 'Did the Dalradian slides originate as extensional faults?' Nature, 307, 357-60.

STOKER, M.S. 1983. 'The stratigraphy and strcture of the Moine rocks of Eastern Ardgour'. Scott. J. Geol., 19, 369-85.

STRACHAN, R.A. 1982. 'Tectonic sliding within the Moinian Loch Eil Division near Kinlocheil, W. Inverness-shire.' Scott. J. Geol., 18, 187-203.

_____ 1985. 'The stratigraphy and structure of the Moine rocks of the Loch Eilt area, west Inverness-shire.' Scott. J. Geol., 21, 9-22.

STURT, B.A. 1961. 'The geological structure of the area south of Loch Tummel'. Q. J. Geol. Soc. London, 117, 131-56.

SWETT, K. 1981. 'Cambro-Ordovician strata in Ny Friesland, Spitsbergen and their palaeotectonic significance.' Geol. Mag., 118, 225-336.

_____ & SMIT, D.E. 1972. 'Palaeogeography and depositional environments of the Cambro-Ordovician shallow-marine facies of the north Atlantic.' Geol. Soc. Am. Bull., 83, 3223-48.

TALBOT, C. 1984. 'Microdiorite sheet intrusions as incompetent time- and strain-markers in the Moine assemblage NW of the Great Glen fault, Scotland.' Trans. R. Soc. Edinburgh, 74, 137-52.

TANNER, P.W.G. 1970. 'The Sgurr Beag Slide - a major break within the Moinian of the western Highlands of Scotland.' Q.J. Geol. Soc. London, 126, 435-63.

_____, JOHNSTONE, G.S., SMITH, D.I. & HARRIS, A.L. 1970. 'Moinian stratigraphy and the problem of the Central Ross-shire inliers.' Bull. Geol. Soc. Am., 81, 299-306.

THOMAS, P.R. 1979. 'New evidence for a Central Highland Root Zone.' IN: HARRIS, A.L., HOLLAND, C.H. & LEAKE, B.E. (eds.), The Caledonides of the British Isles - Reviewed. Spec. Publ. Geol. Soc. London, 8, 205-12.

TREAGUS, J. 1974. 'A structural cross-section of the Moine and Dalradian rocks of the Kinlochleven area, Scotland.' J. Geol. Soc. London, 130, 525-44.

TREAGUS, J.E. & KING, G. 1978. 'A complete Lower Dalradian

succession in the Schiehallion district, Central Perthshire.' Scott. J. Geol., 14, 157-66.

UPTON, P.S. in press. 'A structural cross-section of the Moine and Dalradian rocks of the Braemar area'. Rep. Br. Geol. Surv.

VAN BREEMEN, O., PIDGEON, R.T., & JOHNSON, M.R.W. 1974. 'Precambrian and Palaeozoic pegmatites in the Moines of northern Scotland.' J. Geol. Soc. London, 130, 493-507.

―――――――― & BLUCK, B.J. 1981. 'Episodic granite plutonism in the Scottish Caledonides'. Nature, 291, 113-7.

―――――――― & PIASECKI, M.A.J. 1983. 'The Glen Kyllachy Granite and its bearing on the nature of the Caledonian Orogeny in Scotland.' J. Geol. Soc. London, 140, 47-62.

―――――――― & ―――――――― 1983. 'The Glen Kyllachy Granite and its bearing on the nature of the Caledonian Orogeny in Scotland.' J. Geol. Soc. London, 140, 47-62.

VOLL, G. 1964. 'Deckendau und fazies im Schottischen Dalradian.' Geol. Rdsch., 21, 590-612.

WATSON, J. 1984. 'The ending of the Caledonian orogeny in Scotland.' J. Geol. Soc. London, 141, 193-214.

SUBJECT INDEX

Aberarth 53
Aberdeenshire (Younger) gabbros 118
Abereiddi 41-42, 45
Abereiddi Bay 43-44
Aberystwyth 53-55
Aberystwyth Grits 53-55
Abington 90-92
accretionary prism 100
Afton Formation 93, 216
Afton Greywackes 90
Afton Water 92-95
American faunas 80
American faunas of Storen nappe 240
Anglas Bay 52
Anglesey 4-5, 29, 60-63, 67-70
Appin Group (see Dalradian)
Appin Limestone 129, 133 (Dalradian)
Appin Phyllites 129, 133 (Dalradian)
Appin Phyllites and Limestone 130 (Dalradian)
Appin Quartzite 129-130, 133 (Dalradian)
Appin syncline 130-131, 133
Appin Transition Group 132 (Dalradian)
Applecross 148, 155
Aran Volcanic Group 29, 55-56, 191
Archaean-to-Proterozoic gneisses 18
Ardgour granitic gneiss (or orthogneiss) 21-22, 116-117, 119
    135, 137, 139-140, 142-143, 304, 310
Ardmillan Group 101
Ardnamurchan 20-21
Ardrishaig anticline 317
Ardwell Formation 107
Ardwell Group 104
Arenig (mid-) acid volcanism 197
Arenig angular unconformity 5
Arenig island arc 17
Arenig island arc tholeiites 192
Arenig obduction 197
Arenig ophiolite 196
Arenig-to-Llandeilo volcanism 191
Arenig transgression 191-192
Arenig turbidites 192
Arenig-Llanvirn rhyolitic ignimbrites 197
Arfon Group 4
Argyll Group (see Dalradian)
Armorican massif 3
Arnabol thrust 253, 265
Arnaboll sheet 156, 266-267
Arran 196

Arrochar diorite 127
Ashgill alkali gabbros and rhyolites 194
Ashgill and Lower Llandovery basic lavas and tuffs 198
Assynt 156-159, 161, 168-171, 178, 259, 228-267, 272-274
Assynt sheet 265
Assynt thrust 261
Auchtertyre 146
Aultbea 155
Avalon (or Avalonian) terrane 31, 62, 188, 281, 283, 287-288
Avalonian 4
Back of Keppoch 141
Badcallian 18, 154
Bail Hill Volcanic Group 210
Bala fault 55, 288
Balclatchie Group 104
Ballachulish 112
Ballachulish Limestone 128-132 (Dalradian)
Ballachulish Slates 129-130, 132 (Dalradian)
Ballachulish slide 129, 131-132
Ballachulish syncline 130-132
Ballantrae 79-80, 99, 105, 196
Ballantrae complex 17, 80-81, 100, 103, 105, 107, 197-198, 290
Ballantrae complex lavas 101
Ballantrae ophiolite 86, 187
Ballantrae ophiolite obduction 199
Ballantrae Volcanic Group 17
Balloch 125
Balmacara 146-150
Balmacara nappe 146
Balmacara thrust 146, 148
Balmacara thrust plane 150
Balmaha 121-126
Baltica 213
Bannisdale Slates 83
Barr Group 81, 101
Barrovian index minerals 22
Barrovian metamorphism 25, 118, 320
base metal mineralization 6
Bedded Pyroclastic Formation 59
Ben Attow 144
Ben Lawers Schist 127 (Dalradian)
Ben Lawers synform 127
Ben Lomond 108
Ben Lui Schist 127 (Dalradian)
Ben More sheet 156, 176, 263
Ben More thrust 175, 261, 264
Ben More thrust sheet 174
Ben Nevis 142-143
Ben Vurich granite 117-119, 304
Benan Conglomerate 107
Bennane 105-106
Bennane Head 81, 99, 101, 107

bentonites 198
Bergen arc 229, 231
Berw fault 63, 67-69, 288
beryl-bearing pegmatite 140
bimodal basalt-rhyolite association 191
Binnein Quartzite 129-130 (Dalradian)
Binnein Schists 129-130 (Dalradian)
Binsey-Eycott volcanics 192
Birkhill Cottage 78
Birkhill Shale 88-89
Birkhill shales metabentonite 198
Black Neuk 100
Blackcraig Formation 93, 95, 212, 216
Bloody Bluff-Clinton-Newberry fault system 254
Blue Ridge 249, 254
Blue Ridge fault 250
blueschists 63, 69
Bonney's Dyke 102
Borrolan igneous complex 169, 175 (see also Loch Barrolan)
borrolanite 176
Borrolan igneous intrusion 260
Borrow Bridge 83
Borrowdale (Volcanic) Group 8, 11-12, 78, 82-84, 193
Borrowdale Volcanics 192, 289
Borrowdale volcanoes 193
Brathay Flags Formation 11
Brevard 247
Brevard fault 219, 250, 255
Brevard fault zone 253
British Institutions Reflection Profiling syndicate (BIRPS) 7, 284, 287-288
Buchaille Etive Mhor 127
Buchan 25, 118
Buchan metamorphism 320
Bute 196
Caddroun Burn Beds 216
Cader Idris 55
Caerfai Group 40
Cailleach Head 155
Cairnsmore of Fleet granite 13
calc-silicate mineral assemblages 313
Caledonian deformation front 24
Caledonian foreland 146, 148
Caledonian granite vein complexes 138
Caledonian microdiorite suite 139
Caledonian pegmatite 139
Caledonian shear zones 156
Caledonian (Caledonide) suture 187-188, 199
Caledonide suture 199
Caledonides of Wales 28
Cambrian (Basal) Quartzite (Scotland) 155, 165-167, 173-175, 266
Cambrian (late) ocean-floor basalts 192

Cambrian of foreland 168
Cambrian quartzites 155, 161, 164, 169, 174
Cambrian volcanism 195
Cambro-Ordovician sequence 19, 155
Cambro-Ordovician sediments 153
Canadian Shield 213
Canisp porphyry 156
Capel Curig 56-58, 60
Capel Curig Volcanic Formation 58, 191
Capel Horeb 36
Caradoc palaeogeography 56
Caradoc-to-Llandovery Birkhill metabentonites 198
Caradocian volcanic climax 193
Cardiff 30,33
Carn Chuinneag granite 21-22, 117, 119, 304, 310, 313
Carrock Fell igneous complex 8
Carrovarre nappe 228
Carsphairn granite 13
Cashell Burn 125
cauldron-subsidence 128
Celtic 5
Cemaes Bay 68
Central Highland Division (see Moine)
Central Highland Granulites (Moine) 303, 306
Central Highlands 23, 25, 118
Central Penmynydd zone 67
Channel Islands 3
Charlie fracture zone 283
Charlotte belt 250
Charlotte belt and Carolina slate belt 249
Charnian 3
Chauga belt 249
Cheviots calc-alkaline andesitic to dacitic lavas 198
Church Stretton fault 29
Clew Bay 196
Cluanie granite 143
Cockburnland 212
Coed-y-Brenin porphyry copper 6
Coedana complex 63, 66-67, 69
Coedana granite 67, 69
Collector anomaly 283
Coniston Grits 11, 83
Coniston Limestone Group 8, 12, 79, 83
Coniston-Howgill Fells 7
continent/ocean boundaries 187
Corsewall Group 216
Corsewall-Glen App 96
Cowal antiform 122
Criffel granite 13
Crincoed Point 53
Cross Fell inlier 8, 10

SUBJECT INDEX

Cross Fell volcanics 192
Cruachan granite 127
Cumbria 8
Cumbrian Mountains 7
Cwm Eigiau Formation 58
Cwm Idwal 58-60
Dalradian (Grampian) nappe complex 26
Dalradian 1, 19, 112-114, 116, 118-119
Dalradian basement 17
Dalradian basin 25
Dalradian formations (listed separately by name)
Dalradian groups
    Appin Group 19, 21, 24, 116, 304-305, 314-316
    Argyll Group 19, 21, 24-25, 116, 122, 304-305, 314-315
    Grampian Group/Division 19, 21, 24-25, 116, 304-306,
    309, 314-316
    Southern Highland Group 19, 21, 24, 116, 119, 122, 195,
    304-305, 314-315
Dalradian Supergroup 24, 114, 118, 125
Deep Park complex 194
deformed eclogites 152
deformed pegmatites 140
Devonian plutonism 26
Diabaig Formation 155
diopside-nodules 145
Dobb's Linn 78-79, 81, 87-89
Dornie 145, 147-148, 151
double unconformity 19
Dover fault 283
Downan Point 102
Downan Point lavas 100-101, 105
Drumindarroch 141
Drymen 108
Durness Limestone 19, 156, 165-167, 169, 171, 173-174, 176,
   252, 262, 274
Dyfed 41
Eastern Lewisian 144-145
eclogite 144-145
Ediacaran fauna 36
Eigg 141
Eilde Flags and Quartzite 128 (Dalradian)
Eilde Quartzite 129-130 (Dalradian)
Eilde Schist 129-130 (Dalradian)
Eilean Dubh 156
Eilean Dubh white dolomite 178 (Cambrian)
English Midlands 3
Eriboll 159, 262, 264, 265
Eriboll Quartzite 19 (see also under Cambrian basal quartzite)
Etive complex 127
eulysites 144
extensional tectonics 190

Eycott Group 8, 11-12
Eycott Lavas 82
Fannich Forest 311
Fassfern 138
Faunal Provincialism in the Scandinavian Caledonides 239
Finnmarkian orogenic belt 235
Finnmarkian orogeny 221-222, 225, 227, 232, 238, 241
Fishguard 41-53
Fishguard volcanic complex 191
Fishguard Volcanic Group 29, 41, 48
Five Sisters of Kintail 143, 145
Flannan thrust 272
forsterite (-diopside) marble 144-145
Fort William 133, 142
Fort William slide 129-130
Foyers granite 21, 27
Fucoid Beds 156, 163, 165-167, 169, 171-173, 262, 274 (Cambrian)
Gader complex 60
Gala Group 89, 216
Galdenoch Group 216
Galway granite 198
Gander terrane 283
Geophysics in the Caledonides 281
Ghrudaidh Formation 156 (Cambrian)
Ghrudaidh Limestone 172 (Cambrian)
Girvan 13, 14, 17, 79-81, 89, 96, 98-100, 104-105, 108
Girvan-Ballantrae 16, 78, 100-101
Glen Afton 79, 96
Glen App fault 100
Glen App Group 100-102
Glen Dessary syenite 116, 119, 135, 137, 140, 310, 312
Glen Garry 143
Glen Shiel 142-143
Glencoe 112, 115, 199
Glencoe cauldron subsidence 127, 130
Glencoe fault-intrusion 127-128, 132
Glencoe Quartzite 129-130, 132 (Dalradian)
Glencoul 115, 156-157, 168, 267
Glencoul fault 174
Glencoul nappe 154, 161
Glencoul sheet 154, 174, 259, 263
Glencoul surge 160
Glencoul thrust 160, 253, 261, 264-269, 272
Glenelg 147-148, 309
Glenelg (Lewisian) inlier 142, 144-146, 149-152
Glenfinnan 139
Glenfinnan Division/Group (see Moine)
Glenkiln Shales 87-88
Goodwick (Volcanic) Formation 48, 50
Grampian deformation and metamorphism 119

Grampian Division/Group (see Dalradian)
Grampian Highlands 114, 200
Grampian orogeny 1, 25, 114, 195, 323
Grampian slide 306, 309, 311-312, 316
Great Glen fault 18, 20, 24-25, 27, 112, 120, 133-134, 142, 195, 198, 282-283, 291-292, 303, 308-309, 311-312, 314, 318, 322-323
Grenville 114, 116-117, 213, 217, 304, 306
Grenville Province 212
Gruadaidh Dolomite 178 (Cambrian)
Gula nappe 236
Gwna Group (see Monian)
Gwna melange 4, 68-69 (Monian)
Gwna volcanics 63, 68 (Monian)
Harlech dome 55, 191, 288
Hartfell Shales 89
Hawick line 290
Hawick rocks 213, 216
Hebridean craton 18-20
Helgeland nappe complex 225, 229-230, 233
Highland Border 112, 118-122
Highland Border series 303
Highland Border steep belt 120, 122, 125
Highland Border zone 27
Highland Boundary complex 19, 26, 117-120, 123, 187, 196-197, 201, 303-305, 321
Highland Boundary fault 1, 23, 26-27, 108, 114, 118, 195, 199, 209, 217, 282, 284, 291-292, 303, 312, 318
Highland Boundary fracture zone 122, 124
Holland Arms 68
Holland Arms complex 60
Holyhead Quartzite Formation 65 (Monian)
Humber terrane 283
Humberian orogeny 217
Iapetus lithosphere 188, 190
Iapetus margin volcanism 198
Iapetus northern (continental) margin 195, 197
Iapetus Ocean 1, 7, 11, 14, 16, 18, 26, 63, 81, 221
Iapetus suture 88, 195, 282, 284
Inchnadamph 174
Inner Piedmont 249-250
Inner Piedmont thrust sheet 254
Invergarry 142-143
Inverian 18, 154
Irish Sea horst 192
Islay 24
Islay anticline 317
Isle of Man 7, 62
Kalak nappe complex 222-223, 228, 241
Karelian gneisses 222
Karmoy 230, 233

Karmoy igneous complex 225
Karmoy ophiolite 238
Kennedy's Pass 100, 107
Kiln Formation 210
Kinlochewe 163
Kinlochewe thrust 164, 253, 272
Kinlochewe thrust sheet 265
Kinlochewe-Kishorn sheets 157
Kintail 147, 311
Kiokee belt 249
Kirkholm Group 216
Kishorn nappe 146-147
Kishorn sheet 259, 265
Kishorn syncline 164
Kishorn thrust 146, 148, 162, 164, 253, 261
Knochan 173, 175-178, 272-273
Knoydart thrust 20, 311
Knoydartian 116 (Morarian)
Kyle of Lochalsh 134, 142, 147-149, 161
Lake Char 247
Lake Char fault 254
Lake Char-Honey Hill fault 255
Lake District 1, 7-12, 62, 78-79, 81, 83, 85, 191, 289, 291
Lake District anticline 12
Lake District basement 12
Lake District sedimentary cover 12
Lake District volcanic arc 187
Lake District/Leinster 200-201
Laksefjord nappe 238
Laurentian craton 145
Laxford Bridge 154
Laxford front 266
Laxfordian 18-19, 154, 260, 262
Laxfordian fold structures 164
Leadhills imbricate zone 90-92
Ledmore Junction 173, 175-176
Leinster 7
Leinster-Lake District zone 29
Leinster-Lake District volcanic island arc 192
Lendalfoot 99, 102, 107
Leny Grits 125 (Dalradian)
Leny Limestone 118 (Dalradian)
Leven Schists 128-132 (Dalradian)
Lewisian 1, 19, 24, 114, 143, 147, 151, 154, 162, 164-165, 166, 169, 259, 261, 303
Lewisian basement 20, 22, 24, 119, 274
Lewisian complex 18, 145, 153
Lewisian gneiss(es) 116, 146, 148-149, 153-155, 169, 173-174, 260
Lewisian inliers 24, 116
Lewisian mylonite 149

SUBJECT INDEX

Lewisian slices 311
Lingula Flags 40
Lintrathen ignimbrite 199
LISPB 274, 285, 290, 292
Llandeilo calc-alkaline volcanics 193
Llandeilo-Caradoc Bail Hill volcanics 197
Llandovery Skomer volcanics 192
Llandovery submarine lavas 190
Llanrian Volcanic Group 41
Llewelyn 191
Llewelyn Volcanic Group 56
Llyn Peninsula 4, 7, 60, 62, 70
Llyn Traffwll 66
Loch Ailsh 156, 263, 267
Loch Alsh 142, 146-147, 150-151
Loch Assynt 19, 158
Loch Awe syncline 23, 317-319
Loch Borrolan 156, 253, 263
Loch Borrolan (igneous) complex 174, 274 (see also Borrolan)
Loch Borrolan syenite 313
Loch Cluanie 115, 143
Loch Doon granite 13
Loch Duich 141, 144, 151
Loch Eil 115
Loch Eilt 139
Loch Garry 143
Loch Glencoul 174
Loch Iain Oig 149
Loch Kishorn 164
Loch Leven 129
Loch Lomond 109, 115, 120-126
Loch Lomond clastic unit 123
Loch Loyne 143
Loch Maree 164
Loch Maree Group 262 (Lewisian)
Loch More klippe 266
Loch Quoich line 22-23, 114, 116, 134-135, 137, 139, 143, 153, 308, 311-314, 318
Loch Tay 123, 127
Loch Tay Fault 27
Loch Tay Limestone 127 (Dalradian)
Lochailort Pelite 136 (Moine)
Lochalsh fold 265
Lochalsh syncline 148-149
Locheil (Loch Eil) Division/Group (see Moine)
Lochnager complex 199
Longford-Down Inlier 198
Longmyndian 3
Lorne andesitic lavas 199
Lough Nafooey island arc 196
Lower Allochthon 227 (Scandinavia)

Lower Crafnant Volcanic Formation 56
Lower Hartfell Shales 87-88
Lower Morar Psammite 136 (Moine)
Lower Old Red Sandstone 122-123, 151, 190, 212
Lower Ordovician plate motion 236
Lower Palaeozoic volcanic rocks 187-188
Lower Rhyolitic Tuff Formation 56, 59-60
Lyngen nappe 222, 229, 231, 233
Maen Jaspis 50-51
mafic microdiorite (sheets) 140-141, 143
Main ring-fault 128, 131
Mallaig 133
Malvern Hills 3
Marchburn Group 100
Meguma terrane 283
melange 62
Menai Straits 33
Menai Straits fault 3
Menai Straits fault system 29
Middle Allochthon (Scandinavia) 225, 227-228, 241-242
Midland platform 31
Midland Valley 16, 27, 81, 107, 120, 187, 195, 197, 200-201, 209, 217, 291, 303
Midland Valley continental basement 18
Midland Valley inliers 17
Midland Valley magmatic arc 196
Migmatitic rocks of Sutherland 21 (?Moine)
Moffat 87, 91
Moffat Shales 87, 89
Moine (Moines) (Moinian) 1, 20, 22, 24, 114, 119, 133, 146 148, 150, 165, 167, 174, 178, 303   Moine formations listed by name
Moine (divisions) (groups) 19, 134
    Central Highland Division 19, 21, 24-25, 305-306, 308-309, 316-317
    Glenfinnan Division/Group 19-20, 22, 24, 114, 116-117, 134-136, 138-140, 142-143, 268, 304-306, 308-311, 313
    Locheil (Loch Eil) Division/Group 19-20, 22, 114, 117, 119, 134-139, 142, 305-306, 308, 310
    Loch Eil/Glenfinnan Division contact 137
    Morar Division/Group 19-22, 24, 114, 116-117, 134-136, 142-144, 268, 304-306, 308-311, 313
Moine and Outer Isles Seismic Traverse (MOIST) 270, 274, 295
Moine basal conglomerates 165
Moine Basal Pelite 136
Moine mylonites 164
Moine nappe 147, 149
Moine schists 153-154, 159
Moine thrust 23, 24, 116, 119, 146, 156-157, 160, 162, 164, 166, 167-169, 175-178, 247, 249, 251-255, 259-260, 262, 265-267, 270, 272, 274-275, 282, 292, 295, 303, 311-312

SUBJECT INDEX

Moine thrust plane 149-150
Moine thrust zone 19-20, 113-114, 134, 142, 147-148, 150,
    153, 159, 161-162, 169, 174
Mona complex 4, 29, 60
Monian (supergroup) 1, 3-5, 60, 62-63, 67, 70 (formations
        listed by name)
    Gwna Group 62-63, 68
    New Harbour Group 62-67
    South Stack Group 62-65
Monian gabbros and serpentinites 65
Monian terrane 7
Morar 113, 133, 137, 139
Morar anticline 140
Morar Division/Group (see Moine)
Morarian 116-117, 304, 306 (Knoydartian)
mylonitised Lewisian (gneiss) 148-150
mylonitised Moine (schists) 150, 166
mylonitised Torridonian sandstones 148, 150
Navar/Swordly slide 20, 24
Naver slide 23, 249, 268, 310, 312
NE Basic Complexes 117, 304
New Harbour Group (see Monian)
Newer Granites 117, 322
North American shelly faunas 17
Northern Appalachians 217
northern flank of the Iapetus 187
Northern Highlands 23, 114, 118, 142
Northern Highlands steep belt 22
Old Red Sandstone 33, 114, 117, 119, 142, 209, 270, 304
Old Red Sandstone lavas 198
Old Red Sandstone molasse 36
Old Red Sandstone volcanic rocks 195
Older granites 322
olistostrome 69, 100, 103
ophiolite complex 26
ophiolite complex of Shetland 21
ophiolite slice 197
Ordovician inliers 190
Ordovician (late) marine transgression 5
Ossian steep belt 317-319
Outer Hebrides 23
Outer (Isles) Hebrides thrust 20, 23, 267, 270, 272, 282,
    284, 295
Padarn Tuff Formation 4
Parys Mountain 68
Parys Mountain copper-lead-zinc deposits 6
Pebidian 40
Pembrokeshire 5, 29, 36, 41, 191
Pen Anglas 52
Pen Caer 48, 49
Penchrise Burn Beds 216

Penfathach 51
Penmynydd 68-69
Penmynydd zone 62, 69
Pentevrian 3
Pinbain 102
Pinbain Beach 100-107
Pipe Rock (Lower Cambrian) 156, 163, 165, 167, 171-175, 177, 253, 262, 265-267
Pitlochry 118
Pitt's Head ash-flow tuff 58
Plock of Kyle 148
Pontesford-Church Stretton fault 3, 5
Pontesford-Linley fault 29
Port Askaig Tillite 116 (Dalradian)
Porth Clais 38
Porth Maen 52
Porth Maen Melyn 47-48
Porth Maen Melyn Volcanic Formation 46, 48
Porth Trecastell 67
Porth Wen 68
Portpatrick Formation 216
post-Finnmarkian nappes 222
pre-Caledonian granite 139
pre-Torridonian Lewisian topography 164
Precambrian volcanics 40
Ptarmigan Lodge 126
pyroxene-granulites 146
Queen's View 108
Ramsey Island 41
Ramsay Island rhyolites 191
Rannoch Moor 113, 127
Ratagan 142
Ratagan granite 144, 151
Rhobell Fawr Volcanic Group 6, 29
Rhoscolyn 63-64
Rhoscolyn anticline 63-64
Rhoscolyn Formation 65 (Monian)
Rhum 141
Riccarton Group 213
Rockall Bank 217
Ross of Mull 308
Ross of Mull granite 260, 274
Rubha Mhor 126
S Britain platform 1, 3
S Britain (platform) terrane 5, 7
Saddle 143
Samnanger complex 236
Sandaig 145
Sarn granite 60
Sarn nappe 229
Scandian motions 236

Scandian nappes 242
Scandian thrusts 236
Scandinavian Caledonides 221-222
Scandinavian nappe complex 223
Scar greywackes 90
Scardroy sheet 311
Scourian 18, 154, 260, 262
Scourian gneiss 172
Scourian gneissic banding 174
Scourian granulites 266
Scourian migmatites 166
Scourian (late) structures 174
Scourie dykes 18-19, 150-151, 166, 260, 266
Seiland igneous province 228, 233
serpentinite(s) 123, 125, 145
serpentinite-clast conglomerates 123
Serpulite Grit (Lower Cambrian) 156, 165, 169, 171-173, 176-177, 262
Sgurr Beag slide (ductile thrust) 20, 22-24, 114, 116, 119, 134-135, 141-143, 152, 268, 306, 309, 311-314, 318
Shap 85
Shap Blue Quarry 85
Shap granite 82, 84-85, 289
Shap granite Pink Quarry 85
Shetland 21, 23, 270, 292
Shetland andesites and dacites 198
Silurian bentonite 192
Silurian proximal volcanism 198
Silurian volcanics 190
Skerries Group 62 (Monian)
Skiddaw Eycott Group 12
Skiddaw Group 8, 12
Skiddaw Slate (Group) 78, 82
Skillefjord nappe 228
Skolithus 156
Skomer Island 6
Skomer volcanics 192
Skye 141, 145, 147-148
Snowdon (Snowdonia) 5-7, 29, 55-56, 58, 60, 70, 83, 191-192
Snowdon Volcanic Group 56, 192
Sole thrust 20, 153, 168-169, 175-177, 267, 272
Solway basin 284
Soroy nappe 228
Soroy succession 222
South Irish Sea lineament 7, 284
South Mayo basin 201
South Stack Formation 63, 65-66 (Monian)
South Stack Group (see Monian)
South West Approaches Traverse SWAT 287
Southern Appalachians 249
Southern Highland Group (see Dalradian)

Southern Highlands 23, 118
Southern Uplands 1, 11, 12-14, 16-17, 26, 78-81, 85-86, 89,
    96, 98, 195, 207, 209, 217, 289, 290
Southern Uplands belts
    Central Belt 14-15, 207, 209
    Northern Belt 14-15, 207, 209
    Southern Belt 14-15, 207, 209
Southern Uplands fault 13, 89, 91-92, 100, 105, 195, 197,
    208-209, 217, 282, 289-291
Southern Uplands ocean floor volcanics 197
Southern Uplands (accretionary) prism 17, 100, 105, 187, 198,
    201
Southern Uplands tracts 211-214, 216
Southern Uplands/S Connemara 200-201
Spango granite 13
Spothfore Formation 210
St Davids 38
Stack of Glencoul 174
Stanner-Hunter rocks 3
Stinchar fault 99-101, 105, 208, 211
Stinchar fault complex 89
Stinchar Limestone 81, 101
Stobs Castle Beds 216
Stockdale Rhyolite 78
Stoer Group 19-155
Storen nappe 239-240, 242
Storen ophiolite 236
Strath Halladale 119
Strath Halladale granite 21, 24, 117, 119, 304, 310
Strathconon fault 144
stratiform composite volcanoes 193
Strontian granite 21, 27
Strumble Head 41
Strumble Head (Volcanic) Formation 46, 48, 50
Stuckivoulich 126
subduction-related magmatism 187, 194
subduction-related volcanism 191
submarine welding (of ashflows) 191
sulphide mineralisation 193
Sutherland migmatite complex 24, 249
Tappins Group 100-101
Tarbet 127
Tarskavaig thrust 253
Tay nappe (complex) 25, 118, 122, 318-319, 324-325
Tayvallich lavas 315 (Dalradian)
Tayvallich Limestone 116, 118 (Dalradian)
Tayvallich volcanics 195 (Dalradian)
Tollie antiform 164
Torridon 115
Torridon Group 19, 155

Torridonian 19-20, 24, 119, 146, 148-149, 153, 155-156, 164-169, 172, 174, 178, 252, 259, 262, 265
Torridonian-Lewisian unconformity 164
Towaliga 247
Towaliga fault 254-255
Towaliga-Goat Rock-Barletts Ferry fault system 254
Townsend tuff 190, 192
Towy 'anticline' 36
Traeth Lligwy 68
Trecastell 67-69
Trefgarn Volcanic Group 29, 190
Tremadoc arc-tholeiites 190
Tremadoc subduction-related magmatism 196
Tremadoc/Arenig marine transgression 190
Trondheim nappe complex 229, 231, 235-236, 239, 241
trumpet pipes 172
Tweeddale lavas 212
Tyrone igneous complex 196
Ullapool 115, 157, 165-167, 178, 270, 272
Unst (Shetland) Ophiolite 188
Upper Allochthon (Scandinavia) 222, 225, 227-229, 233, 235-236, 241-242
Upper Arnaboll thrust 159
Upper Birkhill sandstones 216
Upper Hartfell sandstones 216
Upper Hartfell Shale 88
Upper Llanvirn volcanoes 190
Upper Morar Psammite 136, 141 (Moine)
Upper Old Red Sandstone 123
Uppermost Allochthon (Scandinavia) 222, 227-229, 233, 235-236, 241-242
Vaddas nappe 222, 229, 231, 233, 235
Valley and Ridge province 254
Valley Fault complex 91
Varanger tillite 314
volcanics north of the suture 194
volcanics south of the suture 188
Wales marginal basin 31
Welsh back-arc basin 199
Welsh back-arc ensialic marginal basin 187
Welsh basin 1, 3-7, 33, 41, 70, 190-192, 286, 288
Welsh Borderland 3, 33, 190, 192, 198
Welsh Borderland continental foreland 199
West Finnmark nappes 238
West Karmoy igneous complex 240, 242
West Orkney basin 270-272
Western Lewisian 144-145
Whitesand Bay 40
WINCH 288, 292, 295
Woodland Point 100
Y Penrhyn 52

Yarlside Rhyolite 78
Younger Basic suite 320
Younger gabbros 119